Advances in Nano-Electrochemical Materials and Devices

Advances in Nano-Electrochemical Materials and Devices

Xuyuan Chen
Mei Wang
Nabin Aryal

Basel • Beijing • Wuhan • Barcelona • Belgrade • Novi Sad • Cluj • Manchester

Xuyuan Chen
Institute of Laser Spectroscopy
Shanxi University
Taiyuan
China

Mei Wang
Institute of Laser Spectroscopy
Shanxi University
Taiyuan
China

Nabin Aryal
Department of Process, Energy
and Environmental Technology
University of
South-Eastern Norway
Porsgrunn
Norway

Editorial Office
MDPI AG
Grosspeteranlage 5
4052 Basel, Switzerland

This is a reprint of articles from the Special Issue published online in the open access journal *Nanomaterials* (ISSN 2079-4991) (available at: www.mdpi.com/journal/nanomaterials/special_issues/advances_nanoelectrochemical_materials_devices).

For citation purposes, cite each article independently as indicated on the article page online and using the guide below:

Lastname, A.A.; Lastname, B.B. Article Title. *Journal Name* **Year**, *Volume Number*, Page Range.

ISBN 978-3-7258-1828-0 (Hbk)
ISBN 978-3-7258-1827-3 (PDF)
https://doi.org/10.3390/books978-3-7258-1827-3

© 2024 by the authors. Articles in this book are Open Access and distributed under the Creative Commons Attribution (CC BY) license. The book as a whole is distributed by MDPI under the terms and conditions of the Creative Commons Attribution-NonCommercial-NoDerivs (CC BY-NC-ND) license (https://creativecommons.org/licenses/by-nc-nd/4.0/).

Contents

About the Editors vii

Preface ix

Mei Wang, Xuyuan Chen and Nabin Aryal
Advances in Nano-Electrochemical Materials and Devices
Reprinted from: *Nanomaterials* **2024**, *14*, 712, doi:10.3390/nano14080712 1

Alessio Mezza, Mattia Bartoli, Angelica Chiodoni, Juqin Zeng, Candido F. Pirri and Adriano Sacco
Optimizing the Performance of Low-Loaded Electrodes for CO_2-to-CO Conversion Directly from Capture Medium: A Comprehensive Parameter Analysis
Reprinted from: *Nanomaterials* **2023**, *13*, 2314, doi:10.3390/nano13162314 5

Zanhe Yang, Siqi Zhou, Xiangyu Feng, Nannan Wang, Oluwafunmilola Ola and Yanqiu Zhu
Recent Progress in Multifunctional Graphene-Based Nanocomposites for Photocatalysis and Electrocatalysis Application
Reprinted from: *Nanomaterials* **2023**, *13*, 2028, doi:10.3390/nano13132028 15

Xindan Zhang, Xiao Tan, Ping Wang and Jieling Qin
Application of Polypyrrole-Based Electrochemical Biosensor for the Early Diagnosis of Colorectal Cancer
Reprinted from: *Nanomaterials* **2023**, *13*, 674, doi:10.3390/nano13040674 57

Xinghua Liang, Yujuan Ning, Linxiao Lan, Guanhua Yang, Minghua Li, Shufang Tang and Jianling Huang
Electrochemical Performance of a PVDF-HFP-LiClO$_4$- Li$_{6.4}$La$_{3.0}$Zr$_{1.4}$Ta$_{0.6}$O$_{12}$ Composite Solid Electrolyte at Different Temperatures
Reprinted from: *Nanomaterials* **2022**, *12*, 3390, doi:10.3390/nano12193390 84

Bing Li, Tingting Xiang, Yuqi Shao, Fei Lv, Chao Cheng, Jiali Zhang, et al.
Secondary-Heteroatom-Doping-Derived Synthesis of N, S Co-Doped Graphene Nanoribbons for Enhanced Oxygen Reduction Activity
Reprinted from: *Nanomaterials* **2022**, *12*, 3306, doi:10.3390/nano12193306 94

Xinghua Liang, Xingtao Jiang, Linxiao Lan, Shuaibo Zeng, Meihong Huang and Dongxue Huang
Preparation and Study of a Simple Three-Matrix Solid Electrolyte Membrane in Air
Reprinted from: *Nanomaterials* **2022**, *12*, 3069, doi:10.3390/nano12173069 103

Xinghua Liang, Yu Zhang, Yujuan Ning, Dongxue Huang, Linxiao Lan and Siying Li
Quasi-Solid-State Lithium-Sulfur Batteries Assembled by Composite Polymer Electrolyte and Nitrogen Doped Porous Carbon Fiber Composite Cathode
Reprinted from: *Nanomaterials* **2022**, *12*, 2614, doi:10.3390/nano12152614 115

Jifeng Gu, Jiaping Zhang, Yun Su and Xu Yu
Aramid Fibers Modulated Polyethylene Separator as Efficient Polysulfide Barrier for High-Performance Lithium-Sulfur Batteries
Reprinted from: *Nanomaterials* **2022**, *12*, 2513, doi:10.3390/nano12152513 129

Xu Yu, Zhiqiang Pan, Zhixin Zhao, Yuke Zhou, Chengang Pei, Yifei Ma, et al.
Boosting the Oxygen Evolution Reaction by Controllably Constructing FeNi$_3$/C Nanorods
Reprinted from: *Nanomaterials* **2022**, *12*, 2525, doi:10.3390/nano12152525 138

Wenlei Zhang, Hongwei Yin, Zhichao Yu, Xiaoxia Jia, Jianguo Liang, Gang Li, et al.
Facile Synthesis of 4,4′-biphenyl Dicarboxylic Acid-Based Nickel Metal Organic Frameworks with a Tunable Pore Size towards High-Performance Supercapacitors
Reprinted from: *Nanomaterials* **2022**, *12*, 2062, doi:10.3390/nano12122062 . . . . **149**

Deyu Mao, Zirui He, Wanni Lu and Qiancheng Zhu
Carbon Tube-Based Cathode for Li-CO_2 Batteries: A Review
Reprinted from: *Nanomaterials* **2022**, *12*, 2063, doi:10.3390/nano12122063 . . . . **159**

Yifei Ma, Jiemin Han, Zhaomin Tong, Jieling Qin, Mei Wang, Jonghwan Suhr, et al.
Porous Carbon Boosted Non-Enzymatic Glutamate Detection with Ultra-High Sensitivity in Broad Range Using Cu Ions
Reprinted from: *Nanomaterials* **2022**, *12*, 1987, doi:10.3390/nano12121987 . . . . **177**

About the Editors

Xuyuan Chen

Prof. Xuyuan Chen is currently working at the Institute of Laser Spectroscopy, Shanxi University, China. He received his Ph.D. in Semiconductor Materials and Devices from the Eindhoven University of Technology, the Netherlands, in 1997. After that, he joined the Microelectronics Lab at the School of Engineering Science, Simon Fraser University, Canada, where he worked on silicon microelectronic devices and III-V DFB laser diodes. From 1997, he worked at the Physics Department at the University of Tromso, Norway, as an Associate Professor, and he continued at this institution as a Professor (2001) until 2002. During that time, his research topics were microelectronic devices and optoelectronics. Dr. Xuyuan Chen joined Microsystems and Nanotechnology, SINTEF ICT, Norway, at the end of 2002, where he started as a Senior Research Scientist. In 2004, he was appointed as a Professor of the Department of Microsystems at the University of South-Eastern Norway (previous Vestfold University College). In 2023, he joined Shanxi University. He is currently working on advanced materials and micro- and nano-technology, as well as their applications in electrochemical energy storage systems, MEMs, and so on. He is the author and co-author of more than 250 scientific publications in international journals and conferences, invited talks, and overview papers in the area of semiconductor technology, micro/nano-systems, and electrochemical technology.

Mei Wang

Prof. Mei Wang received her Ph.D. degree in Energy Science from Sungkyunkwan University, South Korea, in 2014. After conducting post-doctorate research at Sungkyunkwan University for 3 years, she joined the State Key Laboratory of Quantum Optics and Quantum Optics Devices, Institute of Laser Spectroscopy, Shanxi University, as a Professor in 2017. Prof. Mei Wang has been engaged in the precise control of material microstructures and the design of material structures for applications for 14 years, including supercapacitors, electrochemical sensors, photocatalysis, strain/sensors, etc. Prof. Mei Wang has published more than 50 research papers and has been awarded expert contracts by the Shanxi Provincial Party Committee, and she was recognized as a young outstanding talent in Shanxi Province, China.

Nabin Aryal

Nabin Aryal is an Associate Professor at the Department of Process, Energy and Environmental Technology, Porsgrunn Campus, University of South-Eastern Norway (USN), Norway. He received his MSc in Environmental Technology and Engineering from Ghent University, Belgium, with an Erasmus Mundus fellowship from the European Union and then a PhD. from the Technical University of Denmark (DTU) in Environmental Engineering with a fellowship from Novo Nordisk Foundation, Denmark. Before joining USN, Dr Aryal was engaged in joint industrial postdoctoral research at Aarhus University and the Danish Gas Technology Center, Denmark. He has more than a decade of work experience from Norway, Denmark, Belgium, the Czech Republic, India, and Nepal. His current research focuses on bioelectrochemical systems, microbe–material interaction, biofilm-based bioprocess development, and anaerobic technology. He is an active reviewer of more than 50 International peer reviewed journals, an Editorial Board Member of the *International Journal of Environmental Science and Technology*—Springer, and a Lead Editor of several Special Issues for international journals. He has co-authored over 50 scientific publications and edited two books on biogas upgrading and microbe–material interaction that received an h-index of 18 (>1500 citations; Google Scholar).

Preface

This Special Issue focuses on recent advances in the design, manufacture, and application of nanoelectrochemical materials and devices. Twelve papers were included in this Special Issue, where nine original research papers highlighted novel research progress on advanced nanomaterials and nanosystems for oxygen reduction reactions, batteries, supercapacitors, and sensors. In addition, recent achievements in photocatalysis and electrocatalysis, biosensors, and Li-CO_2 batteries were comprehensively summarized in three review papers. We are confident that this Special Issue will attract a high degree of attention in the field of electrochemical materials and devices and promote the development of advanced functional nanomaterials and nanotechnologies.

Xuyuan Chen, Mei Wang, and Nabin Aryal
Editors

Editorial

Advances in Nano-Electrochemical Materials and Devices

Mei Wang [1,*], Xuyuan Chen [1,*] and Nabin Aryal [2,*]

[1] State Key Laboratory of Quantum Optics and Quantum Optics Devices, Institute of Laser Spectroscopy, Collaborative Innovation Center of Extreme Optics, Shanxi University, Taiyuan 030006, China

[2] Department of Process, Energy and Environmental Technology, Porsgrunn Campus, University of South-Eastern Norway, 3918 Porsgrunn, Norway

* Correspondence: wangmei@sxu.edu.cn (M.W.); xuyuan.chen@usn.no (X.C.); nabin.aryal@usn.no (N.A.)

Nano-electrochemical materials and devices are at the frontier of research and development, advancing electrochemistry and its applications in energy storage, sensing, electrochemical processing, etc. The synergy of nanotechnology and electrochemistry has led to advances in nanostructured electrode materials [1,2]. Nanotechnology is supported to advance nanomaterials with high spatial surface area, nanosized and porous-induced physical effects, and multi-dimensional structure construction, which boosts prominent properties bordering its potential applications in developing electrochemical devices.

The recent critical developments in nano-electrochemical materials are devices that focus on developing nanostructured electrode material [3], nano-electrochemical sensors [4], nanomaterials for energy storage and conversion [5,6], etc. Developing a novel material synthesis process enables smaller-scale prototypes for application, optimization, and innovation experimentation. Therefore, researchers must explore novel concepts and apply synergy from multiple disciplines to explore fundamental science. That could build a strong foundation for further industrial applications, technology upscaling and commercialization. The research and development on laboratory-scale applications required controlled environmental research with various parameters to bridge the gap between concepts and laboratory-scale innovation testing. In this Special Issue, various advanced nano-scale materials and novel processing techniques are reported, including material synthesis, structure design, interface engineering, and characterization methods.

Demonstrating and developing new nanomaterials and devices are not only academic exercises for teaching but also include heavy practical applications. However, practical technical information on the innovation and application of electrochemical materials and devices is still limited. To overcome such a gap, this Special Issue was aimed at providing technical details on nano-electrochemical materials and devices. The Special Issue was focused on summarizing the advances in the design, development, manufacture, and application of nano-electrochemical materials and devices. In the Special Issue, recent advances in nanomaterial development, particularly graphene-based, polypyrrole-based, and carbon tube-based material developments, were critically reviewed. The critical review summarized the applications of carbon-based materials in electrochemical sensor device applications, photocatalyst and electrocatalyst applications, and biosensors for early disease diagnosis. Experts wrote the individual scientific manuscripts published in this Special Issue, consolidating the most recent state-of-the-art and focusing on innovation.

In the field of energy conversion and storage, novel electrode material synthesis, especially N, S, and Co-doped graphene (Contribution 1), and constructing $FeNi_3/C$ nanorods (Contribution 2) to enhance oxygen reduction reactions have been tested. A low-loaded silver-based electrode is also fabricated via the sputtering deposition technique for CO_2-to-CO conversion directly from the capture medium (Contribution 3). Moreover, solid-state batteries have boosted the development of solid-state electrolytes. In this Special Issue, the electrochemical performances of PVDF-HFP-$LiClO_4$-$Li_{6.4}La_{3.0}Zr_{1.4}Ta_{0.6}O_{12}$ composite

Citation: Wang, M.; Chen, X.; Aryal, N. Advances in Nano-Electrochemical Materials and Devices. *Nanomaterials* **2024**, *14*, 712. https://doi.org/10.3390/nano14080712

Received: 4 April 2024
Accepted: 15 April 2024
Published: 18 April 2024

Copyright: © 2024 by the authors. Licensee MDPI, Basel, Switzerland. This article is an open access article distributed under the terms and conditions of the Creative Commons Attribution (CC BY) license (https://creativecommons.org/licenses/by/4.0/).

solid-state electrolytes were researched, which is essential for the development of high-energy density all-solid-state lithium-ion batteries (Contribution 4). A three-matrix solid electrolyte membrane in air was also tested and published (Contribution 5). The critical scientific information from this research could be an essential reference for further developing all-solid-state batteries. In addition, solid-state lithium-sulfur batteries have recently received widespread attention for energy storage due to their high current density, performance, and economics [7]. Therefore, composite polymer electrolytes and nitrogen-doped porous carbon fiber composite cathodes were synthesized (Contribution 6), and their electrochemical performance was tested to advance the technology. Advances in the separator also promote improving the performance of lithium-sulfur batteries. An aramid fiber-modulated polyethylene separator was reported as an efficient polysulfide barrier for high-performance lithium-sulfur batteries, providing a facile way to fabricate the separator for inhibiting polysulfides in the lithium-sulfur battery (Contribution 7).

Nowadays, due to the increasingly severe greenhouse effect, the treatment and utilization of carbon dioxide (CO_2), which is the biggest contributor to greenhouse gases (GHGs), has become a worldwide hot topic. Current utilization methods for CO_2 include chemical conversion, photoreduction, electrochemical reduction [8], bioconversion, etc. In this Special Issue, two review papers were published, summarizing recent progress in photoreduction and electrochemical reduction of CO_2. One paper reviewed the carbon tube-based cathode for Li-CO_2 batteries (Contribution 8). As an electrochemical device that can capture, fix, and convert CO_2, electric energy can be stored for energy utilization in various applications. The other review paper reviewed the multifunctional graphene-based nanocomposites for photocatalysis and electrocatalysis applications, including photocatalytic hydrolysis, pollutant degradation, and the photocatalytic reduction of CO_2 (Contribution 9).

Recent laboratory research has shown that metal-organic framework (MOF) materials have demonstrated superior performances because of their high-tunable conductivity and their structure's pore size [9]. One of the research papers published in this Special Issue is the synthesis of 4,4′-biphenyl dicarboxylic acid-based nickel metal-organic frameworks for supercapacitor applications (Contribution 10). The research evidence published in the manuscript will be beneficial as reference information to develop high-performance hybrid MOF composites for future electrochemical energy storage applications.

Employing electrochemical devices as sensors represents a groundbreaking capability in identifying environmental traces, significantly contributing to the monitoring and restoration of natural ecosystems. This sensor technology has also propelled advancements in medical diagnostics and industrial processes. Apart from electrochemical devices, nanomaterials have also gained broader application in medical applications for disease detection. Towards such aims, one of the manuscripts developed a non-enzymatic electrochemical sensor for the detection of glutamate based on an advanced porous carbon electrode, and impressively, a novel detection mechanism based on Cu ions was first published (Contribution 11). Recently, conductive polymers have also drawn great attention in the field of electrochemical sensors. A review of polypyrrole-based electrochemical biosensors for the early diagnosis of colorectal cancer is published, reporting the properties, synthesis techniques, and applications of the biosensors (Contribution 12). These achievements demonstrated the advances and innovation of electrochemical devices for medical and biological applications, such as microbial fuel cells and amicrobial electrochemical systems.

In conclusion, nanometer-scale electrochemical materials and devices have made gratifying progress in various fields in recent years. This Special Issue reported several novel research works on advanced nanomaterials and nano-systems for oxygen reduction reactions, batteries, supercapacitors, and sensors. Recent achievements in photocatalysis and electrocatalysis, biosensors, and Li-CO_2 batteries were comprehensively summarized in three review papers. We strongly believe that this issue will draw wide attention in the field of electrochemical materials and devices and promote the development of advanced functional nanomaterials and nanotechnologies.

Funding: This research received no external funding.

Conflicts of Interest: The author declares no conflicts of interest.

List of Contributions

1. Li, B.; Xiang, T.; Shao, Y.; Lv, F.; Cheng, C.; Zhang, J.; Zhu, Q.; Zhang, Y.; Yang, J. Secondary-Heteroatom-Doping-Derived Synthesis of N, S Co-Doped Graphene Nanoribbons for Enhanced Oxygen Reduction Activity. *Nanomaterials* **2022**, *12*, 3306. https://doi.org/10.3390/nano12193306
2. Yu, X.; Pan, Z.; Zhao, Z.; Zhou, Y.; Pei, C.; Ma, Y.; Park, H.S.; Wang, M. Boosting the Oxygen Evolution Reaction by Controllably Constructing $FeNi_3$/C Nanorods. *Nanomaterials* **2022**, *12*, 2525. https://doi.org/10.3390/nano12152525
3. Mezza, A.; Bartoli, M.; Chiodoni, A.; Zeng, J.; Pirri, C.F.; Sacco, A. Optimizing the Performance of Low-Loaded Electrodes for CO_2-to-CO Conversion Directly from Capture Medium: A Comprehensive Parameter Analysis. *Nanomaterials* **2023**, *13*, 2314. https://doi.org/10.3390/nano13162314.
4. Liang, X.; Ning, Y.; Lan, L.; Yang, G.; Li, M.; Tang, S.; Huang, J. Electrochemical Performance of a PVDF-HFP-$LiClO_4$-$Li_{6.4}La_{3.0}Zr_{1.4}Ta_{0.6}O_{12}$ Composite Solid Electrolyte at Different Temperatures. *Nanomaterials* **2022**, *12*, 3390. https://doi.org/10.3390/nano12193390
5. Liang, X.; Jiang, X.; Lan, L.; Zeng, S.; Huang, M.; Huang, D. Preparation and Study of a Simple Three-Matrix Solid Electrolyte Membrane in Air. *Nanomaterials* **2022**, *12*, 3069. https://doi.org/10.3390/nano12173069
6. Liang, X.; Zhang, Y.; Ning, Y.; Huang, D.; Lan, L.; Li, S. Quasi-Solid-State Lithium-Sulfur Batteries Assembled by Composite Polymer Electrolyte and Nitrogen Doped Porous Carbon Fiber Composite Cathode. *Nanomaterials* **2022**, *12*, 2614. https://doi.org/10.3390/nano12152614
7. Gu, J.; Zhang, J.; Su, Y.; Yu, X. Aramid Fibers Modulated Polyethylene Separator as Efficient Polysulfide Barrier for High-Performance Lithium-Sulfur Batteries. *Nanomaterials* **2022**, *12*, 2513. https://doi.org/10.3390/nano12152513
8. Mao, D.; He, Z.; Lu, W.; Zhu, Q. Carbon Tube-Based Cathode for Li-CO_2 Batteries: A Review. *Nanomaterials* **2022**, *12*, 2063. https://doi.org/10.3390/nano12122063
9. Yang, Z.; Zhou, S.; Feng, X.; Wang, N.; Ola, O.; Zhu, Y. Recent Progress in Multifunctional Graphene-Based Nanocomposites for Photocatalysis and Electrocatalysis Application. *Nanomaterials* **2023**, *13*, 2028. https://doi.org/10.3390/nano13132028
10. Zhang, W.; Yin, H.; Yu, Z.; Jia, X.; Liang, J.; Li, G.; Li, Y.; Wang, K. Facile Synthesis of 4,4′-biphenyl Dicarboxylic Acid-Based Nickel Metal Organic Frameworks with a Tunable Pore Size towards High-Performance Supercapacitors. *Nanomaterials* **2022**, *12*, 2062. https://doi.org/10.3390/nano12122062
11. Ma, Y.; Han, J.; Tong, Z.; Qin, J.; Wang, M.; Suhr, J.; Nam, J.; Xiao, L.; Jia, S.; Chen, X. Porous Carbon Boosted Non-Enzymatic Glutamate Detection with Ultra-High Sensitivity in Broad Range Using Cu Ions. *Nanomaterials* **2022**, *12*, 1987. https://doi.org/10.3390/nano12121987
12. Zhang, X.; Tan, X.; Wang, P.; Qin, J. Application of Polypyrrole-Based Electrochemical Biosensor for the Early Diagnosis of Colorectal Cancer. *Nanomaterials* **2023**, *13*, 674. https://doi.org/10.3390/nano13040674

References

1. Nanotechnology for Electrochemical Energy Storage. *Nat. Nanotechnol.* **2023**, *18*, 1117. [CrossRef] [PubMed]
2. Gao, Y.; Zao, L. Review on Recent Advances in Nanostructured Transition-Metal-Sulfide-Based Electrode Materials for Cathode Materials of Asymmetric Supercapacitors. *Chem. Eng. J.* **2022**, *430*, 132745. [CrossRef]
3. Yan, J.; Li, S.H.; Lan, B.B.; Wu, Y.C.; Lee, P.S. Rational Design of Nanostructured Electrode Materials toward Multifunctional Supercapacitors. *Adv. Funct. Mater.* **2020**, *30*, 1902564. [CrossRef]
4. Mahato, K.; Wang, J. Electrochemical Sensors: From the Bench to the Skin. *Sens. Actuat. B Chem.* **2021**, *344*, 130178. [CrossRef]
5. Simon, P.; Gogotsi, Y. Perspectives for Electrochemical Capacitors and Related Devices. *Nat. Mater.* **2020**, *19*, 1151–1163. [CrossRef] [PubMed]
6. Du, X.; Ren, X.; Xu, C.; Chen, H. Recent advances on the manganese cobalt oxides as electrode materials for supercapacitor applications: A comprehensive review. *J. Energy Storage* **2023**, *68*, 107672. [CrossRef]
7. Li, J.; Xie, F.; Pang, W.; Yang, X.; Zhang, L. Regulate Transportation of Ions and Polysulfides in All-Solid-State Li-S Batteries using Ordered-MOF Composite Solid Electrolyte. *Sci. Adv.* **2024**, *10*, eadl3925. [CrossRef]

8. Ross, M.B.; De Luna, P.; Li, Y.; Dinh, C.-T.; Kim, D.; Yang, P.; Sargent, E.H. Designing Materials for Electrochemical Carbon Dioxide Recycling. *Nat. Catal.* **2019**, *2*, 648–658. [CrossRef]
9. Wu, H.B.; Lou, X.W. Metal-Organic Frameworks and Their Derived Materials for Electrochemical Energy Storage and Conversion: Promises and Challenges. *Sci. Adv.* **2017**, *3*, eaap9252. [CrossRef] [PubMed]

Disclaimer/Publisher's Note: The statements, opinions and data contained in all publications are solely those of the individual author(s) and contributor(s) and not of MDPI and/or the editor(s). MDPI and/or the editor(s) disclaim responsibility for any injury to people or property resulting from any ideas, methods, instructions or products referred to in the content.

Article

Optimizing the Performance of Low-Loaded Electrodes for CO_2-to-CO Conversion Directly from Capture Medium: A Comprehensive Parameter Analysis

Alessio Mezza [1,2,*], Mattia Bartoli [1], Angelica Chiodoni [1], Juqin Zeng [1,2], Candido F. Pirri [1,2] and Adriano Sacco [1,*]

1 Center for Sustainable Future Technologies @Polito, Istituto Italiano di Tecnologia, Via Livorno 60, 10144 Torino, Italy; mattia.bartoli@iit.it (M.B.); angelica.chiodoni@iit.it (A.C.); juqin.zeng@polito.it (J.Z.); fabrizio.pirri@polito.it (C.F.P.)
2 Department of Applied Science and Technology, Politecnico di Torino, Corso Duca degli Abruzzi 24, 10129 Torino, Italy
* Correspondence: alessio.mezza@polito.it (A.M.); adriano.sacco@iit.it (A.S.)

Abstract: Gas-fed reactors for CO_2 reduction processes are a solid technology to mitigate CO_2 accumulation in the atmosphere. However, since it is necessary to feed them with a pure CO_2 stream, a highly energy-demanding process is required to separate CO_2 from the flue gasses. Recently introduced bicarbonate zero-gap flow reactors are a valid solution to integrate carbon capture and valorization, with them being able to convert the CO_2 capture medium (i.e., the bicarbonate solution) into added-value chemicals, such as CO, thus avoiding this expensive separation process. We report here a study on the influence of the electrode structure on the performance of a bicarbonate reactor in terms of Faradaic efficiency, activity, and CO_2 utilization. In particular, the effect of catalyst mass loading and electrode permeability on bicarbonate electrolysis was investigated by exploiting three commercial carbon supports, and the results obtained were deepened via electrochemical impedance spectroscopy, which is introduced for the first time in the field of bicarbonate electrolyzers. As an outcome of the study, a novel low-loaded silver-based electrode fabricated via the sputtering deposition technique is proposed. The silver mass loading was optimized by increasing it from 116 µg/cm^2 to 565 µg/cm^2, thereby obtaining an important enhancement in selectivity (from 55% to 77%) and activity, while a further rise to 1.13 mg/cm^2 did not provide significant improvements. The tremendous effect of the electrode permeability on activity and proficiency in releasing CO_2 from the bicarbonate solution was shown. Hence, an increase in electrode permeability doubled the activity and boosted the production of in situ CO_2 by 40%. The optimized Ag-electrode provided Faradaic efficiencies for CO close to 80% at a cell voltage of 3 V and under ambient conditions, with silver loading of 565 µg/cm^2, the lowest value ever reported in the literature so far.

Keywords: carbon capture and utilization; CO_2 valorization; bicarbonate electrolyzer; electrochemical impedance spectroscopy

1. Introduction

During the previous decades, human activities have increased atmospheric CO_2 concentrations, stimulating the scientific community toward the development of less-carbon intensive technologies and depleting the use of fossil fuels. However, renewable energy use requires time to be assessed, and it is of capital relevance to reduce carbon emissions during the energy transition. As a consequence, the renewable-energy-powered electrochemical reduction reaction of CO_2 (eCO_2RR) into added-value chemicals and fuels (e.g., syngas and methane) has attracted strong interest as a solution for a close carbon cycle [1]. The possibility to obtain carbon-based products from eCO_2RR at high rates has already been deeply investigated in gas-fed electrolyzers [2], where a stream of pure CO_2

needs to be delivered to the cathode. The perspective to employ such technology in an industrial setting requires coupling between the electrolyzer and CO_2 separation from the other components of flue gasses (e.g., O_2, N_2, and H_2O) emitted by a point source (e.g., an industrial plant). As an example, alkaline solutions (e.g., KOH) are able to capture gaseous CO_2 from flue gasses thanks to reactions that form (bi)carbonates [3]. Since it is known that CO_2 may be extracted from bicarbonate through energy-intensive processes [4], once it has been pressurized, it can be exploited for further valorization through electrolytic conversion. [5] In such kinds of platforms for carbon capture and utilization (CCU), since common gas-fed electrolyzers exhibit low single-pass utilization, around 80% of delivered CO_2 exits from the platform as unreacted gas [6]. In this framework, liquid-fed bicarbonate (HCO_3^-) electrolyzers have arisen as a new, groundbreaking technology to integrate the capture and conversion of CO_2 (ICCU) [7–9] into CO. These reactors introduced the chance to eliminate all the energy-demanding processes (capture/stripping and pressurization) necessary to feed a classical gas-fed eCO_2RR system. The electrolysis of carbonate solution, i.e., the capture media, is possible using a cation exchange membrane (CEM) or a bipolar membrane (BPM) that, providing an acidic local environment, makes gaseous CO_2 available in proximity to the catalyst for electroreduction. The utilization of a BPM instead of a CEM is now benchmarked since it allows for the employment of an inexpensive nickel anode and prevents products' cross-over [10,11]. The BPM, together with anodic and cathodic catalysts, constitutes a membrane electrode assembly (MEA), which is the benchmark configuration in bicarbonate electrolyzers. In the MEA, the presence of H^+ (produced in the BPM by water splitting) at the membrane/catalyst interface is responsible for the in situ acidification and thus extraction of CO_2 (i-CO_2) from (bi)carbonate, which is converted into CO (Equations (1) and (2)), thanks to the eCO_2RR catalyst.

$$H^+ + HCO_3^- \leftrightarrow H_2O + CO_2 \text{ (g)} \qquad (1)$$

$$CO_2\text{(g)} + H_2O + 2e^- \rightarrow 2OH^- + CO \qquad (2)$$

Since OH^- is a product of the CO_2RR as well, the original alkaline capture solution is regenerated, making this system able to implement a closed cycle where CO_2 is sequentially captured and converted. The presence of a MEA ensures a very high local concentration of CO_2 at the electrocatalyst interface without the need of supplying the reactor by a stream of gaseous CO_2 in stoichiometric excess, as happens with gas-fed electrolyzers. This also means that CO_2RR products are generated at higher concentrations [12]. The gas diffusion electrode (GDE) employed in this kind of system has to ensure the efficient transport of carbon feedstock (i.e., HCO_3^-) at the BPM/catalyst interface. Therefore it must be engineered differently with respect to the GDEs used in gas-fed electrolyzers, which usually exhibit hydrophobic properties to avoid the accumulation of water and to mitigate the hydrogen evolution reaction (HER) [13], the competing reaction of the CO_2RR.

Despite the promising advantages, research conducted on this technology so far is still limited compared to more well-known gas-fed electrolyzers. Therefore, a deep investigation into every aspect of the system is needed. Among the first works, Y. C. Li et al. [8] reported a bicarbonate electrolyzer able to keep the high pH of the capture solution for 145 h by using a carbon composite silver electrode, but the highest Faradaic efficiency (FE) toward CO (FE_{CO}) was ~35%. T. Li et al. [7] with a silver nanoparticle-coated carbon support obtained impressive FE_{CO} at a low current density and showed how the employment of an anion exchange membrane (AEM) is detrimental to the electrolyzer's performance. The same conclusion was reported by C. Larrea et al. [14], whereby, although it was responsible for a large ohmic drop between the two electrodes, the necessity to use a BPM in order to have appreciable FE was proven. Z. Zhang et al. [15] showed how the increase in porosity of a silver foam, employed as a cathode, enables more efficient CO_2 conversion; however, even avoiding the utilization of composite carbon electrodes, the FE_{CO} achieved at ambient conditions is around 60%. In addition, they illustrated how higher pressure and higher temperature promote the CO_2RR. Y. Kim et al. [16] underlined the importance of a trade-off between the active surface and the permeability of the GDE in order to guarantee both a high CO_2RR rate and efficient transport of bicarbonate (i.e., i-CO_2 generation). E. W. Lees et al. [17] reported important information on the spray coating of a silver catalyst in order to have an efficient GDE in terms of Nafion content and Ag nanoparticle loading. By adding a preliminary deposition step by sputtering physical vapor deposition (PVD) before the spray coating, they reached a very good FE_{CO} (around 82%) using a high silver loading of 2 mg/cm^2.

The sputtering technique serves as a rapid and reproducible method for manufacturing nanostructured Ag-GDEs with a high surface area in a single step, offering precise control over catalyst loading, layer thickness, and homogeneity. Our research group has already explored and established the reliability of this approach [18]. In this study, we further optimized the sputtering process to fabricate a GDE specifically designed for bicarbonate electrolyzers. To investigate the electrode's performance, we tested different commercial carbon supports with distinct characteristics such as gas diffusion layers (GDLs). This allowed us to delve into the GDL's role and its impact on the FE_{CO} and CO_2 utilization, representing the extent of CO_2 conversion compared to the unreacted CO_2. During the analysis, we explored the influence of several structural and morphological properties of the cathode on the electrochemical performance. These properties encompassed different catalyst distributions on the GDL, GDL hydrophobicity and permeability. By understanding the significance of these factors, we gain critical insights into optimizing the GDE's design and performance for bicarbonate electrolyzers.

Moreover, electrochemical impedance spectroscopy (EIS) has already demonstrated its efficacy as an efficient tool for studying the charge transfer and transport processes involved in typical systems for CO_2 reduction reactions (CO_2RRs) [19]. Its utility has made it a valuable technique for the characterization of materials and reactors in this field [20,21]. Despite this, to the best of our knowledge, it has never been employed in studies involving bicarbonate electrolyzers. Remarkably, our paper presents a pioneering application of EIS to investigate the performance of GDEs employed in bicarbonate electrolyzers. This novel approach represents a significant advancement in the field, as previous research has primarily focused on using EIS for other CO_2RR systems. Through careful modeling of the GDE/electrolyte interface using an equivalent electrical circuit, this paper successfully elucidates the underlying factors influencing the activity and Faradaic efficiency trends of the GDEs.

2. Materials and Methods

2.1. Preparation and Morphological Characterization of Ag GDEs

DC Sputtering (Quorum Technologies Ltd., Lewes, UK, Q150T) was used to prepare the Ag electrodes. Three commercial carbon papers (GDL, Ion Power) of 5 cm^2 characterized by different permeability and wettability (Table S1) were used as substrates, with a silver disk (99.999%, Nanovision, Brugherio, Italy) as the target. The deposition current was fixed at 50 mA, while the deposition time was varied to control the silver mass-loading (100 s, 300 s, and 600 s). A total of 6 GDE samples (A–F) were prepared, whose properties are reported in Table S2. All of the samples were prepared by depositing silver on both faces of the carbon papers, except for sample A. The mass loading was determined by weighing the sample before and after the silver deposition and then by dividing the weight difference by the geometric area of the GDL. The morphology of the commercial carbon-based supports and Ag-GDE samples was investigated by field emission scanning electron microscopy (FESEM, Zeiss Auriga, Oberkochen, Germany).

2.2. Electrochemical Tests and Product Analyses

The electrochemical screening was performed in a bicarbonate electrolyzer (Scribner, Cell Fixture) placed in a vertical position, whose schematic representation is reported in Figure 1. A more detailed description of the system is provided in the Supporting Information. A 5 cm^2 MEA was employed in the electrolyzer, and it was made by a bipolar membrane (FumaSep FBM, FumaTech, Bietigheim-Bissingen, Germany) sandwiched between a Nickel foam (99.5%, GoodFellow, Huntingdon, UK) and a Ag GDE. KHCO$_3$ (99.5%, Sigma-Aldrich, St. Louis, MO, USA) 2 M was used as a catholyte and KOH (Sigma-Aldrich) 1 M was used as an anolyte by dissolving 200 g and 56 g, respectively, in 1 L of ultra-pure water. A peristaltic pump was used to continuously recirculate 60 mL of bicarbonate solution and 40 mL of potassium hydroxide at a flow rate of 5 mL/min. Electrolysis was carried out at ambient temperature and pressure by applying a constant cell voltage (V_{cell}) of 3 V (Potentiostat, BioLogic VSP, Seyssinet-Pariset, France). Gas-phase products were delivered to a microgas chromatograph (μGC, Fusion, INFICON) by a N$_2$ 35 mL/min stream (Bronkhorst, EL-FLOW select) and analyzed on-line throughout the entire duration of the experiment. The microgas chromatograph, which is preceded by a mass flow reader (Bronkhorst, Ruurlo, The Netherlands, EL-FLOW prestige), is composed of two channels with a 10 m Rt-Molsieve 5A column and an 8 m Rt-Q-Bond column, and each channel has a microthermal conductivity detector. Two tests were conducted per set of experiments, and the results are reported as average values (the error bars correspond to the absolute error). Additional details on the calculation of the CO partial current density, FE, CO$_2$ utilization, mass activity, and partial mass activity are provided in the Supporting Information.

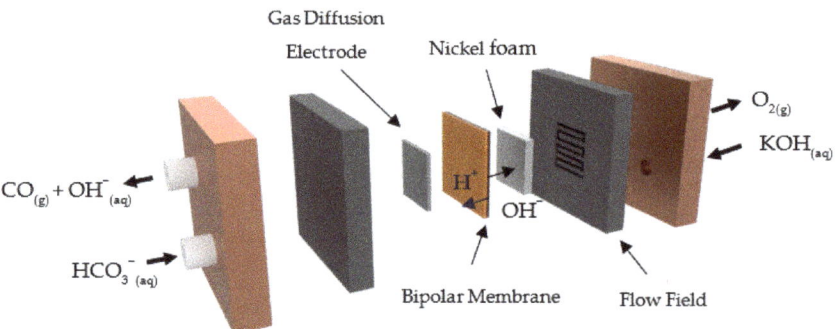

Figure 1. Schematic representation of the bicarbonate MEA electrolyzer.

2.3. Electrochemical Impedance Spectroscopy

EIS measurements were performed in a three-electrode single compartment cell at room temperature with a Biologic VSP electrochemical workstation. The working electrode was a Ag GDE with a geometric area of 0.4 cm^2. A Pt wire was used as the counter electrode, and Ag/AgCl (3 M Cl$^-$) was used as the reference, with both purchased from ALS. The electrolyte was a CO$_2$-saturated 2 M KHCO$_3$ (99.5%, Sigma-Aldrich) aqueous solution. The analysis was performed at a potential of -1 V vs. a reversible hydrogen electrode (RHE) with an AC signal with 10 mV of amplitude and a 0.1–10^5 Hz frequency range.

3. Results and Discussion

As the first step of GDE optimization, it was investigated as to whether it is more convenient to deposit the silver only on one side of the carbon support or on both of them (Figure 2a). Therefore, keeping the same sputtering parameters and carbon support, two samples were made. On the first one (sample A), silver was sputtered only on the face in contact with the bipolar membrane, while on the second one, the sputtering process was replicated on the opposite side as well (sample B), the one facing the graphite flow field. As shown in Figure 2b, sample A exhibits relatively good activity and FE$_{CO}$. This implies that the most active interface is the one facing the bipolar membrane, namely the region with the highest concentration of i-CO$_2$, since it is in proximity to the BPM. However, sputtering the silver on the other side of the GDL as well boosted the FE$_{CO}$ from 55% to 77%. Considering the results of this experiment, silver was deposited on both faces of the GDE samples tested from then on.

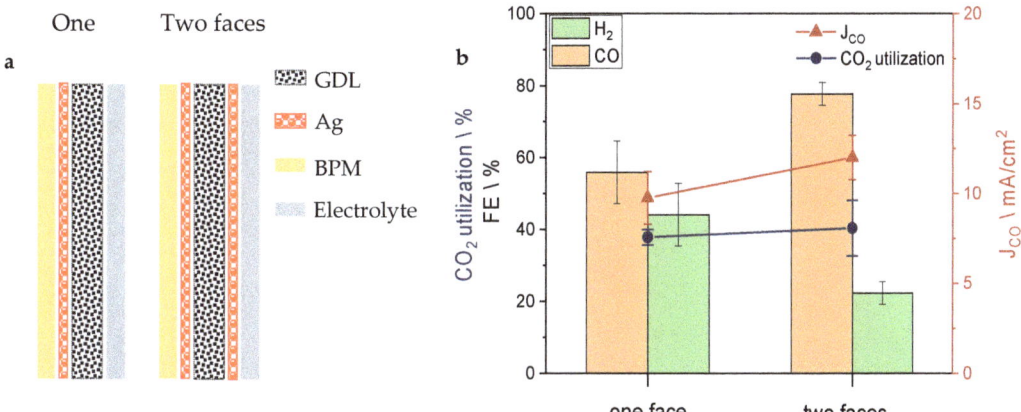

Figure 2. (**a**) Schematic representation of the GDEs with silver sputtered on one face (sample A) and both (sample B) faces. (**b**) FE, CO_2 utilization and CO partial current density obtained by the two GDEs samples.

The carbon support used to obtain the results mentioned just above was not treated with polytetrafluoroethylene (PTFE) nor did it include a microporous layer (MPL). These two characteristics are fundamental if this electrode had been used in a common gas-fed CO_2RR reactor [18]. In gas-fed reactors, the MPL and the hydrophobic treatment produce water repellent properties that can prevent the carbon fiber backing from flooding. In bicarbonate electrolyzers, the hydrophobic feature inhibits the transport of bicarbonate from the flow field toward the BPM, where the low-pH region is located. In this way, the production of i-CO_2 drastically decreases, hence also the FE_{CO}, J_{CO}, and the CO_2 utilization (Figure 3). By using a hydrophobic carbon support (sample C), the FE_{CO} decreases to 23%, while J_{CO} and the overall activity (J_{tot}) (Figure S1a) are significantly affected as well.

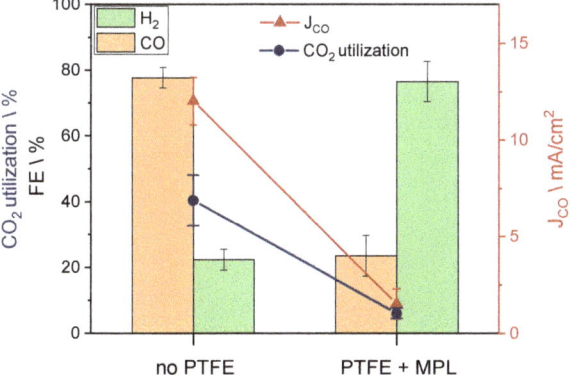

Figure 3. FE, CO_2 utilization, and CO partial current density obtained with GDEs with two different GDLs: one has not been treated with PTFE and does not include an MPL (sample B), while the other has a strong hydrophobic feature thanks to the PTFE and MPL (sample C).

Once the importance of using a GDE with no hydrophobic treatment and presence of a catalyst on both its faces has been confirmed, the silver mass loading was optimized by modulating the sputtering time. The performances of GDEs with silver mass loading of 116 µg/cm² (sample D), 565 µg/cm² (sample B), and 1.13 mg/cm² (sample E) were explored by carrying out electrolysis in the flow cell and an EIS analysis in a three-electrodes set-up. As reported in Figure 4a, the sample with 116 µg/cm^2 of silver shows the lowest FE (55%), while the other two samples with higher mass loading exhibit better selectivity toward CO. However, since the 1.13 mg/cm^2 sample did not provided any improvement in selectivity with respect to 565 µg/cm^2, the latter was identified as optimal catalyst loading since it achieved an FE_{CO} value of 77%, namely, to the best of our knowledge, the lowest loaded silver-based GDE reported in the literature so far (Table 1). Most probably, the great amount of material deposited in sample E lowered the permeability of the GDE, inhibiting the mass transport of bicarbonate and affecting the selectivity. The performance in terms of CO_2 utilization followed a similar trend to the Faradaic efficiency: it was doubled by increasing the silver loading from 116 µg/cm^2 to 565 µg/cm^2, while, with sample E, the increase to 1.13 mg/cm^2 of silver loading did not further enhance the CO_2 utilization. J_{CO} and J_{tot} (Figure S1a) increased with higher loadings as confirmed by the EIS analysis (Table S3 and Figure S2). Indeed, the increasing trend of activities observed during bicarbonate electrolysis could be related to the value of the charge transfer resistance (R_{ct}). This parameter describes the catalyst's ability to exchange electrons with the reactants, applicable to both the CO_2RR and HER. R_{ct} decreases from 1.42 Ω cm^2 to 0.92 Ω cm^2 when augmenting the amount of silver from 116 µg/cm^2 to 565 µg/cm^2 (Figure 4b). A further decrease (0.38 Ω cm^2) was experienced with the highest loaded sample (E). Since the electrochemical surface area (ECSA) is considered to be proportionally associated to the double layer capacitance C_{dl} derivable from the EIS analysis (Table S3) [22], the intrinsic activity of various materials can be compared by investigating the C_{dl} normalized current densities (Figure S3) [23]. This investigation confirmed 565 µg/cm^2 as the optimal mass loading since it showed the highest C_{dl} normalized current density, hence the largest presence of active sites for the CO_2RR to CO. However, the higher mass activity obtained with sample D (87.1 mA/mg$_{Ag}$) compared to samples B and E highlights the excellent performance of this type of GDE even at very low mass-loading (Figure 4c).

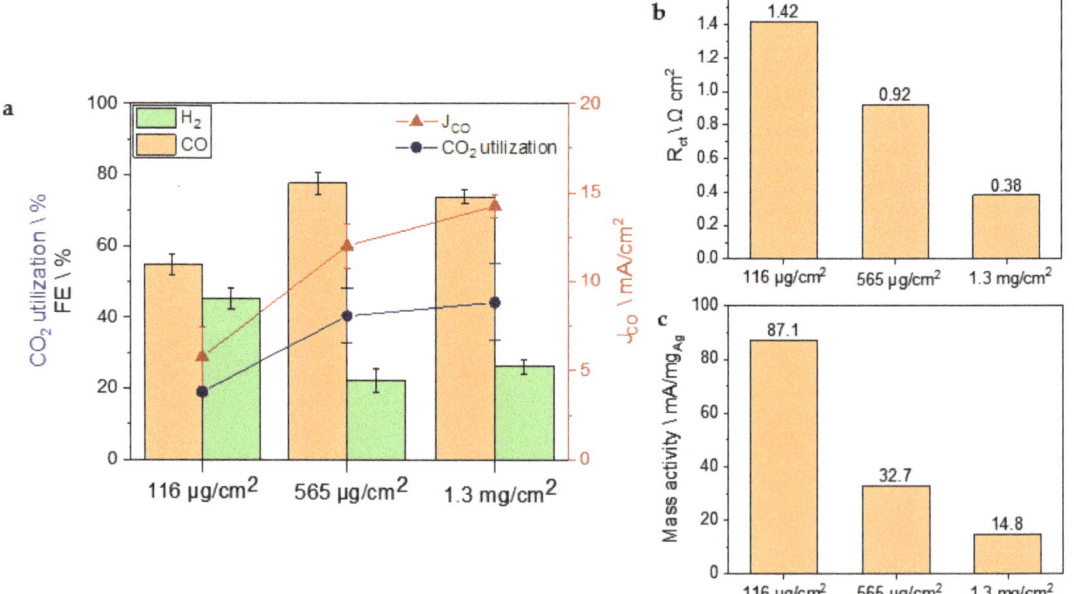

Figure 4. (a) FE, CO_2 utilization, and CO partial current density for different silver mass-loadings: 116 µg/cm^2 (sample D), 565 µg/cm^2 (sample B), and 1.13 mg/cm^2 (sample E). (b) Values of R_{ct} that emerged from the EIS analysis. (c) Values of mass activity for each sample.

Table 1. The reported state-of-the-art silver GDEs' performance for liquid-fed bicarbonate electrolyzers.

Ag Mass Loading (mg/cm^2)	Deposition Technique	Feedstock [KHCO$_3$ (M)]	FE$_{CO}$ (%)	Cell Potential (V)	J$_{CO}$ (mA/cm^2)	Partial Mass Activity (mA/mg$_{Ag}$)	Reference
13 *	Spray coating	3	80	3	20	2 *	[7]
2	PVD + spray coating	3	25	3.5	25	13	[8]
2	Spray coating	2	58	3	14	7	[14]
Foam **	Free standing electrode **	3	60	3.7	60	-	[15]
3	Electrodeposition	3	70	3.5	70	23	[16]
2	PVD + spray coating	3	82	3.6	82	41	[17]
0.565	PVD	2	77	3	13	25	This work
0.565	PVD	2	58	3	22	40	This work
0.116	PVD	2	55	3	6	48	This work

* This is the nominal loading; the experimental one was not reported by T. Li et al. [7]. ** Loading not present since a silver foam was used as a free-standing GDE.

As already mentioned, the structural characteristics of the carbon composite electrode are crucial in the determination of the catalytic behavior of the GDE in bicarbonate electrolyzers. In particular, the choice of the GDL is critical, as its permeability to the bicarbonate solution directly impacts the i-CO$_2$ production efficiency. It is known that the catalyst's selectivity toward CO tends to increase when the system is more proficient in producing i-CO$_2$ [24].

This observation was further confirmed by comparing the performance of the same GDE (sample B) using a less concentrated bicarbonate solution. When the concentration is halved from 2 M to 1 M, the carbon feedstock is poorer and the i-CO$_2$ generated drops. This introduces a mass transport limitation, causing a decrease in FE$_{CO}$ from 77% to 55%, while the CO$_2$ utilization drastically increased from 40% to 83%, as reported in Figure 5.

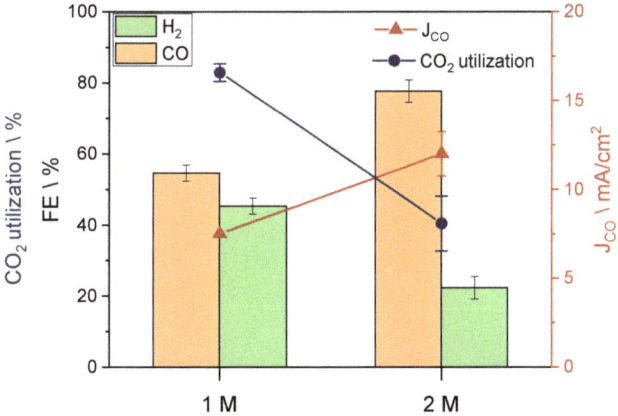

Figure 5. FE, CO$_2$ utilization, and CO partial current density when varying the concentration of the bicarbonate solution. GDE sample: B.

Increasing the permeability of the GDL would have a similar effect to using a higher electrolyte concentration. This improvement allows for enhanced flow of bicarbonate through the GDE, reaching the BPM, and consequently, the low-pH region becomes capable of producing a larger amount of i-CO$_2$. Therefore, sample B was compared, whilst keeping the same mass loading (565 μg/cm^2), to a GDE (sample F) whose GDL has a permeability that is four times higher. The effect of permeability is evident in Figure 6a, which shows the amount of CO$_2$ released inside the reactor as a function of the GDE's permeability. The graph presents the total i-CO$_2$ produced, which was calculated by summing the concentrations of CO$_2$ and CO detected at the electrolyzer outlet during electrolysis.

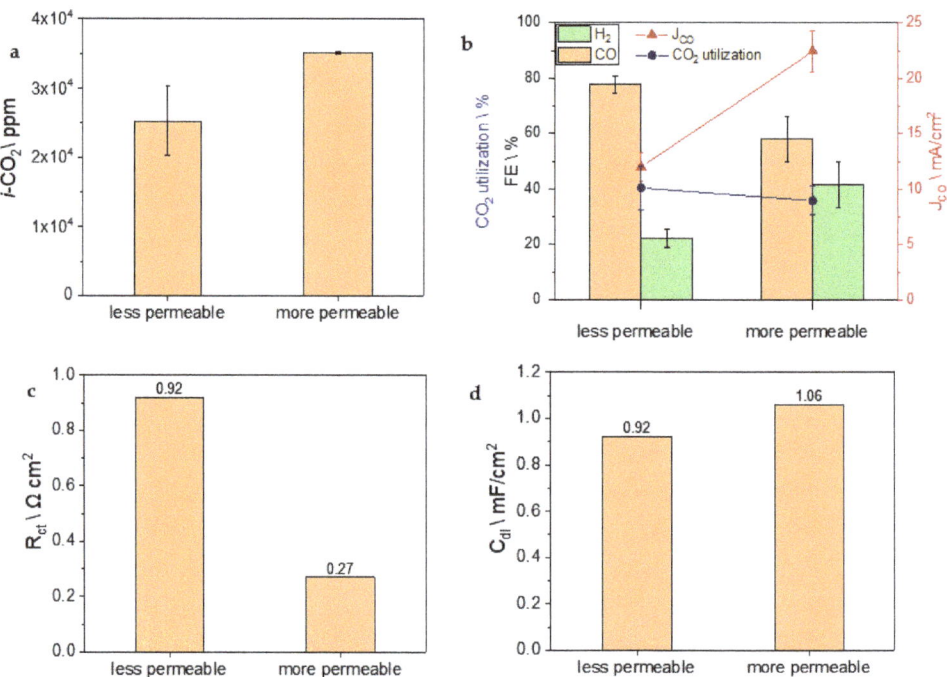

Figure 6. (**a**) Concentration of *i*-CO_2 produced by the electrolyzer using carbon supports with different permeability: less permeable (sample B) and more permeable (sample F). (**b**) FE, CO_2 utilization, and partial current density for CO. (**c**) Charge transfer resistance values and (**d**) double-layer capacitance values obtained by EIS analysis.

However, as shown in Figure 6b, the improvement in *i*-CO_2 production given by the high permeability of sample F did not provide an enhancement of selectivity; in fact, FE_{CO} dropped from 77% in sample B to 58%. Most probably, having a very open structure (see FESEM micrographs in Figure S4), which allows it to be more permeable to the bicarbonate, introduces a problem of mass transportation of *i*-CO_2 toward the active sites, affecting the FE_{CO}. The R_{ct} provided by the EIS analysis in sample F is around three times lower (Figure 6c), meaning that it includes a higher number of active sites for catalysis, either for CO_2RR or HER, as evidenced by the double-layer capacitance and displayed in Figure 6d. The slightly larger value of the C_{dl} of the most permeable GDE, 1.06 mF/cm^2 compared to 0.92 mF/cm^2, indicates a higher ECSA and confirms the presence of a larger number of active sites. The R_{ct} and C_{dl} values account for the high values of J_{tot} (Figure S1a), partial mass activity (Figure S1b), and J_{CO} observed in sample F, ensuring a good CO_2 utilization percentage even with a lower FE_{CO} and increased *i*-CO_2 production. In fact, the partial mass activity for CO was found to be 40 mA/mg$_{Ag}$ (Figure S1b), surpassing values reported in the literature (Table 1). It is important to emphasize that despite the lower Faradaic efficiency, the significantly high J_{CO} achieved, explained by the higher ECSA, makes sample F likely the most suitable GDE for industrial purposes in syngas production.

4. Conclusions

In this work, novel high-performance Ag electrodes for bicarbonate electrolyzers were fabricated via a simple and scalable sputtering method. Silver thin films were deposited on commercial carbon supports and used as free-standing gas diffusion electrodes without any post-treatment. Thanks to the highly repeatable deposition technique, GDEs with different carbon substrates and silver mass loadings were reliably tested at V_{cell} = 3 V to understand their effect in terms of activity and selectivity in CO_2-to-CO conversion. The final result of this investigation presents Ag-GDEs with a FE_{CO} close to 80%, which is comparable to the state-of-the-art achievement with a mass loading of 565 µg/cm^2 (sample B). This mass loading is significantly lower compared to the well-performing Ag-GDEs reported in the literature. Moreover, increasing the permeability of the carbon GDL significantly enhanced the activity and, consequently, the mass-activity. As a result, sample F exhibited remarkably high partial mass activity compared to the values reported in the literature for bicarbonate electrolyzers. The new Ag electrode reported respectable results in terms of CO_2 utilization, which turned out to be around 40%, while, when the bicarbonate concentration was halved to 1 M, it reached 83%. Additionally, our research marks a significant advancement in the field of GDE development for bicarbonate electrolyzers by introducing the application of electrochemical impedance spectroscopy. This innovative technique provided us with a valuable opportunity to delve deeper into the underlying factors that influenced the GDE's performance within the reactor. In fact, the charge transport resistances and the double-layer capacitances derived from the fitting of the experimental Nyquist plot provided an effective explanation for the different behaviors of the GDEs during bicarbonate electrolysis. Therefore, by using EIS as a powerful characterization tool for GDEs in bicarbonate electrolyzers, this work contributes to the growing body of knowledge in this emerging field of research.

The profound insights gained from this study offer a comprehensive understanding of the intricate electrochemical processes taking place within GDEs during bicarbonate electrolysis. Based on the results obtained by this work, the herein-proposed Ag GDEs demonstrate exceptionally promising potential for low-cost electrodes in the future industrial implementation of integrated carbon capture and conversion through bicarbonate electrolyzers.

Supplementary Materials: The following supporting information can be downloaded at: https://www.mdpi.com/article/10.3390/nano13162314/s1. Tables S1 and S2 on the features of commercial carbon papers and Ag-GDEs; Figure S1 reporting the total density current and the partial mass activity; details on the reactor set-up; formulas used to calculate Faradaic efficiencies, partial density current, i-CO_2, CO_2 utilization, mass activity, and partial mass activity; details on and results (Table S3 and Figure S2) of the EIS analysis; Figure S3 reports the values of J_{tot}/C_{dl}; Figure S4 displays the FESEM images of the GDEs. Ref. [25] is cited in the Supplementary Materials.

Author Contributions: Conceptualization, A.M., A.S. and J.Z.; methodology, A.M. and A.S.; validation, A.M. and A.S.; investigation, A.M., A.S., M.B. and A.C.; resources, C.F.P.; writing—original draft preparation, A.M.; writing—review and editing, A.M., A.S., M.B., A.C. and J.Z.; supervision, A.S. and C.F.P.; funding acquisition, C.F.P. All authors have read and agreed to the published version of the manuscript.

Funding: This research received no external funding.

Data Availability Statement: Data are available upon request.

Conflicts of Interest: The authors declare no conflict of interest.

References

1. Pardal, T.; Messias, S.; Sousa, M.; Machado, A.S.R.; Rangel, C.M.; Nunes, D.; Pinto, J.V.; Martins, R.; da Ponte, M.N. Syngas production by electrochemical CO_2 reduction in an ionic liquid based-electrolyte. *J. CO2 Util.* **2017**, *18*, 62–72. [CrossRef]
2. Lees, E.W.; Mowbray, B.A.W.; Parlane, F.G.L.; Berlinguette, C.P. Gas diffusion electrodes and membranes for CO_2 reduction electrolysers. *Nat. Rev. Mater.* **2021**, *7*, 55–64. [CrossRef]
3. Peng, Y.; Zhao, B.; Li, L. Advance in Post-Combustion CO_2 Capture with Alkaline Solution: A Brief Review. *Energy Procedia* **2012**, *14*, 1515–1522. [CrossRef]
4. Iizuka, A.; Hashimoto, K.; Nagasawa, H.; Kumagai, K.; Yanagisawa, Y.; Yamasaki, A. Carbon dioxide recovery from carbonate solutions using bipolar membrane electrodialysis. *Sep. Purif. Technol.* **2012**, *101*, 49–59. [CrossRef]
5. Mezza, A.; Pettigiani, A.; Monti, N.B.D.; Bocchini, S.; Farkhondehfal, M.A.; Zeng, J.; Chiodoni, A.; Pirri, C.F.; Sacco, A. An Electrochemical Platform for the Carbon Dioxide Capture and Conversion to Syngas. *Energies* **2021**, *14*, 7869. [CrossRef]
6. Dinh, C.-T.; Li, Y.C.; Sargent, E.H. Boosting the Single-Pass Conversion for Renewable Chemical Electrosynthesis. *Joule* **2018**, *3*, 13–15. [CrossRef]
7. Li, T.; Lees, E.W.; Goldman, M.; Salvatore, D.A.; Weekes, D.M.; Berlinguette, C.P. Electrolytic Conversion of Bicarbonate into CO in a Flow Cell. *Joule* **2019**, *3*, 1487–1497. [CrossRef]

8. Li, Y.C.; Lee, G.; Yuan, T.; Wang, Y.; Nam, D.-H.; Wang, Z.; de Arquer, F.P.G.; Lum, Y.; Dinh, C.T.; Voznyy, O.; et al. CO_2 Electroreduction from Carbonate Electrolyte. *ACS Energy Lett.* **2019**, *4*, 1427–1431. [CrossRef]
9. Agliuzza, M.; Mezza, A.; Sacco, A. Solar-driven integrated carbon capture and utilization: Coupling CO_2 electroreduction toward CO with capture or photovoltaic systems. *Appl. Energy* **2023**, *334*, 120649. [CrossRef]
10. McDonald, M.B.; Ardo, S.; Lewis, N.S.; Freund, M.S. Use of Bipolar Membranes for Maintaining Steady-State pH Gradients in Membrane-Supported, Solar-Driven Water Splitting. *ChemSusChem* **2014**, *7*, 3021–3027. [CrossRef] [PubMed]
11. Li, Y.C.; Yan, Z.; Hitt, J.; Wycisk, R.; Pintauro, P.N.; Mallouk, T.E. Bipolar Membranes Inhibit Product Crossover in CO_2 Electrolysis Cells. *Adv. Sustain. Syst.* **2018**, *2*, 1700187. [CrossRef]
12. Welch, L.M.; Vijayaraghavan, M.; Greenwell, F.; Satherley, J.; Cowan, A.J. Electrochemical carbon dioxide reduction in ionic liquids at high pressure. *Faraday Discuss.* **2021**, *230*, 331–343. [CrossRef] [PubMed]
13. Park, G.; Hong, S.; Choi, M.; Lee, S.; Lee, J. Au on highly hydrophobic carbon substrate for improved selective CO production from CO_2 in gas-phase electrolytic cell. *Catal. Today* **2019**, *355*, 340–346. [CrossRef]
14. Larrea, C.; Torres, D.; Avilés-Moreno, J.R.; Ocón, P. Multi-parameter study of CO_2 electrochemical reduction from concentrated bicarbonate feed. *J. CO2 Util.* **2022**, *57*, 101878. [CrossRef]
15. Zhang, Z.; Lees, E.W.; Habibzadeh, F.; Salvatore, D.A.; Ren, S.; Simpson, G.L.; Wheeler, D.G.; Liu, A.; Berlinguette, C.P. Porous metal electrodes enable efficient electrolysis of carbon capture solutions. *Energy Environ. Sci.* **2022**, *15*, 705–713. [CrossRef]
16. Kim, Y.; Lees, E.W.; Berlinguette, C.P. Permeability Matters When Reducing CO_2 in an Electrochemical Flow Cell. *ACS Energy Lett.* **2022**, *7*, 2382–2387. [CrossRef]
17. Lees, E.W.; Goldman, M.; Fink, A.G.; Dvorak, D.J.; Salvatore, D.A.; Zhang, Z.; Loo, N.W.X.; Berlinguette, C.P. Electrodes Designed for Converting Bicarbonate into CO. *ACS Energy Lett.* **2020**, *5*, 2165–2173. [CrossRef]
18. Monti, N.B.D.; Fontana, M.; Sacco, A.; Chiodoni, A.; Lamberti, A.; Pirri, C.F.; Zeng, J. Facile Fabrication of Ag Electrodes for CO_2-to-CO Conversion with Near-Unity Selectivity and High Mass Activity. *ACS Appl. Energy Mater.* **2022**, *5*, 14779–14788. [CrossRef]
19. Sacco, A. Electrochemical impedance spectroscopy as a tool to investigate the electroreduction of carbon dioxide: A short review. *J. CO2 Util.* **2018**, *27*, 22–31. [CrossRef]
20. Zhao, H.-Z.; Chang, Y.-Y.; Liu, C. Electrodes modified with iron porphyrin and carbon nanotubes: Application to CO_2 reduction and mechanism of synergistic electrocatalysis. *J. Solid State Electrochem.* **2013**, *17*, 1657–1664. [CrossRef]
21. Kleiminger, L.; Li, T.; Li, K.; Kelsall, G.H. Syngas ($CO-H_2$) production using high temperature micro-tubular solid oxide electrolysers. *Electrochim. Acta* **2015**, *179*, 565–577. [CrossRef]
22. Zeng, J.; Castellino, M.; Bejtka, K.; Sacco, A.; Di Martino, G.; Farkhondehfal, M.A.; Chiodoni, A.; Hernández, S.; Pirri, C.F. Facile synthesis of cubic cuprous oxide for electrochemical reduction of carbon dioxide. *J. Mater. Sci.* **2020**, *56*, 1255–1271. [CrossRef]
23. Zeng, J.; Castellino, M.; Fontana, M.; Sacco, A.; Monti, N.B.D.; Chiodoni, A.; Pirri, C.F. Electrochemical Reduction of CO_2 with Good Efficiency on a Nanostructured Cu-Al Catalyst. *Front. Chem.* **2022**, *10*, 931767. [CrossRef] [PubMed]
24. Gutiérrez-Sánchez, O.; de Mot, B.; Bulut, M.; Pant, D.; Breugelmans, T. Engineering Aspects for the Design of a Bicarbonate Zero-Gap Flow Electrolyzer for the Conversion of CO_2 to Formate. *ACS Appl. Mater. Interfaces* **2022**, *14*, 30760–30771. [CrossRef] [PubMed]
25. Zeng, J.; Rino, T.; Bejtka, K.; Castellino, M.; Sacco, A.; Farkhondehfal, M.A.; Chiodoni, A.; Drago, F.; Pirri, P.C.F. Coupled Copper–Zinc Catalysts for Electrochemical Reduction of Carbon Dioxide. *ChemSusChem* **2020**, *13*, 4128–4139. [CrossRef]

Disclaimer/Publisher's Note: The statements, opinions and data contained in all publications are solely those of the individual author(s) and contributor(s) and not of MDPI and/or the editor(s). MDPI and/or the editor(s) disclaim responsibility for any injury to people or property resulting from any ideas, methods, instructions or products referred to in the content.

Review

Recent Progress in Multifunctional Graphene-Based Nanocomposites for Photocatalysis and Electrocatalysis Application

Zanhe Yang [1], Siqi Zhou [1], Xiangyu Feng [1], Nannan Wang [1,*], Oluwafunmilola Ola [2,*] and Yanqiu Zhu [1]

[1] State Key Laboratory of Featured Metal Materials and Life-Cycle Safety for Composite Structures, School of Resources, Environment and Materials, Guangxi University, Nanning 530004, China; 2039200206@st.gxu.edu.cn (Z.Y.); 2039200234@st.gxu.edu.cn (S.Z.); 2139200335@st.gxu.edu.cn (X.F.); y.zhu@exeter.ac.uk (Y.Z.)

[2] Advanced Materials Group, Faculty of Engineering, The University of Nottingham, Nottingham NG7 2RD, UK

* Correspondence: wangnannan@gxu.edu.cn (N.W.); oluwafunmilola.ola1@nottingham.ac.uk (O.O.)

Abstract: The global energy shortage and environmental degradation are two major issues of concern in today's society. The production of renewable energy and the treatment of pollutants are currently the mainstream research directions in the field of photocatalysis. In addition, over the last decade or so, graphene (GR) has been widely used in photocatalysis due to its unique physical and chemical properties, such as its large light-absorption range, high adsorption capacity, large specific surface area, and excellent electronic conductivity. Here, we first introduce the unique properties of graphene, such as its high specific surface area, chemical stability, etc. Then, the basic principles of photocatalytic hydrolysis, pollutant degradation, and the photocatalytic reduction of CO_2 are summarized. We then give an overview of the optimization strategies for graphene-based photocatalysis and the latest advances in its application. Finally, we present challenges and perspectives for graphene-based applications in this field in light of recent developments.

Keywords: graphene composites; photocatalysis; electrocatalysis; pollutant degradation; CO_2 fixation

1. Introduction

With the acceleration of urbanization, mankind in the 21st century is facing the two problems of global warming and energy shortage. Among the abundant gases in the atmosphere, CO_2 is the most important factor determining the trend of the greenhouse effect. The excessive development of fossil energy causes energy shortage and also promotes the excessive emission of CO_2. Various environmental pollution problems around the world are becoming more and more serious [1]. There are many studies on the reduction of CO_2, but only a few semiconductors meet the reduction requirements for CO_2 under sunlight irradiation [2–5]. The reason is that effective charge separation is crucial for the photocatalytic reduction of CO_2. A large number of studies have found that the composite material formed by using graphene-doped semiconductors in photocatalysis significantly improves the effect of the photocatalytic reaction and greatly improves the reduction efficiency [6–10]. In particular, graphene can generate specific fuels in the photocatalytic reduction of CO_2 through its unique electron collection and transfer capabilities [11,12].

Water resources are important survival resources. According to statistics, human beings can drink only 3% of the global water resources. With the increase in the world's population, there is a shortage of water resources in some parts of the world. More seriously, the water environment has also deteriorated, making the water resources available for human use significantly reduced [13]. The deterioration of water resources usually comes from domestic sewage, industrial wastewater, mining wastewater, etc. This sewage contains

organic refractory substances, heavy metal ions, etc., which will cause damage to the human body upon drinking and also cause certain damage to the ecological environment [14]. In view of the shortage of water resources and serious pollution [15–19], low-cost and efficient sewage purification methods are of great significance to human beings [20–23]. Sewage purification can be performed via physical or chemical methods, including filtration, adsorption, boiling, distillation, chlorination, electromagnetic radiation, etc.; however, these methods still contain obvious drawbacks [13,24]. For example, the chlorination of purified water will produce many carcinogens during the process, and the hypochlorite of purified water has strong corrosivity, which further increases its treatment cost [25–27].

In recent decades, the use of renewable solar energy technology to remove pollutants has gradually become a research trend [28–30]. This is because photocatalysis can produce reactive oxygen species and sterilize and destroy various organic and inorganic pollutants; furthermore, it is cheap, which also shows that photocatalysis is an important research direction for water purification [31]. However, considering the cost and photocatalytic effect, doping graphene is a good method [32–34]. Graphene is a single-layer two-dimensional structure formed by the arrangement of carbon atoms. It has a high surface area, optical transmittance, and chemical stability [35]. It is not only an ideal choice for pollutant treatment applications, but because of its biocompatibility and efficient use of photocatalytic efficiency, it has also become a leading candidate in the field of photocatalysis [35]. The structure of graphene is formed by arranging carbon atoms one by one. The internal carbon atoms are bonded by "sp^2" hybrid orbitals. The normal "pz" orbital of the layer plane of the cabinet carbon atom can form a large π bond throughout the whole layer. There is a strong "s" bond between the two carbon atoms, which greatly enhances its structural strength and flexibility. Based on this hybridization, the carbon atoms connected by graphene "sp^2" hybridization can be tightly stacked into a single-layer two-dimensional honeycomb lattice structure, with a high specific surface area and good electronic properties [36]. At the same time, the unique band structure also makes graphene have good conductivity and electron mobility [37]. Graphene has a variety of application characteristics. In the field of multifunctional components, it can be used in the tunable terahertz filter/antenna-sensor of graphene metamaterials. The center frequency of the filter and antenna sensor can be adjusted by changing the chemical formula of graphene, and the thickness of the graphene-layered material can be increased to increase the depth of the enhanced resonance effect [38,39].

On the other hand, in the application of three-dimensional absorbers, three-dimensional metamaterial curved ultra-wideband absorbers can exhibit better microwave absorption properties similar to those of reduced graphene oxide (rGO)/MWNTs hybrids [40–43]. Borah et al. prepared the expanded graphite non-metallic flexible metamaterial absorber by using the linear low-density polyethylene as the substrate material to composite the expanded graphite (EG). Compared with copper-based absorbers, the resonance frequencies of expanded graphite-based absorbers are similar (EG = 11.56 GHz, Cu = 11.73 GHz), but the reflection loss of expanded graphite-based absorbers is much lower than that of copper-based absorbers (EG = −24.51 dB, Cu = −7.32 dB) [44]. The EG material also has the characteristics of heat resistance, low thermal expansion, and oxidation resistance, and it has certain application potential in the field of electromagnetic wave absorbers [45,46]. Singhal studied an ultra-wideband infrared absorption device with an absorption rate of more than 90% and an operating bandwidth of more than 74 THz from 6 THz. This device uses a dielectric material such as graphite as a substrate composite of SiO_2. Due to the temperature stability of graphene [47], such devices can be well applied in the terahertz and infrared spectral bands in the future [48]. Norouzi et al. proposed a low-cost, simple, and efficient 3D metamaterial ultra-wideband absorber that is insensitive to the incident angle of 60° in the TE mode (Transverse Electric, no electric field in the direction of propagation) and 90° in the TM mode (Transverse Magnetic, no magnetic field in the direction of propagation), and that is not affected by the polarization of the incident wave. This shows that the device is not sensitive to other electromagnetic wave segments under the initial receiving condition setting, and that it has strong anti-interference or shielding ability [49]. The flexible structural characteristics and broadband-absorption capacity, low thermal-expansion coefficient and antioxidant functional characteristics of graphene-based materials confer them great development potential and application prospects in the fields of medical treatment, imaging, and microwave absorption in the future [50–55].

In addition to the important application of graphene 3D metamaterials in the field of microwave absorption, some graphene aerogel materials also have outstanding performance in electromagnetic wave absorption [56–59]. Wu et al. dispersed graphene oxide (GO) uniformly in the chain formed by a polypyridine gel and reduced GO to rGO by the hydrothermal method. After the purification and drying process, using a large amount of distilled water and ethanol, a sponge-like polypyrrole (S-PPy)/rGO aerogel was obtained. This material not only has the lightweight properties of aerogel materials but also has low reflection loss (−54.4 dB at 12.76 GHz) [60]. The graphene@SiC aerogel composites studied by Jiang et al. also have the characteristics of low reflection loss (−47.3 dB at 10.52 GHz) [61]. The special structure of the aerogel material endows it with compressible characteristics, and it shows obvious electromagnetic sensitivity and strong adsorption performance under certain conditions [62–65]. Wang et al. prepared macroscopic 3D-independent porous all-graphene aerogels with ultra-light density and high compressibility by an in situ self-assembly and thermal annealing process. The ice crystal growth, GO reduction and the restoration of π-conjugation during the freeze-drying process will give the material a 3D structure, thereby obtaining good recoverable compressibility and a strain level of up to 75%, which gives the aerogel a highly sensitive strain response characteristic in volume conductivity. At the same time, the high-temperature-stable graphene composition and large porosity aerogel structure can quickly remove heat during the combustion process, reflecting a certain flame retardancy [66]. Li et al. made full use of the adsorption of aerogels and the hydrophobicity of graphene sheets to synthesize hydrophobic aerogels with high porosity, which can absorb different organic liquids or be used to separate and absorb organic pollutants from water [67]. Hong et al. also studied the selective adsorption of aerogels. On the basis of non-functionalized graphene aerogels with high porosity and hydrophobicity after surface modification, they introduced fluorinated functional groups into the surface of three-dimensional macroporous graphene aerogels by a one-step immersion method to obtain functionalized rGO (F-rGO). This material has the physical properties of low density (bulk density of 14.4 mg cm^{-3}), high porosity (>87%), mechanical stability (supports at least 2600 times its own weight), and hydrophobicity (contact angle of 144°). At the same time, the team tested the absorption capacity of F-rGO aerogels for various oils and organic solvents such as pump oil, chlorobenzene, tetrahydrofuran, and acetone. The results show that F-rGO aerogel has excellent adsorption efficiency for various oils and organic solvents, and the adsorption capacity is 34~112 times its weight. The absorption capacity depends on the density of organic solvents. Pre-introduction of fluorinated functional groups can be used to selectively remove oil or take away more O atoms [68]. There are also many examples of supercapacitor applications [69–71]. A graphene-based nitrogen self-doped hierarchical porous carbon aerogel was synthesized by Hao et al. using chitosan as the raw material through a carefully controlled aerogel formation–carbonization–activation process. The specific capacitance calculated from charge–discharge measurements using an all-solid-state symmetric supercapacitor was about 197 F g^{-1} at 0.2 A g^{-1} with an excellent capacitance retention of ~92.1% over 10,000 cycles. The energy density reached as high as 27.4 W h kg^{-1} at a power density of 0.4 kW kg^{-1} and 15 W h kg^{-1} at a power density of 20 kW kg^{-1} [72]. In terms of photocatalysis, of the many metal-based materials, an organic-semiconductor has certain applications [73,74], with a long service life and high water decomposition efficiency, but its effect and economy can still be improved. The doping of graphene and graphene-based materials provide a good basis for forming heterojunctions. The doping of graphene brings its own high specific surface area, high carrier migration speed, high conductivity and other characteristics to the doped composite materials, which can effectively solve the problems of material environmental protection and the rapid recombination of photogenerated electron holes and significantly improve the overall efficiency and resource utilization of photocatalysis [75,76]. With regard to the selection of dopants, these are generally divided into metal and nonmetal. Metal can act as an 'electronic warehouse' for releasing electrons, fix on some sites of graphene to improve the photocatalytic effect [77,78], or fix on graphene to provide appropriate adsorption and activation to improve the activity of the whole material [79]. Metal doping is quite a good idea, but because the price of metal materials is slightly expensive and some metals are not friendly to the environment, the development of metal-anchored graphene composites is still challenging. Non-metallic doping, single doping or multi-doping have significantly improved the trend of photocatalysis, and can also achieve the same effect as metal doping [80]. N and S doping is quite typical of non-metallic doping; when these non-metallic material co-catalysts are anchored in the active site of graphene, they can limit some metal-atom couplings. In terms of dimensional stacking, the extension of linear one-dimensional conjugated polymers to two- or three-dimensional polymers can significantly enhance exciton dissociation, effectively producing free electrons and abundant reaction sites.

This paper reviews the mechanism of the photocatalytic production of various products and also shows the optimization and promotion of graphene in photocatalysis, as well as the latest research progress of graphene in this field. In the following chapters, we will focus on the effect and mechanism of graphene doping and its different dimensional applications on the improvement of the photocatalytic effect, including photocatalytic water splitting to produce hydrogen, the reduction of carbon dioxide, the degradation of pollutants, etc. Finally, we try to provide an understanding of the current progress, future trends, and challenges of graphene photocatalysis.

2. Photocatalytic Water Splitting for Hydrogen Production and Electrocatalysis

In the context of energy shortage and environmental degradation, hydrogen, as a clean energy that can be stored in large quantities, is considered to be the main carrier of future energy and has attracted more and more attention in various fields [81–86]. In 1972, the University of Tokyo in Japan used the n-type TiO_2 semiconductor as an anode and Pt as a cathode to produce a solar photoelectrochemical cell, which realized the photodecomposition of water to produce hydrogen [87]. According to the characteristics of semiconductor light-excitation electron transition and photocatalytic reaction, researchers have developed a device for hydrogen production by the photocomplexation catalytic decomposition of water. Through a series of coupling processes, that is, artificially simulating the process of water decomposition by photosynthesis, the efficiency of hydrogen production is generally 6% [88–91]. In order to use hydrogen energy to solve the problem of energy shortage or to provide better commercial benefits, technical breakthroughs and more effective materials are needed. At present, graphene-based materials are the better choice, because graphene can be used as a transfer carrier of electrons, reducing the requirement for electrons to pass through the valence band and significantly enhancing the photocatalytic reaction effect [92]. In the next chapter, the mechanism of the photocatalytic reaction and the beneficial optimization strategy of graphene-based photocatalysis will be introduced, mainly from the dimension structure of doping and graphene.

2.1. Mechanism

According to the light quantum theory, when the semiconductor receives a light quantum energy $h\upsilon$ higher than the band gap, the photoelectron overcomes the escape work and escapes, producing free electrons in the conduction band and positive holes in the valence band (Figure 1) [93]. The electron produced in this process exhibits a strong reducing ability in the electron-donor reaction, and the produced holes exhibit an oxidizing ability in the electron-acceptor reaction. In the reduction process, the electron can undergo a molecular oxygen reaction with superoxide anions, and in the oxidation process, electrons in water or hydroxyl ions can be supplemented into holes to produce hydroxyl radicals. The superoxide anions and hydroxyl radicals produced in these two processes can degrade organic matter, and microorganisms and bacteria will be eliminated by OH⁻ and $O_2^{·-}$ to complete the entire photocatalytic process [94–96]. The overall water decomposition reaction and the complete water decomposition reaction are as follows.

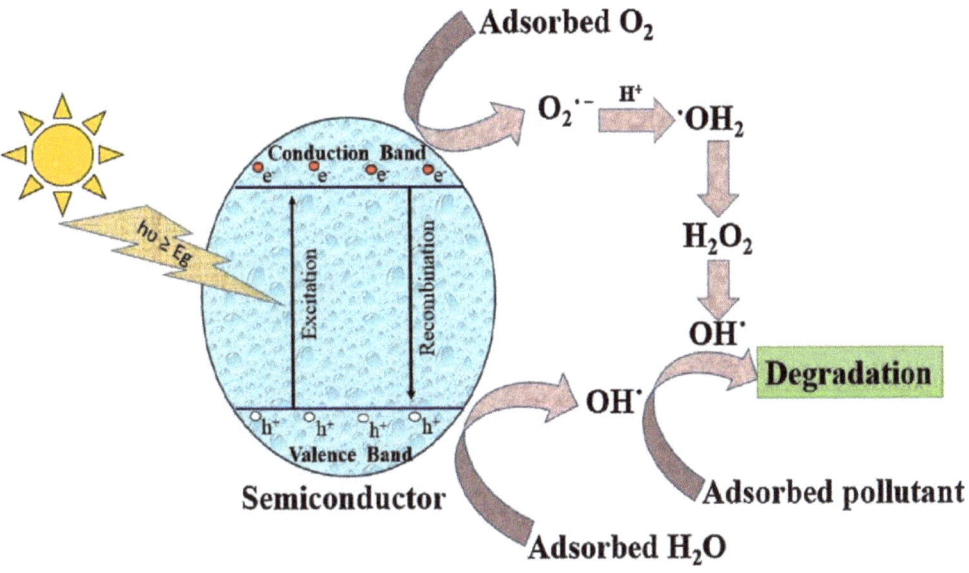

Figure 1. Semiconductor−based photocatalysis perspective. Reprinted with permission from Ref. [93]. Copyright©2018 Production and hosting by Elsevier B.V. on behalf of King Saud University.

Overall water decomposition reaction:

$$H_2O \xrightarrow{hv} \frac{1}{2}O_2 + 2H^+ \tag{1}$$

$$2H^+ + 2e^- \rightarrow H_2 + \frac{1}{2}O_2 \tag{2}$$

Complete water decomposition reaction:

$$2H_2O \xrightarrow{hv} O_2 + 4H^+ \tag{3}$$

$$4H^+ + 4e^- \rightarrow 2H_2 + O_2 \tag{4}$$

During the catalytic splitting of water into H_2, the CO_2 that is generally reduced to CO can also be reduced to methane or ethane using special materials [97].

The photocatalytic reaction of metal oxides contains the above general photocatalytic reaction mechanism. The difference is that metal oxide photocatalysis mainly involves a catalyst containing transition metals. The whole cycle includes four main steps: oxidation, reduction, addition, and removal. The main mechanism is that the d-orbital electrons in the transition metal has the characteristics of easy separation or addition, which makes it easier to carry out redox reactions. At the same time, the requirement of d-orbital bonding and bond-angle matching is low and the bonding energy is not high, which is beneficial for the formation of low-orbital bonding of reactants. In some of the transition metal catalytic systems, after the introduction of platinum, palladium, rhodium, silver, and other precious metals, these transition metals can be used as photon acceptors to produce photoelectrons, which is conducive to the formation of active compounds, quantum light, and effective hole separation [98,99].

2.2. Graphene Optimization Method for the Photocatalytic Effect

2.2.1. Graphene Matrix Composites and Graphene-Doped-Metal Matrix Composite

Graphene can be doped with B, Se, N, P, O, S, F, Cl, Br, I, etc. [100–107]. Titanium-based materials are widely used in purification and disinfection, building materials, and agriculture because of their non-toxicity, high reaction efficiency, and economy. They were considered to be the frontier materials in the field of photocatalysis in the previous stage [107,108]. TiO_2-composite graphene nanomaterials have been proved to have better photoelectrochemical activity and photocatalytic activity [109–112]. The reason is that GO can fully act as an electron collection library to accept excited electrons from TiO_2. At the same time, a Ti–O–C bond is formed that introduces the TiO_2 bandgap intermediate state to promote a visible light reaction. Unfortunately, TiO_2 also has the limitation of a light wave frequency. It is only suitable for the ultraviolet band, while ultraviolet only accounts for 4% of the solar spectrum, and a large number of electromagnetic waves in the band cannot be used [113]. In order to further develop photocatalytic energy efficiency and maximize the use of solar clean energy, various types of TiO_2 photocatalytic derivatives of new materials have emerged ceaselessly and have been developed into a variety of composite types of metal oxides, organic photocatalytic semiconductors, etc. An organic photocatalytic semiconductor is a good alternative material. Its efficient photoelectric conversion efficiency and longer service life have a profound impact on solving the increasingly tense resource and environmental crisis and energy shortage. However, an unsatisfactory aspect is that there are fewer active sites in the material, the two carriers excited by the energy given by the light have a short interval, and the overall photocatalytic efficiency is not high, which make the development of this material still have great limitations [114,115].

In addition to metal-based materials, graphene-based composite conductive materials are also promising research directions. The main problem of this graphene-based material is that the free electrons and holes diffuse from the conduction band and valence band and then the free electrons and holes recombine quickly, losing the conductivity. This phenomenon affects photochemical stability and ultimately leads to a decrease in photocatalytic efficiency. The macroscopic performance shows that the material has high resistance. At present, there are a large number of strategies to solve the problem of the recombination of free electrons and holes excited by the optical quantum [116–119]. Now the more efficient solution for photocatalysis is based on the doping of graphene. The main reason for doping is that doping other elements can improve the absorption electromagnetic wave region of photocatalysis, change the graphene semiconductor from a broadband gap to a narrow band gap, reduce the bandgap energy, improve the photochemical stability, work efficiently in the absorption band, and further improve the photocatalytic efficiency [120]. Due to the special layered structure of graphene, it can carry and immobilize other materials, and it has obvious composite material characteristics, making the composite material composed of graphene and various semiconductors a new generation of more promising photocatalytic materials [121,122].

In addition to the above mentioned, graphene can be used as a photocatalytic material substrate and as a co-catalyst. In order to improve the photocatalytic effect of graphene materials, doped N has particularly prominent derivatives [80,123]. At the same time, nitrogen-doped (N-doped) graphene has been shown to be a high-performance co-catalyst that can effectively increase hydrogen production [124]. Liu et al. anchored Ni onto an N-doped graphene (NG) material as a co-catalyst for SrTiO(Al), effectively carrying the Ni-based material to improve the photocatalytic water splitting effect. In this experiment, the N on graphene can be used as the anchor for a single metal atom, and the atomic coupling of the single metal atom, Ni, is limited [125]. In this way, the effect of precious metals as co-catalysts can be achieved without the high-cost limits of precious metals. Liu's team provided a good direction for photocatalytic water splitting, and in the past two years, Yang and Li's team, advancing further in this direction, have proposed an N and S co-doped approach to obtain a nitrogen and sulfur co-doped graphene that has a better effect than single-doped N [126,127].

2.2.2. Structure Design

Graphene's ability to function as a photocatalytic substrate material and auxiliary material has a lot to do with its structure [128]. It is well known that monolayer nanomaterials are classified as two-dimensional structures, with graphene being the most common. Multilayer graphene stacked by chemical bonds is classified as a three-dimensional structure.

The enhanced catalytic effect of high-dimensional graphene can be attributed to the fact that, compared with one-dimensional graphene, two-dimensional and three-dimensional graphene can reduce the Coulomb binding energy of electron holes and increase the specific surface area to solve the shortcomings of the loose substrate contact and fewer active sites on graphene. At the same time, fixing metals at specific sites can also avoid the aggregation of metal oxides [129]. In particular, graphene materials with a three-dimensional porous structure have the synergistic effect of light absorption and size-stacking to form a higher specific surface area, which is conducive to multiple reflections of light and improves the light capture ability. They are also widely used in capacitors [130,131].

Graphitic Carbon Nitride (g-C_3N_4) is a two-dimensional material that has been intensively studied in ozone oxidation and in composites with rhodium phosphide [132]. Han reported a porous g-C_3N_4 material with heterostructural defects. They observed enhanced water-splitting hydrogen-production efficiency and photocatalytic H_2-evolution activity by the heterojunction and the construction of defective ultrathin two-dimensional materials, respectively [133]. Liu et al. improved the structure of g-C_3N_4 and divided it into three groups: g-$C_{2.52}N_4$, g-$C_{1.95}N_4$ and g-$C_{1.85}N_4$. These three groups were also doped with Ag–Pd to form a two-dimensional composite material that strongly promoted photocatalysis. They controlled the ratio of carbon and nitrogen, and the removal rate of NO_2^- and NO_3^- was used as a reference for the degree of photocatalytic reaction. The XRD pattern showed that g-$C_{1.95}N_4$ had the best photocatalytic activity. After 3 h of reaction, the removal rate of NO_3^- was 87.4%, and the removal rate of NO_2^- was 61.8% [134]. This shows that by changing or utilizing the material structure of traditional photocatalytic materials, the catalytic hydrogen evolution performance can be significantly improved. At the same time, two-dimensional graphene semiconductor materials have lamellar stacking, and the intrinsic Dirac band structure and performance are not fully utilized. This is an obvious disadvantage of two-dimensional graphene, which is not mentioned in the above articles.

The photocatalytic reaction of TiO_2 composite photocatalytic materials is a good means to reduce the greenhouse effect gas, CO_2. The efficiency of the photocatalytic conversion of CO_2 into available chemical fuels mainly depends on the adsorption and diffusion of CO_2 by the material. Wang et al. developed a porous composite structure by the in situ weaving of hyper-crosslinked polymers (HCPs) on TiO_2-functionalized graphene (TiO_2-FG) without adding precious metal co-catalysts. The porous HCP–TiO_2 graphene composite structure was used as a photocatalytic material to achieve a CO_2 absorption capacity of 12.87 wt% and a CH_4 yield of 27.62 µmol g^{-1} h^{-1}. The principle of effective CO_2 absorption in this process is to introduce the material into the HCP layer (Figure 2). This method can increase the micropore volume and significantly increase the specific surface area of TiO_2 graphene, thereby improving the overall energy efficiency of photocatalysis. This is a good example of a porous material application that offers a research direction for solar fuels [135].

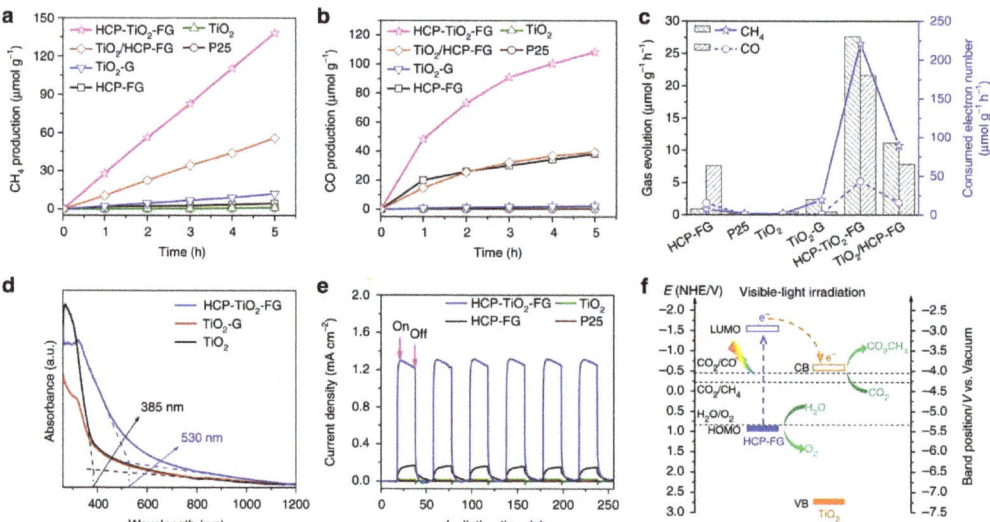

Figure 2. The photocatalytic performance of CO_2 reduction, optical and photoelectrical properties, and mechanism of the charge transfer pathway. Time-dependent production of (**a**) CH_4 and (**b**) CO in photocatalytic CO_2 reduction with different catalysts under visible light ($\lambda \geq 420$ nm). The photocatalytic reactions were carried out in a batch system under standard atmospheric pressure. The partial pressure of CO_2 and H_2O were constant, with the water content below the scaffold-loading photocatalyst. Under visible-light irradiation, the temperature of the water was measured to be about 50 °C. (**c**) Average efficiency of photocatalytic CO_2 conversion with different catalysts during 5 h of visible-light ($\lambda \geq 420$ nm) irradiation. (**d**) UV–Vis absorption spectra of TiO_2, TiO_2-G, and HCP–TiO_2-FG catalysts. (**e**) Amperometric I–t curves of samples under visible-light ($\lambda \geq 420$ nm) irradiation. (**f**) Proposed mechanism of charge separation and transfer within the HCP–TiO_2-FG composite photocatalyst under visible-light ($\lambda \geq 420$ nm) irradiation. Reprinted with permission from Ref. [135]. Copyright©2023 Springer Nature Limited.

Lee et al. provided a heterostructural treatment of graphite carbon three-dimensional nanomaterials. The team used hydrothermal treatment and the simultaneous reduction of GO and TiO_2 crystals, and the absorption tail of GO-coated amorphous TiO_2 was red-shifted (2.80 eV) [136]. Through hydrothermal treatment, during the conversion of TiO_2 into amorphous TiO_2, Ti atoms are rearranged, GO is reduced, and unpaired electrons are formed. The unpaired electrons and Ti atoms reduce the conduction-band level of TiO_2, thereby reducing the band gap and improving the conductivity and photocatalytic effect. Qiu's team provided a new idea for the treatment of graphene heterostructures by self-doping Ti^{3+} to place GO and TiO_2 on the surface of graphene. GO-cracking into small size graphene; the smaller the size of the graphene, the more binding surface with TiO_2 there is in the structure to promote the efficiency of photocatalysis. Moreover, this process also generates Ti–O–C bonds to form shallow surface defects, and the band gap is slightly reduced and red-shifted to 2.98 eV [137].

Many functional catalysts are used to enhance two-dimensional graphene phase carbonitrides [138–142]. The use of functional catalysts on three-dimensional porous graphene phase carbonitrides (g-C_3N_4/GO (p-CNG)) is also a good idea. Li's team fixed Au, Pd, Pt, and other precious metals on the three-dimensional p-CNG skeleton. By monitoring the positive correlation between the metal content and the diffraction intensity of the three-dimensional porous-graphene-phase carbon nitride skeleton, it was determined that the three metals were successfully anchored on the three-dimensional porous-graphene-phase carbon nitride body. Subsequently, the performance, photostability and thermoelectrochemical performance of the hydrogen evolution reaction (HER) were tested. Because precious metals can be used as electron acceptor sites, the effect of electron-hole separation is significantly improved, and it has strong HER activity under simulated solar light (SSL). The test results show that the 3D p-CNG–Pt composite catalyst exhibits better HER activity than the 3D p-CNG–Au and 3D p-CNG–Pd composite catalysts under optimal conditions. This is due to the effective charge transfer of platinum and 3D p-CNG skeletons, which shows better HER kinetic curve linearity and lower catalyst dose consumption and results in better durability. In terms of thermoelectrochemistry, the 3D skeleton exhibits a wider absorption boundary (531 nm) than the 2D skeleton, and the anchored Au, Pd, and Pt significantly improve the photoreaction effect of the 3D p-CNG composite catalyst due to the improvement of the electron transition pathway [143].

Wang's team provided a new approach to the photocatalytic charge transfer pathway. The team used the characteristics of the ordered structure of the metal organic layer to easily obtain active sites and prepared a 1.5 nm metal coordination layer with rGO as an electron mediator. The ultrathin two-dimensional metal–organic layers (MOLs) are distributed on the two-dimensional template constructed by rGO. The synergistic effect and electronic mediation of the two compensate for the gap between heterogeneous catalysts and homogeneous antenna molecules, improve the energy efficiency of photocatalysis and CO_2 absorption, and significantly improve the activity of the photocatalytic reduction reaction [144].

Although metal-doped photocatalysts have been extensively studied, as described above, the efficiency and effectiveness of photocatalytic hydrogen production have been significantly improved. However, the use of these doped materials is still challenging in many aspects, especially with regard to their impact on the environment. For example, if the material is applied to too many metals, the cost of subsequent waste recycling will be increased. If not handled properly, this may lead to regional heavy metal pollution. This runs counter to the previous idea of improving global energy tensions and environmental crises through in-depth research on photocatalytic technology. Therefore, research on promoting the photocatalytic effect by relying on the characteristics of different dimensions of graphene-based materials is still advancing. Researchers want to improve the photocatalytic effect of graphene composites through the dimension of materials and the introduction of non-metallic elements, such as carbon, hydrogen, oxygen and phosphorus, to reduce the content of doped metals as much as possible, and even not introduce the whole process. Based on the photocatalytic process of perylene bisimide (PBI) supramolecular materials, the addition of zero-dimensional (or microscopic three-dimensional) graphene quantum dots (GQDs) significantly improve the effect of visible light photocatalysis (Figure 3). The small particle size of GQDs can be used to place high active sites and install PBI layer by layer through electrostatic interaction to form GQD/PBI supramolecular composites. The advantage of these GQD composites is that they can interact with the π-π of PBI to form long-range electron delocalization, and the quantum confinement effect also promotes the transfer of electrons from GQDs to PBI, which improves the reduction ability of PBI and the production characteristics of H_2. This study undoubtedly provides ideas and future research directions for supramolecular organic photocatalysis at the level of quantum modification [145].

The Yan team studied the reduction process of functionalized GQDs to decompose water and absorb CO_2 under visible light. They reported two ways to narrow the band gap of GQD and explained the mechanism. GQD-Anln-OCH_3, GQD-Anln-OCF_3 and GQD-Anln-$OCCl_3$ have a Z-scheme structure, and charge separation can promote coupled photocatalysis (Figure 4). The absorption range of all GQDs is limited to 200–300 nm due to the π-π transition of the sp^2 substrate. Yan also compared the optical absorption capacities of GQD derivatives: GQD-BNPTL > GQD-DNPT18 > GQD-DNPT23 > GQD-OPD > GQD (Figure 5) [146].

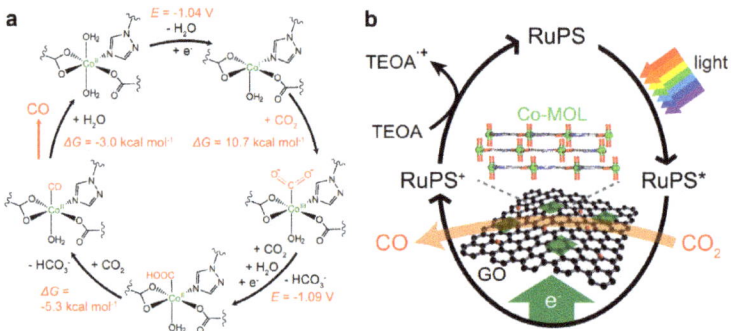

Figure 3. (**a**) Calculated mechanism with the molecular unit of Co−based metal−organic frameworks (MOFs) for photocatalytic CO_2−to−CO conversion, showing the calculated redox potentials and free energy changes. (**b**) Proposed photocatalytic mechanism. Reprinted with permission from Ref. [145]. Copyright©2023 Springer Nature Limited.

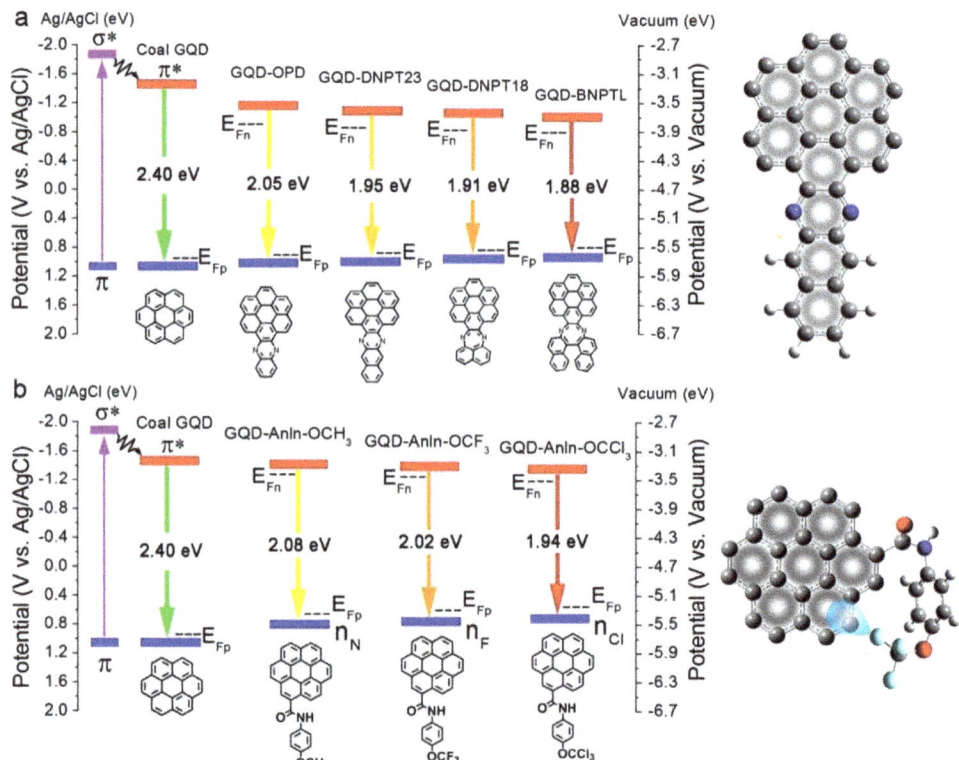

Figure 4. Illustration of energy level diagram and simplified structure of (**a**) coal GQD, GQD−OPD (o−phenylenediamine), GQD−DNPT23, GQD−DNPT18, and GQD−BNPTL and (**b**) coal GQD, GQD−AnIn−OCH$_3$, GQD−AnIn−OCF$_3$, and GQD−AnIn−OCCl$_3$. The Fermi levels for p-type conductivity (E_{Fp}) and n−type conductivity (E_{Fn}) are indicated in the energy diagram. Reprinted with permission from Ref. [146]. Copyright©2023 American Chemical Society.

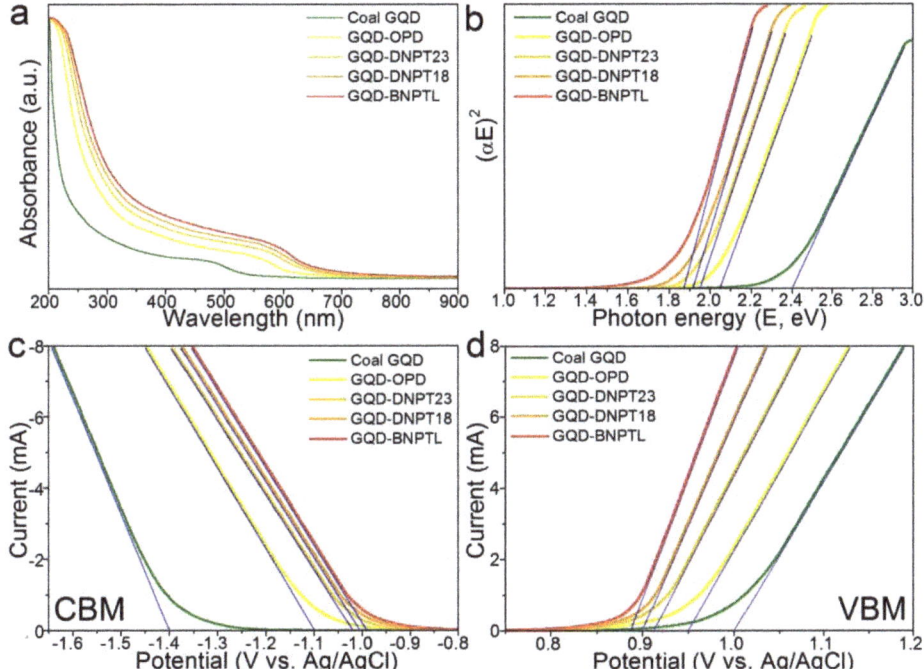

Figure 5. UV–Vis absorbance spectra (**a**) and accordingly obtained plots (**b**) of (αE)2 versus photon energy (E) (where α denotes the absorbance coefficient) of coal GQD, GQD-OPD, GQD-DNPT23, GQD-DNPT18, and GQD-BNPTL. The horizontal intercept of the tangent line in (**b**) indicates the bandgap of each GQD type. (**c**) Cathodic linear sweep voltammetry (**c**) and anodic linear sweep voltammetry (**d**) of coal GQD, GQD-OPD, GQD-DNPT23, GQD-DNPT18, and GQD-BNPTL. The horizontal intercept of the tangent line in (**c**) or (**d**) determines the conduction band minimum (CBM) or the valence band maximum (VBM) of each GQD type, respectively. Reprinted with permission from Ref. [146].Copyright©2023 American Chemical Society.

2.3. Graphene Electrocatalysis

The high specific surface area of graphene can provide more active sites and electron transport channels, which can increase the contact area of reactants and catalysts and the electron transfer between catalysts and electrodes, thereby increasing the catalytic effect [147–150]. At the same time, graphene can also act as a carrier for a stable catalyst. Fixing the catalyst on graphene can improve the dispersion and stability of the catalyst [151–155]. In addition, the functional groups carried on the surface of graphene can also provide additional active sites to regulate the catalytic object or catalytic efficiency. The electrocatalytic mechanism of graphene is based on its high surface area and conductivity, electron transfer ability, catalyst support, and surface functional groups. These factors work together to significantly improve the efficiency and performance of catalytic reactions [74,156–159]. The oxygen reduction reaction (ORR) and the oxygen evolution reaction (OER) are crucial for bifunctional electrocatalysts, which are often used in practical applications of rechargeable metal–air batteries [160].

A 3D nanoporous graphene (np-graphene) bifunctional electrocatalyst is formed by the nitrogen doping and nickel doping of 3D nanoporous graphene, in which nitrogen and nickel have two forms: single atom and cluster. A rechargeable all-solid-state zinc–air battery was prepared by using nitrogen and nickel co-doped np-graphene as a self-made flexible air cathode, PVA gel as an electrolyte, and Zn foil as an anode. As shown in Figure 6 below, the open circuit voltage of this air battery is 1.35 V, and the maximum power density of discharge polarization is 83.8 mW cm^{-2}. Moreover, there is only a slight performance loss after 258 cycles, and the bending of the battery at different angles does not affect its performance, showing good cycle stability and flexibility [78].

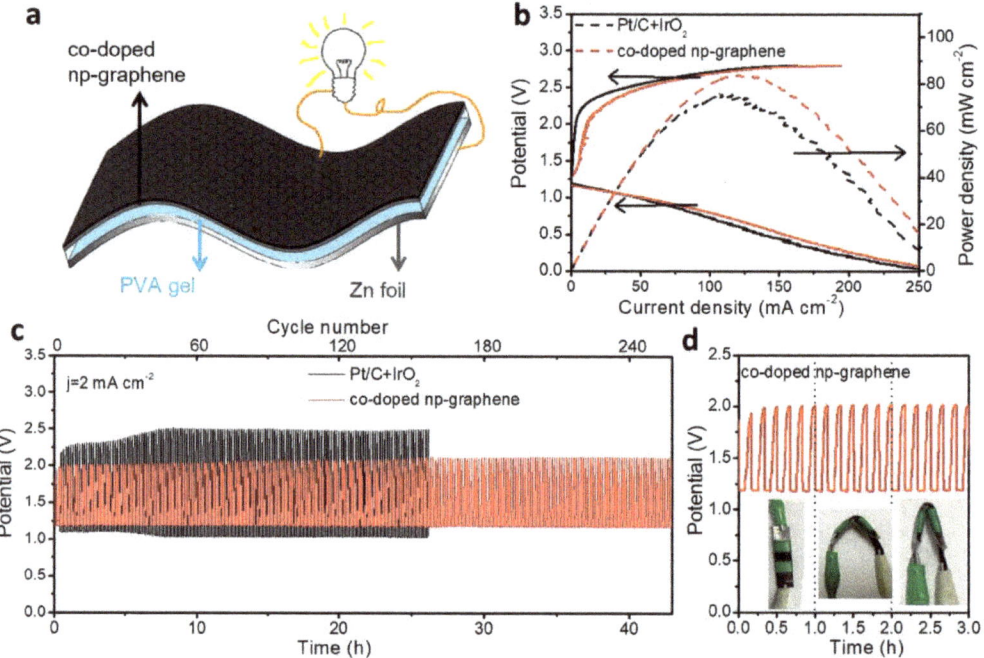

Figure 6. (**a**) Schematic diagram of the co−doped np-graphene−based all−solid-state Zn−air battery. (**b**) Polarization and power density curves of the batteries. (**c**) Discharge/charge cycling curves at 2 mA cm^{-2} and (**d**) discharge/charge curves under different bending states. Reprinted with permission from Ref. [78]. Copyright© John Wiley & Sons, Inc. All rights reserved.

The single atoms of nickel and iron are respectively embedded in the inner and outer walls of graphene hollow nanospheres (GHSs) to form a new type of highly active Ni-N$_4$ and Fe-N$_4$ Janus structure with distributed self-assembly, in which a planar configuration is formed to coordinate with four nitrogen atoms. The Janus structure on the inner and outer walls of the GHSs separates the ORR and OER active sites. The outer Fe-N$_4$ site plays a major role in the ORR activity, and the inner Fe-N$_4$ site plays a major role in the OER. When Ni-N$_4$/GHSs/Fe-N$_4$ is used as the air cathode, the specific capacity of Ni-N$_4$/GHSs/Fe-N4 can reach 777.6 mAh gZn^{-1}, and the energy density can reach 970.4 Wh kgZn^{-1}, showing excellent bifunctional electrocatalytic activity. The Ni-N$_4$/GHSs/Fe-N4-based zinc–air battery still maintains stable electrocatalytic performance after about 200 h of operation, which is considered excellent cycle stability [161].

3. Pollutant Degradation
3.1. Removal of Typical Pollutants
3.1.1. Antibiotics

In recent years, antibiotics have been widely used in human treatment, animal husbandry, and aquaculture [162]. However, antibiotics do not have a good metabolic effect on organisms. The widespread use of antibiotics has caused antibiotic residues to flow into the environment along with raw materials, causing ecological damage. If humans drink water containing antibiotics for a long time, it will also cause damage to human health [163]. Nowadays, adsorption or catalytic oxidation methods, such as the use of natural zeolite, bentonite, activated carbon, carbon nanotubes, biochar, etc., are usually used to remove antibiotics from the water environment [164]. Membrane filtration, oxidation, photocatalysis, and biodegradation can also be used for antibiotic removal [165]. The scientific community has been researching the removal of antibiotics by various nanomaterials such as graphene for a long time [166]. The Ibrahim team studied a new type of on-line potential monitoring technology for antibiotics. It can effectively monitor the degradation or elimination of antibiotics [167].

Antibiotics have a negative impact on plants, aquatic organisms, and microbial communities, causing damage to the ecological environment and affecting human health. The common antibiotics contained in wastewater are tetracycline antibiotics, quinolone antibiotics, β-lactam antibiotics, macrolide antibiotics, sulfonamide antibiotics, etc. [162]. As one of the broad-spectrum tetracycline antibiotics, oxytetracycline (OTC) was identified as having small side effects and was thus widely used in the 1960s and 1970s. However, as OTC became widely used in livestock and aquaculture, people gradually discovered that organisms could not fully absorb these antibiotics. About 90% of OTC enters into the ecological environment [168], which leads to OTC accumulation in the aquatic environment; over time, this will produce resistant bacteria and resistance genes, greatly reducing the effective use of these antibiotics [169]. Another antibiotic, ofloxacin (OFX), one of the third-generation quinolone antibiotics, has been used unrestrainedly and discharged at will. Five years ago, the concentration of antibiotics in groundwater/surface water around the world reached 30 mg/L, which is concerning data [170].

It has been found that the removal efficiency and kinetic constant of rGO-Fe_3O_4 composites formed by adding GO are further improved compared with using Fe_3O_4 alone. The OFX removal rate can even reach 99.9% complete removal. Compared with the alkaline environment, the degradation rate and ability of OFX is more effective under acidic conditions [171]. After 0.5 min visible light irradiation, the degradation rate of ciprofloxacin hydrochloride (CIP) by the perylenetetracarboxylic diimide (PDI)/rGO composite film can reach 94.31% (10 mg/L). This composite film has been found to have two functions: photocatalytic degradation and photothermal conversion. By using rGO as an additive material, the optical absorption range of nano-PDI powder self-assembled using hydrochloric acid can be extended to the near-infrared band. Coupled with the selectivity of the upper band gap, the ions and pores of the upper nano-PDI will relax to the edge of the band, so the excess energy can be converted into heat. The practical significance of studying this PDI/rGO composite membrane shows that there is a visible-light-responsive graphene-based photothermal catalytic material that can achieve two functions. It can degrade antibiotics while recovering pure water in actual water samples to achieve the purification and recovery of wastewater [172].

Using citric acid-modified GO and carboxymethyl cellulose membrane (GO-CMC) to remove antibiotics, the adsorption capacity of OTC, quinic acid (OA), and trimethoprim (TMP) can reach 102.05, 252.68 and 370.93 mg/L, respectively. Antibiotics are deposited on the surface of GO by π-π interactions and cation-π bonds. The citric acid-modified GO-CMC membrane can be reused and maintains stable recyclability after 5 times of recycling, which can remove antibiotics in wastewater [173]. For promoting the degradation of OTC in water, a graphene–TiO_2–Fe_3O_4 nanocomposite plasma has a good degradation effect. Compared with rGO–TiO_2, the addition of Fe_3O_4 enlarged the specific surface area of the composites, accelerated the separation of electron-hole pairs, and increased the magnetic strength, which made the catalyst easy to separate from water. At the best doping amount of Fe_3O_4 of 20 wt%, the removal efficiency can reach 98.1%, which is the highest removal efficiency. Other influencing factors include the catalyst dosage, air-flow rate, peak voltage, and pH value, and their optimal values are 0.24 g/L, 4L/min, 18, and 3.2 kV, respectively. When the above conditions are reached, rGO–TiO_2–Fe_3O_4 has the best removal performance. The rGO–TiO_2–Fe_3O_4 still has high catalytic performance after four uses [174].

Heterogeneous photocatalytic technology based on the TiO_2-based catalytic system is widely considered to remove pollutants in the environment without producing secondary pollutants, which is conducive to environmental protection [175]. For cephalosporin antibiotics, the photocatalytic activity of TiO_2/N-doped porous graphene nanocomposites (TiO_2/NHG) is better than that of bare p25 TiO_2, TiO_2/GO and TiO_2/porous graphene frameworks. The results showed that the oxidative degradation rate of TiO_2/NHG was affected by the catalyst loading, the initial antibiotic concentration, and the presence of H_2O_2. The complete mineralization of 25 mg/L of antibiotics was observed within 90 min of irradiation, and the activity level of the TiO_2/NHG catalyst did not decrease significantly within three repeated cycles. The complete mineralization of antibiotics can be achieved only by sunlight irradiation within a reasonable time span, which means that the oxidative degradation of antibiotics remains efficient even in the absence of H_2O [176].

Magnetic GO/ZnO nanocomposite (MZ) materials offer an excellent adsorption capacity with reusability for tetracycline antibiotics (TCs) in wastewater. The results show that the maximum adsorption capacity of MZ materials for TCs can reach 1590.28 mg g^{-1} at pH = 6.0. At the same time, after four absorption cycles, the adsorption capacity of MZs still did not decrease significantly and maintained a relatively stable adsorption activity. MZ materials have been proved to have the advantages of fast separation speed, strong adsorption capacity, reusability, and simple operation in use [177]. Based on the raw material MXene Ti_3C_2, a graphene-layer anchored TiO_2/g−C_3N_4 (GTOCN) photocatalyst was formed by a one-step in situ calcification method. This method of synthesis can not only act on antibiotics but also on another persistent organic pollutant, namely dyes. Under visible light, such as sunlight in daily environments, GTOCN will have highly oxidizing active ·O_2 and ·OH, hence offering high-efficiency degradation of tetracycline (TC) and ciprofloxacin (CIP) antibiotics and bisphenol A (BPA) and rhodamine B (RhB) dyes [178].

The synergistic catalytic removal of thiamethoxam (TAP) induced by pulsed discharge plasma (PDP) was studied by using graphene–WO_3–Fe_3O_4 nanocomposites. Compared with single WO_3 and rGO–WO_3 without Fe_3O_4, rGO–WO_3–Fe_3O_4 has a larger specific surface area and higher transfer rate of photogenerated carriers. Fe_3O_4 doping does not make the higher the better. With the gradual increase from zero, the TAP removal curve first increased and then decreased. When the Fe_3O_4 doping amount was 24 wt%, the catalyst dosage of 0.23 g/L obtained the best catalytic performance, and the removal rate of TAP could reach 99.3%. Acidic conditions and the presence of O_3, H_2O_2, and ·OH are more conducive to the catalytic degradation of TAP [179].

The three-dimensional (3D) graphene can be designed as an anode to create an enhanced electron and mass-transfer photocatalytic circulation system, which can be used to remove ampicillin in wastewater and to enhance the antibacterial properties of water. It enhances electrons by using a three-dimensional graphene photoelectrode as an anode, a Pt/C air breathing electrode as a cathode, and C_3N_4-MoS_2 loading. Perhaps due to the cleavage of functional groups such as amide bonds and peptide bonds, the removal rate of ampicillin by the photoanode reached 74.6% after the system was fully reacted in sewage for 2 h [180].

The 3D-MoS_2 sponge modified by molybdenum disulfide nanospheres and GO adsorbs organic molecules and provides a multidimensional electron transport path, which has a positive effect on the degradation of advanced oxidation processes (AOPs), especially for aromatic organics. After pilot-scale experiments in 140 L wastewater, it still maintains efficient and stable activity for AOPs. Even after 16 days of continuous experiments, 3D-MoS_2 can still maintain a degradation rate of 97.87% in wastewater containing 120 mg/L antibiotics. This is of great practical and economic interest for industrial applications; if the sponge could be produced industrially in large quantities, the cost of treating a ton of wastewater would be only USD 0.33 in the future [181].

Ion-doped GO nanocomposites can perform photon absorption, electron transfer, and the generation of active species under visible-light irradiation. For example, an iron oxide/hydroxide N-doped graphene-like nanocomposite has been synthesized by a laser-based method to remove antibiotics from wastewater. The method is shown in Figure 7. Moreover, this nanocomposite is still environmentally friendly during the preparation process [182].

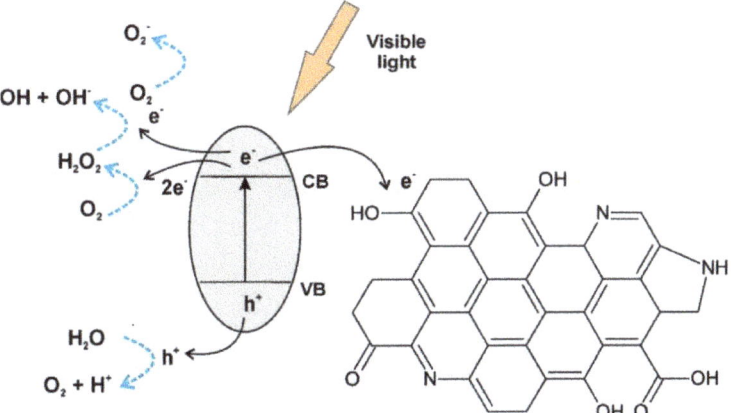

Figure 7. Representation of photon absorption, electron transfer, and generation of reactive species under visible−light irradiation of a Fe oxide/Fe hydroxide/N−rGO nanocomposite. Reprinted with permission from Ref. [182].Copyright©2023 Springer Nature Limited.

3.1.2. Dye

One of the main causes of surface water pollution is from the textile industry, where waste from the printing and dyeing process is directly discharged into the water environment without qualified treatment [183].The increasing concentration of dyes in the water environment seriously affects the refractive index of light, which limits the growth of aquatic plants and further negatively affects the self-purification of the ecosystem. More water resources in barren water resources are no longer suitable for organisms, let alone for household and industrial use [184].

A triphenylmethane cationic dye—Malachite green (MG)—is a common dye in the dyeing process of the textile industry. However, due to the presence of nitrogen (N_2), MG consumption may lead to many serious human health issues, e.g., cancer, etc. [185]. Ozone oxidation, oxidation, membrane filtration, flocculation, biosorption, and electrochemical methods are often used to remove MG dyes on the market. However, these methods are not only costly, but they also have requirements for reaction conditions, such as dissolved oxygen demand. Moreover, due to the problems of high sludge production, short half-life and slow process, the removal efficiency of dyes is not high [186]. As shown in Figure 8, the porous sodium alginate/graphite-based composite hydrogel was modified by the grafting polymerization of acrylic acid on sodium alginate, and graphite powder was loaded to enhance its adsorption capacity. In terms of the effective adsorption for organic pollutants, the maximum adsorption capacity for malachite green dye can reach 628.93 mg g^{-1}. The hydrogel complex showed sustainable usability and could still adsorb 91% of MG after three consecutive dye adsorption–desorption cycles [187].

A novel 3D magnetic bacterial cellulose-nanofiber/GO-polymer aerogel (MBCNF/GOPA) mesoporous structure with a high surface area of 214.75 m^{2}g^{-1} can be used to remove malachite green (MG) dye from aqueous solution with a maximum adsorption capacity of 270.27 mg g^{-1}.When the reaction environment meets the conditions of temperature (25 °C and contact with 30 mg/L MG concentration solution for 25 min, the solution with pH = 7.0 and 5 mg MBCNF/GOPA can have the best performance. The adsorption efficiency of MBCNF/GOPA remained above 62% after 8 times of recovery using 0.1 mol/L acetic acid/methanol in a 1:2 mixing ratio [188].

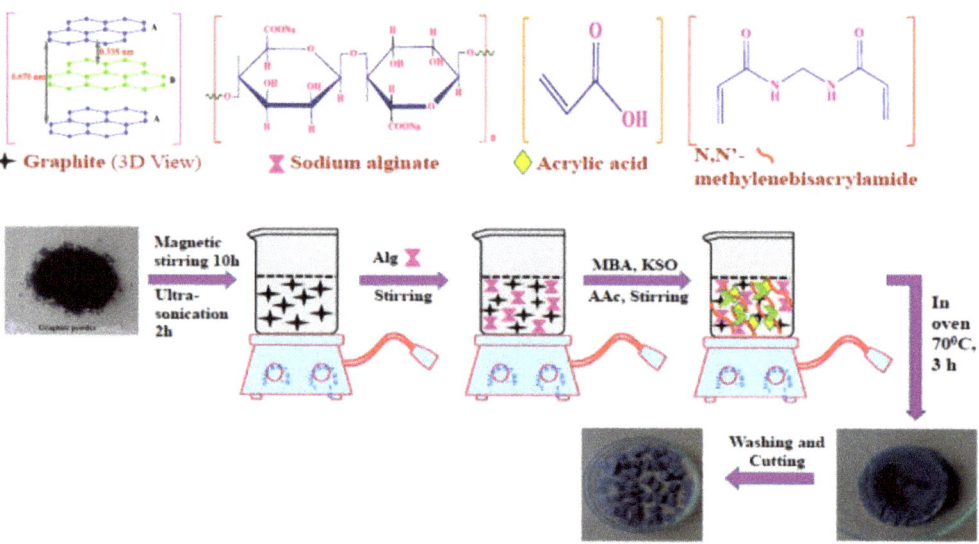

Figure 8. Preparation route diagram of hydrogel composites. Reprinted with permission from Ref. [187]. Copyright©2020 Elsevier B.V. All rights reserved.

A polyvinylidene fluoride (PVDF)–polyaniline (PANI) and GO mixed membrane was prepared by incorporating PANI–GO as a nano-filler, which greatly improved the antifouling performance and solvent content. The pure water flux through the membrane increased from 112 to 454 mL/m^2·h. Compared with other compositions and doping materials, the nanocomposite membrane with 0.1% w/v GO is superior. When operating conditions reach 0.1 MPa operating pressure, the dye rejection rate can reach 98%. After several tests on the membrane, the flux recovery rate of almost all dyes can be stabilized at 94%, while the removal rate of individual dyes such as methyl orange can reach 95%, and that of Allura red can be as high as 98% [189].

A nanocomposite hydrogel (NCH) formed by chitosan (CS) and carboxymethyl cellulose (CMC) crosslinked-modified GO has a significant adsorption effect on methylene blue (MB) and methyl orange (MO). At pH 7, the adsorption rate of 0.4 g/L CS/CMC-NCH for 50 mg/L MB was about 99%. At pH 3, the adsorption rate of 0.6 g/L CS/CMC-NCH for MO was about 82%. The adsorption capacity of CS/CMC-NCH for MO is 404.52 mgdye/gads, and the adsorption capacity for MB can reach 655.98 mgdye/gads. More importantly, the composite hydrogel has a stable adsorption performance for the dye after continuous use of 20 adsorption–desorption cycles. CS/CMC-NCH also has an excellent effect on the removal of anionic and cationic dyes [190]. The novel lysine and ethylenediamine double-crosslinked graphene aerogel (LEGA) exhibited a 3D interconnected porous structure, which greatly increases the adsorption capacity for the MB dye. Compared with other substrate materials, the compression performance of LEGA was significantly improved after adding lysine, and the adsorption capacity of MB could reach 332.23 mg/g [191].

For the removal of the crystal violet (CV) dye, it was found that two kinds of nanocomposite hydrogels, acrylamide-bonded sodium alginate (AM-SA) and acrylamide/GO sodium alginate (AM-GO-SA), can be synthesized by the free radical method, and both of them have good adsorption properties for CV. Compared with other influencing factors, the removal efficiency of AM-SA and AM-GO-SA was more dependent on pH, and the maximum single-layer adsorption capacity could reach 62.07 mg/g and 100.3 mg/g [192].

A novel GO/poly (N-isopropylacrylamide) (GO/PNIPAM) composite system removes organic dyes in water by a similar extraction mechanism and undergoes a reversible sol–gel transition at a temperature higher than the lower critical solution temperature. PNIPAM is anchored on the surface of GO to prevent the reduction of GO and inhibit its aggregation, which greatly improves the stability of GO dispersions. Moreover, the dye can be effectively adsorbed and enriched in the gel phase, which is convenient for its separation from water during the extraction process [193].

3.1.3. Oil and Organic Solvents

The rapid development of modern transportation, petrochemical, and marine engineering has accelerated the release of a large amount of oil into the sea and rivers, causing energy loss, destroying the local ecological environment, and seriously endangering the sustainable development of society [194].The urgent need to solve the problem of oil pollution is self-evident. The public has focused on oil/water separation technology and its ability to treat industrial oily wastewater and oil spill accidents. However, most of the traditional oil/water separation materials are based on activated carbon, polypropylene sponge, and zeolite, and other microporous structure absorbents have the disadvantages of limited absorption capacity and poor wear resistance [195].

Based on polyimide (PI), a novel zeolitic imidazolate framework-8/thiolated graphene (ZIF-8/GSH) nanofiber membrane can be prepared by electrospinning and in situ hydrothermal synthesis. The membrane has superhydrophobicity/superoleophilicity and can effectively purify oily wastewater. For various oil/water mixtures and water-in-oil emulsions, the separation efficiency of oil and water can reach 99.9% through the action of fiber membranes. More importantly, the film maintains superhydrophobicity without requiring the environment to remain under harsh conditions. Under harsh reaction conditions, such as excessive acid and alkali, long-term contact with salt and corrosive organic solvents, high temperature irradiation, mechanical wear, ultrasonic treatment and other simulated environmental conditions, ZIF-8 @ GSH can still exhibit excellent photocatalytic degradation efficiency. This shows that the membrane has self-cleaning and active antibacterial abilities and can maintain its performance through the mechanical and chemical environment, which makes future industrial applications very promising [196].

With the release of a large amount of oil pollution into the water environment, the emergence of reusable superhydrophobic oil adsorption materials are required. A magnetic superhydrophobic polyurethane sponge (Fe_3O_4/OA/GO-PU) was formed based on a 3D microstructure by linking GO and coating with functionalized oleic acid Fe_3O_4 nanoparticles. It can be water repellent, with a contact angle of 158°, and has high selectivity in contact with organic solutions and oils. Theoretically, the microstructure polyurethane (M-PU) sponge has a capacity of 80–160 g/g, and it can undergo 15 adsorption cycles. The oil can be selectively extracted from the wastewater, exhibiting excellent recyclability and the ability to be recovered in a static state using an external magnetic field. Compared with the absence of a magnetic field, the increase of M-PU adsorption capacity seems to be driven by magnetic field exposure due to the enhancement of surface-active sites of the M-PU sponge. The continuous collection of kerosene from the surface water while cleaning wastewater with Fe_3O_4 @ OA @ GO-PU is cost-effective, highly selective, and an excellent recyclable, environmentally friendly oil-spill cleanup option [197].

Using phase inversion technology and the dip coating method, a nanocomposite film was composed of polybenzimidazole (PBI), graGO, rGO, and polydopamine (PDA) in a coated and uncoated manner as shown in Figure 9. When 0.5–1.5 wt% GO was added to the polymer matrix, the antifouling performance of the composite membrane could be improved to the highest extent, and the maximum flux of the membrane reached 91 L/m^2·h·bar. Compared with the original single PBI material, the permeability of the water–oil mixture of the nanocomposite membrane increased by 17%, and the oil removal efficiency also increased from 80% to 100%. After four clean filtration cycles, the water-flux recovery rate (FRR) remained above 90%. Even without any alkaline and acidic cleaning, it exhibits good separation rate for oil-in-water emulsions and stable antifouling and antibacterial properties [198].

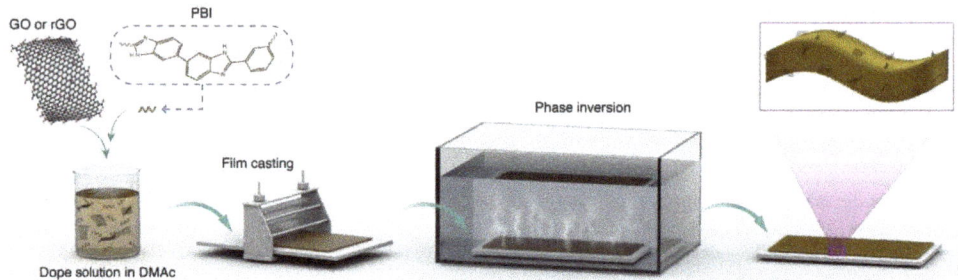

Figure 9. Schematic representation of the membrane fabrication steps. Reprinted with permission from Ref. [198]. Copyright©2020 The Authors. Published by Elsevier B.V.

A novel poly (oxyethylene) graphene oxide-based nanofluid (P-GO-O) prepared as shown in the Figure 10 shows high temperature resistance and high salt tolerance in deionized water, with a potential of 39 mV. The recovery rate of octadecyl-aminated graphene oxide (GO-O) is only 6.7%, while that of P-GO-O is 17.2%, indicating that P-GO-O can improve oil recovery. This is due to the fact that its structural oil–water interfacial tension can be reduced to 12.2 mN/m under the action of P-GO-O, and the oil-wet surface is turned into a water-wet surface. Even under harsh conditions, P-GO-O still exhibits stable properties [199].

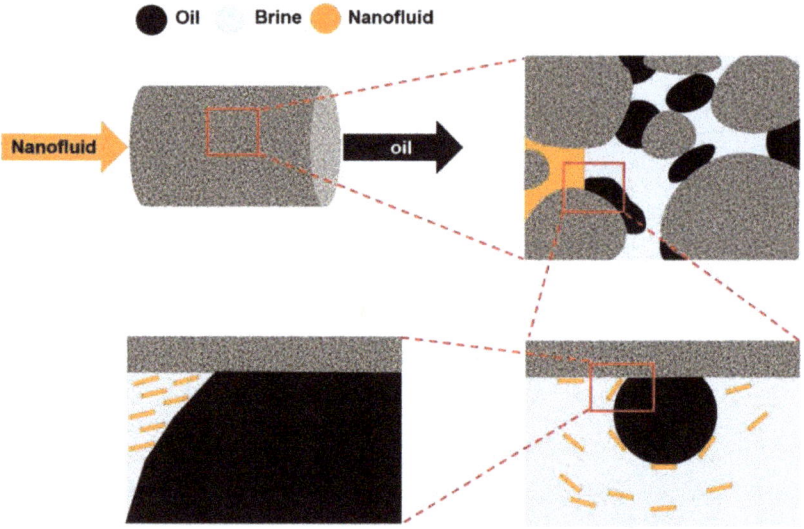

Figure 10. The schematic of displacement mechanism of P-GO-O. Reprinted with permission from Ref. [199]. Copyright© 2021 Elsevier Ltd. All rights reserved.

As shown in Figure 11, a polysulfone (PS) mixed-matrix membrane containing aspartic acid (AA)-functionalized graphene oxide (fGO) has good hydrophilicity, water permeability and oil repellency at a very low GO loading. Thanks to the fluorine GO load, the fluorine GO load performance of the film is much higher than that without addition. The functionalization of AA introduces carboxyl and amino groups, which is beneficial to the performance improvement of the matrix membrane and improves the hydrophilicity and fouling removal rate. After adding very low concentrations of fGO, the incorporation of fGO in the PS membrane had a positive impact on the mechanical properties and antifouling properties of the membrane and enhanced the separation rate of the oil–water emulsion. The affinity of BSA for the membrane surface decreased, which means that the flux recovery of the fGO membrane after bovine serum albumin (BSA) contamination was higher. Compared with the monotonous original membrane, the water permeability of the composite membrane doped with 0.2 wt% fGO increased by 97% and the oil rejection rate reached 97.9% when the 200 mL oil emulsion was filtered [200].

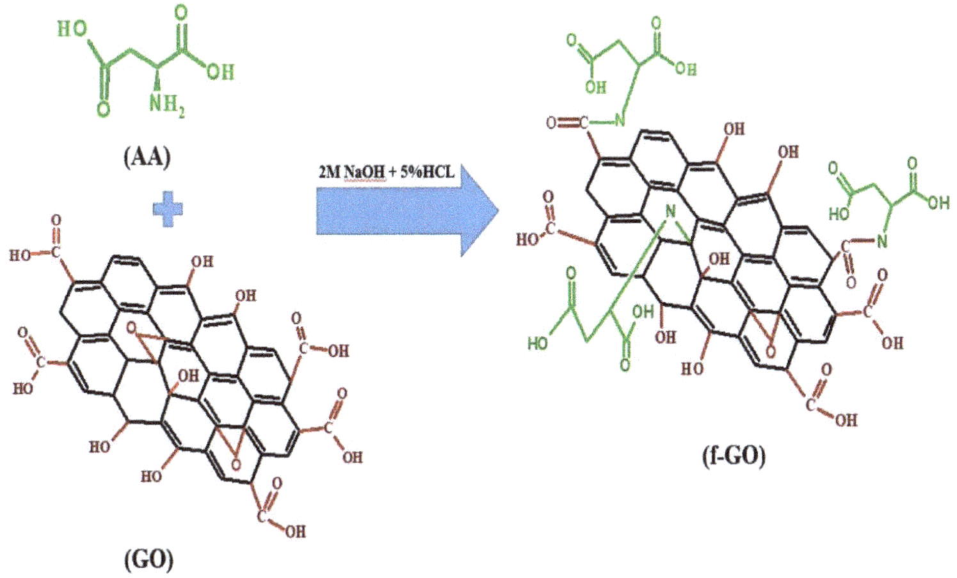

Figure 11. Functionalization of GO with aspartic acid. Reprinted with permission from Ref. [200]. Copyright©2020 The Author(s). Published by Elsevier Ltd.

A small sheet of graphene oxide (SFGO) film for high-performance organic solvent nanofiltration (OSN) applications, using La^{3+} as a crosslinking agent and a spacer layer for insertion; moreover, it stabilizes the SFGO film selection layer and achieves selective molecular transport. The permeability of methanol is 2.9 times higher than that of large-flake graphene oxide (LFGO), and it has high selectivity for three organic dyes. More importantly, the SFGO-La^{3+} film exhibits at least 24 h of stable stability under hydrodynamic stress, which represents real OSN operating conditions. Through the interfacial polymerization (IP) of low-concentration resorcinol on the surface of the graphene quantum dot (GQD)-polyethylenimine (PEI)-modified polyimide substrate, a new nanocomposite (TFN) organic solvent nanofiltration membrane with a sandwich structure was prepared as shown in the Figure 12.The thickness of the IP skin layer of the GQD-interlayer OSN film is about 25 nm, and the average surface roughness is generally less than 2 nm, resulting in an increase in the penetration content of Harnol from 33.5 to 40.3 $Lm^{-2}h^{-1}MPa^{-1}$, and the penetration rate of Rhodamine B from 87.4% to 98.7%. After long-term immersion in pure N, N-dimethylformamide (DMF), it showed superior solubility. After 81 days of storage at room temperature, it was stored at 80 °C for 45 days and then filtered with Bengal Rose (1017 Da) DMF solution at 25 °C for 5 days. There was no scar solute rejection, which proved the antifouling performance during long-term filtration [201].

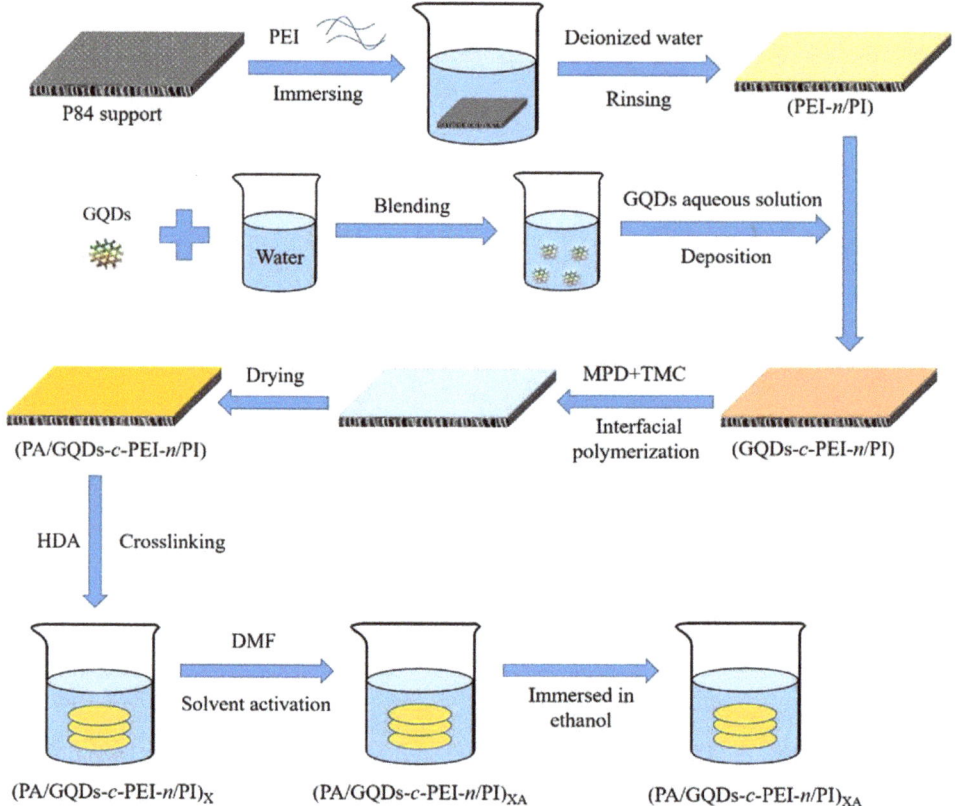

Figure 12. The fabrication process of the GQD-interlayered TFN-OSN membranes. Reprinted with permission from Ref. [201]. Copyright© 2019 Elsevier B.V. All rights reserved.

3.2. Adsorption of Heavy Metals in Sewage

Common heavy metals in sewage, such as Cd, Zn, Pb, Fe, Cu, Hg, Ni, Mn, Co, etc., generally exist in trace amounts. Even in very small amounts, heavy metals are considered to be the most harmful, toxic and most widely distributed components in wastewater due to the mobility of the ions [202].The discharge of heavy metals into the water environment will not only adversely affect the ecological environment but also accumulate in soft tissues after entering the human body, seriously endangering human health and even threatening life and health [203]. For example, copper can lead to liver damage, insufficient blood supply, and night-time insomnia. It can also inhibit the activity of enzymes in the soil and affect the circular development of the ecological environment. Chromium can cause dizziness, headache, nausea, and diarrhea. Excessive inhalation of lead may lead to muscle spasm, even renal failure, and damage to the brain of infants. Mercury can cause rheumatoid arthritis and even threaten the normal work of the human nervous system circulatory system [204]. Due to heavy-metal ions being highly soluble, stable, non-biodegradable, and able to migrate in aqueous media, metal-contaminated wastewater can also cause harmful effects in plants, such as photosynthesis inhibition, and the reduction of seed germination rate, enzyme activity and chlorophyll synthesis [205].

In the past, traditional technologies were used, such as non-destructive processes using resins or adsorbents [206], non-destructive separation using semi-permeable membranes, and solvent separation techniques, which often have disadvantages such as low efficiency, insufficient removal, strict operating conditions, and high prices. Nowadays, high porous nanostructures such as graphene can be used as alternatives to remove heavy metals from contaminated water, e.g., graphite oxide (GO) for Pb^{2+} and Cd^{2+} has excellent adsorption capacity with great potential [134]. Because of its small particle size, large specific surface area and high adsorption efficiency, graphene can break through the limitations of conventional adsorbents and has become a good choice for removing heavy metal ions from water [207].

The preparation of a Bi_2S_3-$BiVO_4$ graphene aerogel (SVGA) requires only a simple hydrothermal method, which can provide effective assistance in both the photogenerated electron transfer and photocatalytic ability of SVGA. The results show that removal rates of Cr (VI) and bisphenol A (BPA) using SVGA materials can be infinitely close to 100% after 40 min of adsorption and 120 min of photocatalysis under visible-light irradiation at 420 nm. The harmful Cr (VI) is preserved as low-toxic Cr (III) after photocatalysis on SVGA, and BPA is degraded into CO_2 and H_2O [208].

Due to the presence of oxygen-containing functional groups and high porosity, 3D magnetic fungal mycelia/graphene oxide nanofibers (MFHGs) can remove Co (II) and Ni (II) from high-salinity aqueous solutions, increasing the ion removal rate. The optimum reaction conditions were as follows: at 323 K and pH = 6.0, MFHGs could remove 97.44 and 104.34 mg/g of Ni (II) and Co (II), respectively, from 2 g/L Na_2SO_4 aqueous solution. Reductive self-assembly (RSA), one of the main materials of MFHGs, is cheap and environmentally friendly, so the cost of MFHGs is not high. They have excellent magnetization and large coercivity and can work normally in high-salinity water [209].

Under aerobic conditions, a novel graphene-like biochar supported trivalent iron (GB/nZVI) to remove Cd (II) and As (III). The main principle of removing As (III) is through oxidation and surface complexation, while the removal of Cd (II) mainly depends on surface complexation. At the same time, the strong synergy between GB and nZVI has a positive effect. The removal ability of GB/nZVI composites was significantly higher than that of pure GB and nZVI under both acidic and neutral conditions. The presence of As (II) significantly promoted the removal of Cd (III) when both ions were present in the same water environment. The maximum removal of As (III) can reach 181.5 mg/g when nZVI is 363 mg/g. The maximum removal capacity of Cd (II) can reach 46.4 mg/g when n ZVI is 92.8 mg/g. It is worth mentioning that the presence of phosphate and humic acid in the coexisting background ions has a reverse inhibitory effect on the removal of Cd (II) and As (III) [210].

GO and highly oxidized graphene oxide (GO_h) with different degrees of oxidation were combined with tridentate terpyridine ligand (Tpy) to form GO, GO_h, and GO–Tpy as shown in Figure 13. Compared with the prepared GO, GO_h, and GO–Tpy, GOh–Tpy has the highest adsorption efficiency for heavy metal ions due to the synergistic effect of GO and Tpy components. The maximum adsorption capacity (q_{max}) of the GOh–Tpy system for Ni (II), Zn (II), and Co (II) reached 462, 421, and 336 mg g^{-1} at pH = 6, respectively, and showed excellent repair performance, which proved that the GOh–Tpy mixture had an ideal cycle stability, reusability, and easy separation operation [211].

3.3. Degradation of Gaseous Pollutants

Exposure to air pollution is one of the five major global human-health risk factors. Photocatalytic oxidation is a promising method for the treatment of environmental pollutants. Nitrogen oxides (NO + NO_2), benzene, and isopropanol are the three major pollutants that can be commonly found outdoors. Titanium dioxide/graphene hybrid nanomaterials were synthesized by the sol–gel method. Under UV-A irradiation, the presence of isopropanol and benzene will form different free radicals, which improves the removal efficiency under UV-A irradiation. The addition of at least 1.0 wt% of granolol to TiO_2 can double the photocatalytic efficiency, and the system exhibits more stability under oxidizing conditions compared with pure TiO [212].

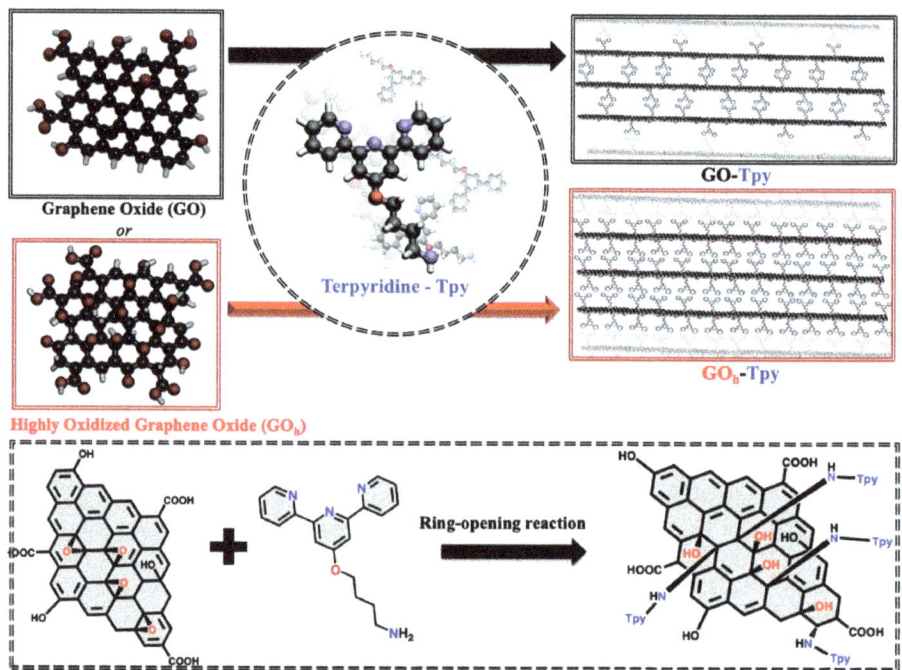

Figure 13. Schematic representation of the functionalization of GO with two levels of oxidation (GO and GO_h) with Tpy using the ring-opening reaction of epoxides yielding GO–Tpy and GO_h–Tpy. Reprinted with permission from Ref. [211]. Copyright©The Royal Society of Chemistry 2021.

A multi-ionic liquid (PIL)/TiO_2 composite material directly degrades pollutants through a free-radical mechanism and promotes the absorption and degradation of composite pollutants. Its photocatalytic degradation is shown in Figure 14. PIL/TiO_2 has high and low concentrations, and its photodegradation rate for benzene and toluene pollutants is also different. When the concentration is high, the photodegradation rate of benzene and toluene pollutants can be obtained at only 59% and 46%, respectively. Meanwhile, the decomposition rates can reach 86% and 74%, respectively, at the low concentration. When the PIL @ TiO_2/modified graphene oxide (m-GO) is at a high concentration, the photodegradation rate of benzene and toluene can reach a 91% oxidation rate, while PIL @ TiO_2/m-GO can reach 97% at a low concentration, and the percentage is obtained within 24 min [213].

A novel high-strength graphene aerogel was prepared by adding tetraethoxysilane (TEOS) to the precursor GO solution, as shown in the Figure 15, which has an ultra-high adsorption capacity for gas pollutants. On this basis, the graphene aerogel modified by $SiO2$ is more sensitive to benzene vapor, and the adsorption capacity of benzene vapor increases from 201.71 mg/g to 809.1 mg/g. In addition, it can also be used for the separation of benzene–toluene mixtures. This graphene aerogel can be reused after heat treatment at 150 °C. More importantly, the strength and adsorption capacity are not significantly reduced. This shows that the graphite aerogel in the field of indoor pollution gas removal has broad application prospects [214].

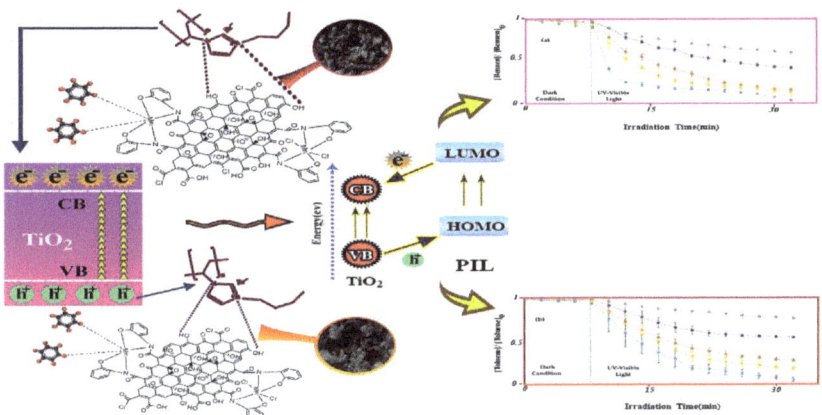

Figure 14. Schematic illustrating photocatalytic degradation with the help of band structures. Reprinted with permission from Ref. [213]. Copyright©2021 by the authors. Licensee MDPI, Basel, Switzerland.

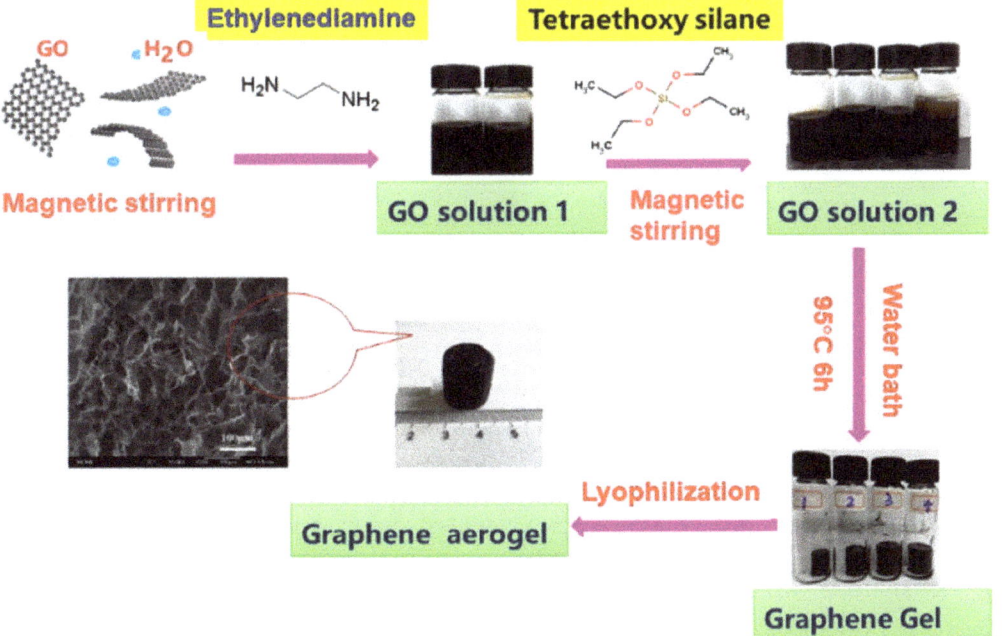

Figure 15. The preparation process of graphene aerogels. Copyright© 2021 The Authors. Reprinted with permission from Ref. [214]. Published by Elsevier Ltd.

4. Photocatalytic Reduction of CO_2

4.1. Mechanism

Solar energy will be an important component of future energy improvements for new sustainable development. Research on the photocatalytic reduction of CO_2 began in 1972 and 1979 when Fujishima and Honda used TiO_2 as a motor to photocatalyze the reduction of CO_2 in water [87,215,216]. The reduction and oxidation potential of CO_2 is matched with the valence band and the conduction-band position of the semiconductor photocatalyst, but most of the semiconductors cannot satisfy this requirement; thus, to achieve CO_2 reduction and emission reduction in photocatalytic reactions under sunlight, it is necessary to find semiconductors that meet the above conditions.

According to the basic reaction principle, the photocatalytic reduction of CO_2 has three main steps: (1) the generation of photogenerated electrons by bandgap engineering; namely, the generation of photogenerated charge carriers and the formation of electron-hole pairs; (2) the transfer of photogenerated electrons—this process can be called charge carrier separation and transport; and (3) the reduction of CO_2 by photogenerated electrons on the surface.

Given that electron-hole pairs combine easily and restrain charge separation to a great extent, effective charge separation is the key to achieve large-scale CO_2 emission reduction. Graphene is used as electron acceptor/transporter due to its high power function and good conductivity.

In the process of photocatalysis, graphene plays the following roles: (i) reduces photogenerated electrons and hole recombination, (ii) promotes CO_2 adsorption through π-π conjugation between graphene and CO_2, (iii) activates CO_2 molecules, (iv) improves corrosion resistance, (v) enhances surface area and light absorption, which subsequently results in higher photocatalytic activity [217].

4.2. Main Optimization Strategies of Graphene

4.2.1. Metal Doping, Non-Metal Doping and Graphene Heterostructure

Since most catalysts are easy to recombine due to the influence of electron-hole pairs in the process of CO_2 reduction, the catalyst surface activity is reduced, and the production efficiency is low [218]. Graphene doping is an effective method to adjust its electrical properties and expand its applications [219]. Graphene with different properties can be obtained by doping [220–222]. Examples include heteroatom doping in graphene, graphene nanosheets (GNS), graphene nanoribbons (GNR), graphene hydrogels, graphene quantum dots (GQD), GO, and rGO [223–227].

Doping is the most feasible and convenient method to adjust the band structure of graphene from semi-metal to p-type or n-type materials [228,229]. Primitive graphene usually shows bipolar characteristics. However, due to different specific needs, we need to make different semiconductors or electronic components, such as those with the above p-type and n-type conductivity, to manufacture logic circuits for industrial applications [230,231].

Graphene doping can be divided into n-type doping, p-type doping and p/n co-doping of single-layer or double-layer graphene. The chemical doping of graphene is realized by attaching heteroatoms to the surface of graphene or by replacing C atoms in graphene. For graphene, it is easier to produce p-type doping through surface adsorption [232]. Exposure of the original graphene in molecules with electron-absorbing groups (H_2O, O_2, N_2, NO_2, PMMA, etc.) will lead to obvious p-type doping. If the heteroatoms are removed from the p-type doping, the p-type doped graphene will return to the state before the original doping [233].

Chemical doping can effectively open the band gap of graphene [234]. The Fermi point on the doped graphene can be moved up and down depending on the different types of dopants, causing charge separation and easier formation of electron-hole pairs. The external factors that determine the band gap of graphene depend on the surface adsorption energy, lattice displacement doping, and so on [235]. X-ray photoelectron spectroscopy (XPS), angle resolved photoemission spectroscopy (ARPES), potential energy surface scanning (PES), and other methods can be used for the characterization of doped graphene and to detect its properties [236].

As a doped atom, graphene can provide a large number of electrons, which is due to the existence of large π bonds. The electrons in its parallel p-orbitals enter the conduction band, leaving a large number of holes in the valence band, forming an electron-hole pair. The Fermi energy level is close to the bottom of the conduction band. This doping method is p-type doping. In general, graphene has a two-dimensional honeycomb structure, so its surface easily adsorbs some small molecules, such as H_2O, N_2, O_2, and CO_2. These small molecules will promote graphene to form p-type doping. We can also open the sp^2 bond of graphene by replacing the position of carbon atoms or other atoms and bonding with carbon atoms to form doping. Because graphene can be prepared in large quantities by chemical vapor deposition (CVD), we can add different reaction sources to it at the end of the production, so that some carbon atoms in graphene can be replaced to form lattice doping. For example, under certain conditions, boron atoms can partially replace carbon atoms, forming p-type graphene. Common p-type doping molecules include fluoropolymer, water, N_2, NO_2, O_2, oxidizing solution, B, Cl, and metal [236]. Khudair and Jappor have proved that the adsorption of CO_2 on B-doped graphene and double B-doped graphite has strong chemisorption. The strong interaction between single- and double-B-doped graphene composites show that B-doped graphene and double B-doped graphene can catalyze or activate, which indicates that B-doped graphene and double B-doped graphene can be used as effective catalysts for CO_2 reduction [237].

Individual metals are easy to dissolve in the reaction environment because of their low bulk-binding energies, which have been extensively calculated by Chen et al. In the ORR performance of the 10 metal-doped graphene (M-G) catalysts (light metal Al; semiconductor Si; 3d-metals Mn, Fe, Co, and Ni; 4d-metals Pd and Ag; and 5d-metals Pt and Au) they studied, the metal bound to the graphene had higher binding energy, which shows that the M-G catalyst can be more stable compared with samples without metal dopants [238].

In the study by Min et al., different kinds of metal-doped graphene (for example, N-doped graphene, Ga-doped graphene and co-doped graphene) were found to have different band gaps [239]. The band gap of N-doped graphene was 0.20 eV, that of Ga-doped graphene was 0.35 eV, and that of N-Ga-doped graphene was 0.49 eV. They also had different electron densities. In N-Ga co-doped graphene, N atoms gained more electrons (-1.027 electrons) than N-doped graphene (-0.6 electrons), and Ga atoms lost more electrons (1.75 electrons) than Ga-doped graphene (1.80 electrons). At the same time, it was also found that different doped elements of graphene also have different electronic densities. For example, in the N-Ga co-doped graphene, the N atom gained more electrons (-0.61 electron) than the N-doped graphene (-0.27 electron), and the Ga atom lost more electrons (1.80 electron) than the Ga-doped graphene (1.75 electron). The analysis of different kinds of doped graphene shows that doped graphene has better photocatalytic performance; this also proves that doping has an important effect on the study of the photocatalysis of graphene [240].

As shown in Figure 16, Pawan Kumar et al. report a novel approach for grafting a copper complex onto N-doped graphene. It shows the reasonable mechanism of reducing CO_2 with a GrN_{700}-CuC catalyst. At the same time, in order to verify that the reduction product methanol is produced by CO_2, they replaced CO_2 with N_2 and found that the methanol content is small enough to be ignored. All the results show that the N-doped graphene catalyst has good chemical stability and can be further applied and developed.

As a material with excellent photosensitivity and conductivity, graphene has a good application prospect in the field of photocatalysis and CO_2 reduction. However, in practical research, it is found that a single-metal material cannot achieve high photocatalytic efficiency, and a graphene heterostructure can effectively solve this problem.

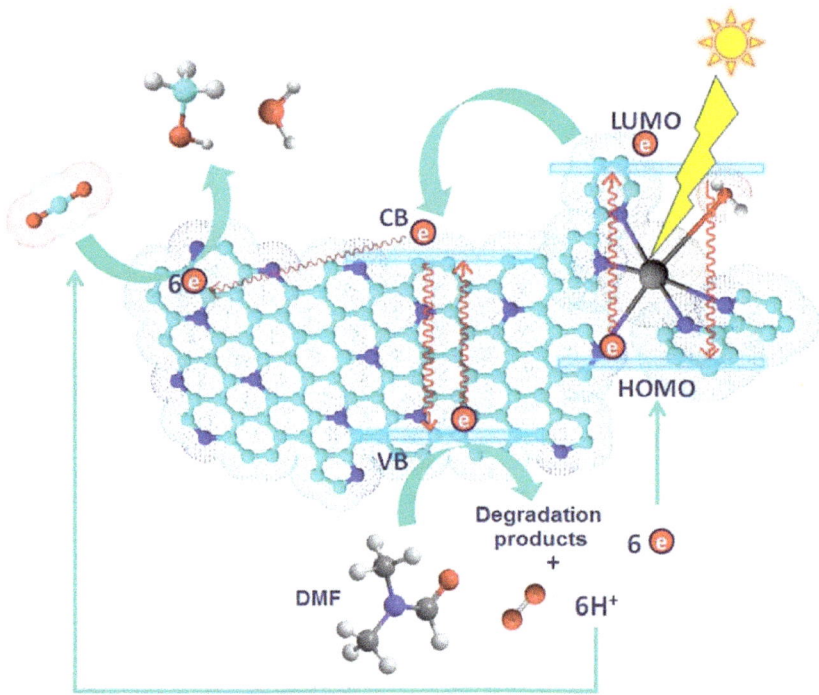

Figure 16. Plausible mechanism of CO_2 reduction by using a GrN_{700}–CuC catalyst. Reprinted with permission from Ref. [241]. Copyright©Royal Society of Chemistry 2023.

First of all, we found that the heterojunction can significantly improve the reaction efficiency in the reduction reaction of CO_2, because the graphene heterojunction effectively reduces the energy band structure of CO_2 and can more easily dissociate the C-O bond (750 kJ·mol^{-1}) [242]. Doping heterostructures into graphene can effectively improve the function of graphene and significantly reduce its energy band structure. For example, a graphene/ZnV_2O_6 heterostructure can reduce its energy band structure to 0.025 eV. This is because heterostructures in graphene can effectively migrate electrons from the interior to the surface. They are excited to graphene through the electrostatic field, promoting the separation of electron-hole pairs and inhibiting the recombination of carriers. In this way, a large number of electrons can be formed on the graphene surface, and a large number of holes can be accumulated on the heterostructure. Such a good combination can greatly promote charge transfer and photocatalytic efficiency [243] (Figure 17).

Zhiling Tang et al. developed the ternary catalyst of rGO-coated Ag/Cu_2O-octahedron nanocrystals (Ag/Cu_2O@rGO). In order to obtain materials with more photoelectrons, Tang et al. coated silver nanoparticles on the materials, according to the characteristics of silver nanoparticles with low Fermi energy, and finally obtained more photoelectrons in Cu_2O. At the same time, because of the high specific surface area and two-dimensional honeycomb structure of GO after sp^2 hybridization, it can adsorb gas better and enhance activation ability by coating a Ag/Cu_2O surface. In the end, they concluded from a number of experiments that the ternary heterojunction can effectively reduce the CO_2 content in photocatalysis and selectively produce CH_4 [242]. Similarly, Fei Li, Li Zhang, and others also proved that when exposed to visible light (λ4400 nm), the use of a single double-sided Cu_2O/graphene/TiO_2 nanotube array (TNA) heterostructure as a separate oxidation and reduction catalyst, which uses anode TNA as the substrate, electrodeposits graphene and Cu_2O in turn. The results of photoelectrochemical measurements show that the ternary heterogeneous materials display the advantages of each component, improve the photocatalytic performance significantly, and show excellent performance in light absorption and the reduction of electron-hole pair coincidence; in other words, ternary heterogeneous materials are excellent catalysts for photocatalytic reactions [244].

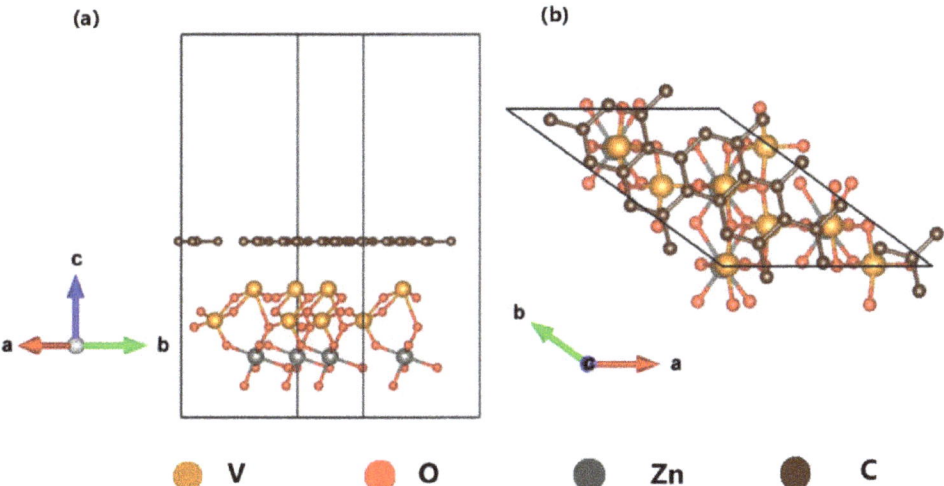

Figure 17. Model for the graphene/ZnV$_2$O$_6$(001) heterostructure viewed from the side (**a**) and the top (**b**). Reprinted with permission from Ref. [243]. Copyright © 2023 Elsevier B.V. or its licensors or contributors.

It is concluded from the above-mentioned optimization strategies that the inhibition of electron-hole-to-ground recombination is the key to enhancing photocatalytic efficiency. Graphene-based photocatalytic engineering is an effective way to reduce CO$_2$ and convert it into fuel because of its superior photocatalytic properties when combined with metal oxides.

Graphene's broad absorption spectrum is one of the main reasons for its excellent photosensitivity, and its excellent performance in light absorption is an effective way to improve the efficiency of photocatalysis. Li and his colleagues reported that the addition of graphene to Pt-TiO enhanced the light absorption. In this process, finding the appropriate doping stoichiometry is the key to improving the optical absorption efficiency.

A great deal of research has been conducted on the preparation of highly efficient heterojunction photocatalysts. Rui Sun et al. obtained the heterojunction of a perylene diimide/Graphene-g-C$_3$N$_4$(PDI/G-CN) nanosheet and demonstrated the potential practicability of an S-type heterojunction in photocatalysis [245].

Radovic et al. have found that the transfer of photogenerated electrons occurs through Valery Karpin sites in graphene after doping and that the surface of the doped graphene has more oxygen vacancies, and realignment occurs. The active sites on the surface provide more sites for CO$_2$ reduction [246].

Khaja Mohaideen Kamal, Rekha Narayan, Narendraraj Chandran, and others studied the synergistic enhancement of plasma gold nanoparticles on a TiO$_2$-modified N-graphene heterostructure catalyst for CO$_2$ reduction for highly selective methane production. Doped graphene itself has excellent photocatalytic performance, and TiO$_2$ also plays an excellent role in synergistic catalysis. Properly designed plasma gold nanoparticles electrodeposited onto TiO$_2$-modified N-doped graphene (ANGT-x) heterostructure catalysts exhibit significant CO$_2$ reduction activity and high methane generation selectivity. Compared with the typical binary Au-TiO$_2$ photocatalyst, the electron consumption rate (Repetron) value of the reduction product of ANGT2 is about 742.39 μ mol g^{-1}h^{-1}, the latter is 60 times more than the former. As far as we know, this is the highest PCO2R rate reported under comparable conditions. These remarkable improvements are attributed to good light selectivity and improved electron transfer dynamics. At the same time, due to the formation of the heterojunction by doping, the components form a seamless interface. Their contact greatly inhibits the recombination of electron-hole pairs and improves the efficiency of charge separation.

4.2.2. Compound Materials

The combination of graphene and semiconductors as a synergistic heterogeneous composite can improve the efficiency of photocatalysis or electrochemical CO_2 reduction. On the premise that the coupling of CO_2 reduction and H_2O oxidation forms a complete photoelectrochemical cycle, graphene can change the conduction-band potential of semiconductors to form graphene-based composites, which leads to the reduction of CO_2. It has been reported that graphene/WO_3 nanobelt composites can reduce CO_2 to CH_4, in which graphene improves the conduction band of WO_3 in the composites, while single WO_3 inherently limits the reduction of CO_2.

The main preparation methods of graphene matrix composites include physical coating, CVD, electrophoresis, and electrodeposition. The physical coating method is to attach the available graphene or graphene-based composite materials to the target materials without changing their physical and chemical properties. The CVD method has been briefly explained in the above section on chemical doping. Electrophoretic deposition (EPD) is a general processing technology for depositing graphene with controllable thickness and uniform structure on a wide range of substrates [247–249].

Herein, we show you the preparation method for TiO_2-rGO nanocomposites developed by Liu et al. They used GO and TiO_2 nanoparticles as starting materials to realize the efficient photocatalysis of GO composites through simple and relatively general methods.

It is difficult to reduce CO_2 efficiently simply by using components as photocatalytic materials. Metal–organic framework (MOF)/GO composites were prepared for CO_2 capture under flue gas conditions by using the method of Mégane Muschi and Sabine Devautour-Vinot, and its performance greatly exceeds that of pure components. Based on the microporous water-stabilized MIL-91(Ti), a series of CO_2-capture composites were prepared by in situ and post-synthesis methods. It was observed that 5 wt% GO in situ composites exhibited semi-conductive behavior, whereas the composite was insulating, even though the GO content was high (20 wt%). Therefore, compared with pure MOFs and post-synthesized materials, this composite material absorbs microwave radiation more effectively. Finally, Muschi and Devautour-Vinot reported that CO_2 desorption under microwave irradiation is faster than direct electrical heating on MOF/GO in situ materials, paving the way for energy-efficient microwave-swing adsorption processes in the future [250].

Graphene itself is a two-dimensional honeycomb material. It is precise because of its structural characteristics of a high specific surface area, which greatly increases its reaction efficiency. Zhuxing Sun and Yun Hang Hu proposed several 3D structure graphene materials that can reduce the accumulation of graphene sheets and help improve the photocatalytic efficiency of graphene-based materials. They found that the exothermic reaction between an alkali metal and inorganic carbon compounds provides an ideal solution for the efficient utilization of CO_2 and the cost-effective production of 3D graphene. In many reactions, in order to solve the problem of building up micro-scale 3D graphene materials, it is necessary to form alkali metals and CO_2 graphene at the same time and use them as original templates. Etching graphene by CO_2 is the key project to control the formation of porous surface structures, so it also provides a good scheme for constructing various excellent 3D graphene materials [251].

The current research on graphene-based composites is aimed at reducing CO_2 emissions. There are many types of materials in this field, but they are mainly based on graphene for the mixing of three or other components. Barbara Szczęśniak et al. have studied three-component GO/ordered mesoporous carbon/metal organic matrix composites to address the need for expensive organic binders in the field of composites, and they found that the three-component composite also contributes to the formation of MOF crystals during air raids on mesoporous networks and CO_2 recovery [252].

4.2.3. Bandgap Engineering

Graphene has a strong s bond between its two carbon atoms due to its "sp^2" hybridization, which greatly enhances its structural strength and flexibility. Based on this hybrid, graphene "sp^2" hybrid-connected carbon atoms can be tightly stacked into a single-layer two-dimensional honeycomb lattice structure. Bandgap engineering can effectively adjust the bandgap of graphene, and it plays a role as a charge-transfer carrier to enhance the efficiency of electron transfer and to increase the reduction rate of CO_2 [144].

The translational symmetry of the original graphene is broken, that is, when two equivalent atoms in the original cell become unequal, a band gap will be generated. The experiment of introducing a band gap into the original graphene includes growing graphene on some substrates, applying an external bias voltage, generating graphene in the form of narrowband graphene, and perforating graphene to form a "graphite nanonet" [253,254].

Through a lot of research on graphene bandgap engineering [255–262], it is found that there are two main methods to change the band gap of graphene: (1) By triggering the quantum confinement effect to open the obvious band gap, which requires greatly reducing the size of graphene to 10 nm through nano-level operations to form nanoribbons. (2) By the adsorption or coating of some substances on the surface of graphene to affect its reaction, thereby changing the electronic structure of graphene. In practice, the first method is not often used in practical applications because it is affected by the nanoribbons and bandgap openings [263,264].

It has been proved by experiments that if nanopores or molecularly modified graphene are introduced into the original graphene, the required translational symmetry can be removed, thus opening the band gap. Doping some elements into graphene can open the band gap. For example, the hydrogenation of graphene can effectively adjust the band gap of graphene. In addition, some graphite intercalation compounds (GIC) are outstanding in inducing superconductivity [265].

At the same time, because the volatile organic compounds adsorbed on the surface of the catalyst combine with the lattice oxygen on the surface of the catalyst, the catalyst surface generates oxygen holes and is reduced, and the catalyst is oxidized by filling the oxygen vacancies with the dissociated adsorbed oxygen. The resulting graphene derivatives, such as GO and rGO, have the characteristics of insulators and similar semiconductors in terms of photocatalysts, resulting in wide band gaps. GO has a limited band gap due to the two-dimensional grid generated by its "sp^2" and "sp^3" hybridization. Therefore, the proportion of operating hybrid-bonding atoms can effectively adjust the broad band gap of GO, transforming it from an insulator to a semiconductor and into a metal similar to graphene. The oxidation functional groups around GO can operate through H_3PO_4. The amount of H_3PO_4 is the key to the operation affecting the oxidation functional groups, which leads to a tuned band gap that is well aligned with the CO_2/CH_3OH redox potential (vs. NHE, pH = 7.0) and shows an improved activity of CO_2 photoreduction to methanol. The conversion of CO_2 to methanol requires stretching the band gap of GO so that there is a CH_4 production orientation in the CO_2 conversion. After much research, it has been found that there are many oxygen groups on the surface of GO, which can effectively stretch the band gap of GO. This stretching regulates the oxidation and reduction potential of CO_2 to match the valence band and the conduction band and promotes the generation and excitation of photogenerated electrons, which can migrate to the surface more effectively and induce CO_2 reduction.

Sokal reported on the bandgap engineering of graphene. We know that photocatalysis produces photogenerated electrons, while doped graphene exhibits better charge separation properties. In his paper, Sokal pointed out that the graphene doped with TiO_2 can separate the electron and hole more effectively and keep them in different positions. Tang and colleagues have also demonstrated that graphene-coated metals exhibit rapid CO_2 reduction over a longer period of time [266].

In order to effectively adjust the band gap structure of graphene, Wu et al. calculated the energy band structure of polycrystalline graphene through the density functional theory (DFT). The results showed that even under the condition of external stress, the grain boundaries (GBs) with symmetrical polycrystalline graphene still have zero band gap, while some asymmetric GBs can also open the band gap, which can be adjusted by external stress (Figure 18). The discovery of this study plays an important role in the further study of graphene bandgap engineering.

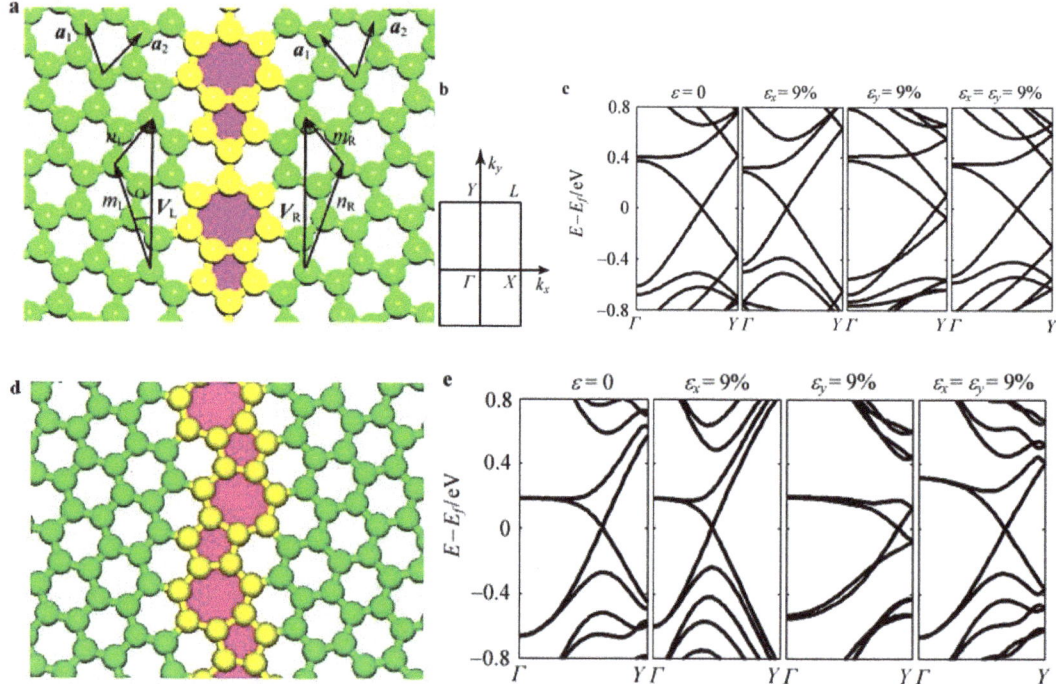

Figure 18. GB structure and band structure of symmetrical polycrystalline graphene. (**a**) Detailed GB structure of armchair-tilt (θ = 21.8°) graphene. (**b**) The first Brillouin zone of all structures and special K points used to calculate the band structures. (**c**) The corresponding band structures for the structure in (**a**) with different types of strains. (**d**) Detailed GB structure of zigzag-tilted boundary with θ = 27.8°. (**e**) The corresponding band structures for the structure in (**d**) with different types of strain. Reprinted with permission from Ref. [267]. Copyright©2023 Springer Nature Switzerland AG.

5. Summary and Outlook

Graphene and its derived materials are functional materials that have emerged in recent years. They have also been involved in a series of breakthrough discoveries in the field of photocatalysis and pollutant degradation and have spawned the creation of many new photocatalytic materials and applications. So far, many of their advantages have been exploited and utilized by people, and there will be more excellent potential applications in the future. However, there is no denying that graphene also has some shortcomings to be overcome, mainly involving the following aspects:

(1) Graphene semiconductor materials have been deeply studied in the field of photocatalysis, and they have shown good photocatalytic material properties when combined with many precious metals. They can reduce the bandgap energy and improve the electron-

hole yield. They are a good choice to reduce the greenhouse gas CO_2 and to obtain chemical fuel–metal composite graphene materials similar to CH_4 that can be used. However, as a new material, at the environmental level, the economic cost and environmental resource cost of using precious metals to improve the photocatalytic effect are still huge. If some metals are not properly treated, this may cause pollution to the environment and indirectly increase the cost of the subsequent treatment of materials. Metal–graphene composites with low cost and high photocatalytic energy efficiency remain to be found. TiO_2 can efficiently utilize ultraviolet light. Although many TiO_2 composite graphene materials can expand the range of the photocatalytic reaction light wave, the effect still has a large room for improvement. Making full use of the visible-light band will greatly improve the photocatalytic energy efficiency. In graphene bandgap engineering, the band gap can be opened by reducing the size of the graphene material [268]. This method does provide a certain bandgap engineering idea, but in view of the edge structure of the bandgap opening and its strong chemical modification sensitivity, the application of this method is still challenging.

(2) By constructing two-dimensional and three-dimensional graphene composites, or by constructing porous materials, the high surface-area ratio, optical properties, and mechanical structure stability of graphene can be maximized. These characteristics can increase the photocatalytic reaction area, CO_2 absorption rate, water decomposition efficiency, etc. and significantly improve the photocatalytic effect. Furthermore, there will be no environmental disadvantages of noble-metal composite graphene photocatalytic materials. However, to be economically viable, researchers need to continue to push for lower material costs so that they can be produced on a large scale and use solar energy efficiently. At the same time, there are also some problems to be addressed in the construction of heterojunction structures by organic non-metallic graphene materials. For example, the existence time of two carriers excited by quantum light is short and the photocatalytic efficiency is low, and it is difficult to avoid agglomeration when graphene nanomaterials are introduced. The reaction site of photocatalysis is not positively correlated with the amount of material input. According to the semiconductor properties of graphene, the development and production of highly active graphene-based photocatalysts can be further studied. For different types of graphene-based materials (open-bandgap graphene materials, zero-bandgap perfect-single-layer graphene, etc.), their electrical properties may have unexpected effects in their applications. At the same time, the application of graphene-based composites in photochromic or electrochromic fields is not mature. Metal oxide-based composite graphene materials or glass-fiber composite graphene may make full use of the semiconductor properties of graphene to participate in the redox process of electrochromic materials and to promote the development of products such as "smart glass". Furthermore, in catalytic reactions, due to the easy recombination effect of electron-hole pairs, we need to introduce heteroatoms to adjust the band structure of graphene and to reduce the recombination rate of electron-hole pairs. For different types of doped graphene, we have made a detailed overview of metal-doped graphene and non-metal-doped graphene and have given certain conclusions in terms of their basic electronic structure, catalytic active site, and morphology. However, the structural characteristics and functions of doped graphene still need further discussion and research.

(3) The use of graphene for pollutant degradation has given rise to more graphene-based composites for environmentally friendly cleaning purposes or to enhance the performance of graphene. However, most forms are just composite materials, composite membranes, composite hydrogels, etc., and few become a green and clean treatment system. They have shown good performance in the laboratory, and some can even achieve a 100% removal rate. They also show good recyclability after an adsorption cycle and will not cause secondary pollution to the environment. However, most of them only have good adsorption for individual pollutants, and the removal efficiency for most of the pollutants not mentioned in the discussion are still unknown. In addition, it is also necessary to consider whether the laboratory simulation environment or sampling can fully match the

real environment. The environment of different regions is also under different conditions. How to carry out the next industrialization or commercialization step is still a question mark. The cost of material manufacturing or cleaning system construction is one of the issues we need to consider. Nonetheless, if there is a better application than the current choice, that would be a small success.

(4) Some defects of graphene-based photocatalyst lead to its low recycling rate, which is rarely considered in most of the current literature. For example, the synthesis of N-doped graphene-based photocatalysts will produce vacancy defects at the same time, which significantly affects the activity of photocatalytic materials. For environmental protection and sustainable development, how to maintain the stability and service life of graphene-based photocatalysts is a direction worthy of study.

(5) Graphene has become an important photocatalytic material in the 21st century. In this article, we discuss in detail the bandgap engineering for constructing graphene, which can fully utilize its role as a charge carrier and improve electron transfer efficiency and CO_2 reduction rate. These methods include adding graphene to the substrate, applying an external voltage, producing graphene in the form of narrowband graphene, and using "graphite nanogrids". The specific graphene band gap provides enormous room for progress in photocatalytic reactions and will be further expanded and developed in the future field of photocatalysis. Single-layer pure graphene sheets lack hydrophilic functional groups, and the preparation of semiconductor/pure graphene composites is extremely challenging. At the same time, as a 2D material, it is still worth exploring how to achieve large-area interface dynamics with other non-2D materials or to load various components to achieve charge separation in space.

(6) Studies suggest that cytotoxicity mainly depends on the physical and chemical properties of nanomaterials. Surface-doped chemical or biological molecules will also affect the expression of properties to some extent. Nowadays, many graphene products exist in the market with smart devices and sensors, which also creates a problem. Graphene-based materials have toxic effects. When put into use, graphene nanoparticles have a potential risk of exposure to air, which will cause certain irritation to the human body and affect health. In order to give full play to the commercial value of graphene nanocomposites in practical market applications, more in-depth research still needs to be conducted in this area.

Author Contributions: Conceptualization, Z.Y.; writing—original draft preparation, Z.Y. and S.Z.; writing—review and editing, O.O.; visualization, X.F.; Funding, N.W. and Y.Z. All authors have read and agreed to the published version of the manuscript.

Funding: This study was supported by grants from the National Natural Science Foundation of China (12205056).

Data Availability Statement: Not applicable.

Conflicts of Interest: The authors declared that they have no conflicts of interest to this work.

References

1. Zhang, S.; Gu, P.; Ma, R.; Luo, C.; Wen, T.; Zhao, G.; Cheng, W.; Wang, X. Recent developments in fabrication and structure regulation of visible-light-driven g-C3N4-based photocatalysts towards water purification: A critical review. *Catal. Today* **2019**, *335*, 65–77. [CrossRef]
2. Freund, H.J.; Roberts, M.W. Surface chemistry of carbon dioxide. *Surf. Sci. Rep.* **1996**, *25*, 225–273. [CrossRef]
3. Sakakura, T.; Choi, J.C.; Yasuda, H. Transformation of carbon dioxide. *Chem. Rev.* **2007**, *107*, 2365–2387. [CrossRef]
4. Feng, J.; Huang, H.; Yan, S.; Luo, W.; Yu, T.; Li, Z.; Zou, Z. Non-oxide semiconductors for artificial photosynthesis: Progress on photoelectrochemical water splitting and carbon dioxide reduction. *Nano Today* **2020**, *30*, 100830. [CrossRef]
5. Ahmad, A.; Ali, M.; Al-Sehemi, A.G.; Al-Ghamdi, A.A.; Park, J.-W.; Algarni, H.; Anwer, H. Carbon-integrated semiconductor photocatalysts for removal of volatile organic compounds in indoor environments. *Chem. Eng. J.* **2023**, *452*, 139436. [CrossRef]
6. Chi, K.; Wu, Z.; Tian, X.; Wang, Z.; Xiao, F.; Xiao, J.; Wang, S. Boosting hydrogen evolution via integrated construction and synergistic cooperation of confined graphene/CoSe2 active interfaces and 3D graphene nanomesh arrays. *Appl. Catal. B Environ.* **2023**, *324*, 122256. [CrossRef]

7. Lin, S.; Lu, Y.; Xu, J.; Feng, S.; Li, J. High performance graphene/semiconductor van der Waals heterostructure optoelectronic devices. *Nano Energy* **2017**, *40*, 122–148. [CrossRef]
8. Lee, S.J.; Theerthagiri, J.; Nithyadharseni, P.; Arunachalam, P.; Balaji, D.; Kumar, A.M.; Madhavan, J.; Mittal, V.; Choi, M.Y. Heteroatom-doped graphene-based materials for sustainable energy applications: A review. *Renew. Sustain. Energy Rev.* **2021**, *143*, 110849. [CrossRef]
9. Hu, Y.; Zhou, C.; Wang, H.; Chen, M.; Zeng, G.; Liu, Z.; Liu, Y.; Wang, W.; Wu, T.; Shao, B.; et al. Recent advance of graphene/semiconductor composite nanocatalysts: Synthesis, mechanism, applications and perspectives. *Chem. Eng. J.* **2021**, *414*, 128795. [CrossRef]
10. Cheng, C.; Liang, Q.; Yan, M.; Liu, Z.; He, Q.; Wu, T.; Luo, S.; Pan, Y.; Zhao, C.; Liu, Y. Advances in preparation, Y., mechanism and applications of graphene quantum dots/semiconductor composite photocatalysts: A review. *J. Hazard. Mater.* **2022**, *424 Pt D*, 127721. [CrossRef]
11. Yan, W.; Xiao, F.; Li, X.; He, W.; Yao, Y.; Wan, D.; Liu, Y.; Feng, F.; Zhang, Q.; Lu, C.; et al. Nickel and oxygen-containing functional groups co-decorated graphene-like shells as catalytic sites with excellent selective hydrogenation activity and robust stability. *Chem. Eng. J.* **2023**, *452*, 139361. [CrossRef]
12. Wang, D.; Dong, S.; Hu, H.; He, Z.; Dong, F.; Tang, J.; Lu, X.; Wang, L.; Song, S.; Ma, J. Catalytic ozonation of atrazine with stable boron-doped graphene nanoparticles derived from waste polyvinyl alcohol film: Performance and mechanism. *Chem. Eng. J.* **2023**, *455*, 140316. [CrossRef]
13. Kumwimba, M.N.; Zhu, B.; Wang, T.; Dzakpasu, M.; Li, X. Nutrient dynamics and retention in a vegetated drainage ditch receiving nutrient-rich sewage at low temperatures. *Sci. Total Environ.* **2020**, *741*, 140268. [CrossRef] [PubMed]
14. Zhang, S.; Li, B.; Wang, X.; Zhao, G.; Hu, B.; Lu, Z.; Wen, T.; Chen, J.; Wang, X. Recent developments of two-dimensional graphene-based composites in visible-light photocatalysis for eliminating persistent organic pollutants from wastewater. *Chem. Eng. J.* **2020**, *390*, 124642. [CrossRef]
15. Lewis, N.S.; Nocera, D.G. Powering the planet: Chemical challenges in solar energy utilization. *Proc. Natl. Acad. Sci. USA* **2006**, *103*, 15729–15735. [CrossRef]
16. Theerthagiri, J.; Senthil, R.; Senthilkumar, B.; Polu, A.R.; Madhavan, J.; Ashokkumar, M. Recent advances in MoS 2 nanostructured materials for energy and environmental applications—A review. *J. Solid State Chem.* **2017**, *252*, 43–71. [CrossRef]
17. Jayaraman, T.; Murthy, A.P.; Elakkiya, V.; Chandrasekaran, S.; Nithyadharseni, P.; Khan, Z.; Senthil, R.A.; Shanker, R.; Raghavender, M.; Kuppusami, P.; et al. Recent development on carbon based heterostructures for their applications in energy and environment: A review. *J. Ind. Eng. Chem.* **2018**, *64*, 16–59. [CrossRef]
18. Theerthagiri, J.; Salla, S.; A Senthil, R.; Nithyadharseni, P.; Madankumar, A.; Arunachalam, P.; Maiyalagan, T.; Kim, H.-S. A review on ZnO nanostructured materials: Energy, environmental and biological applications. *Nanotechnology* **2019**, *30*, 392001. [CrossRef]
19. Theerthagiri, J.; Madhavan, J.; Lee, S.J.; Choi, M.Y.; Ashokkumar, M.; Pollet, B.G. Sonoelectrochemistry for energy and environmental applications. *Ultrason. Sonochem.* **2020**, *63*, 104960. [CrossRef]
20. Bora, A.P.; Gupta, D.P.; Durbha, K.S. Sewage sludge to bio-fuel: A review on the sustainable approach of transforming sewage waste to alternative fuel. *Fuel* **2020**, *259*, 116262. [CrossRef]
21. Liu, X.; Zhu, F.; Zhang, R.; Zhao, L.; Qi, J. Recent progress on biodiesel production from municipal sewage sludge. *Renew. Sustain. Energy Rev.* **2021**, *135*, 110260. [CrossRef]
22. Świerczek, L.; Cieślik, B.M.; Konieczka, P. Challenges and opportunities related to the use of sewage sludge ash in cement-based building materials—A review. *J. Clean. Prod.* **2021**, *287*, 125054. [CrossRef]
23. Zhong, L.; Ding, J.; Wu, T.; Zhao, Y.-L.; Pang, J.W.; Jiang, J.-P.; Jiang, J.-Q.; Li, Y.; Ren, N.-Q.; Yang, S.-S. Bibliometric overview of research progress, challenges, and prospects of rural domestic sewage: Treatment techniques, resource recovery, and ecological risk. *J. Water Process Eng.* **2023**, *51*, 103389. [CrossRef]
24. Lu, F.; Astruc, D. Nanomaterials for removal of toxic elements from water. *Coord. Chem. Rev.* **2018**, *356*, 147–164. [CrossRef]
25. Al-Omari, A.; Muhammetoglu, A.; Karadirek, E.; Jiries, A.; Batarseh, M.; Topkaya, B.; Soyupak, S. A Review on Formation and Decay Kinetics of Trihalomethanes in Water of Different Qualities. *Clean-Soil Air Water* **2014**, *42*, 1687–1700. [CrossRef]
26. Duranceau, S.J.; Smith, C.T. Trihalomethane Formation Downstream of Spray Aerators Treating Disinfected Groundwater. *J. Am. Water Work. Assoc.* **2016**, *108*, E99–E108. [CrossRef]
27. Kinani, S.; Roumiguières, A.; Bouchonnet, S. A Critical Review on Chemical Speciation of Chlorine-Produced Oxidants (CPOs) in Seawater. Part 1: Chlorine Chemistry in Seawater and Its Consequences in Terms of Biocidal Effectiveness and Environmental Impact. *Crit. Rev. Anal. Chem.* **2022**. [CrossRef] [PubMed]
28. Rahman, A.; Farrok, O.; Haque, M. Environmental impact of renewable energy source based electrical power plants: Solar, wind, hydroelectric, biomass, geothermal, tidal, ocean, and osmotic. *Renew. Sustain. Energy Rev.* **2022**, *161*, 112279. [CrossRef]
29. Daaboul, J.; Moriarty, P.; Honnery, D. Net green energy potential of solar photovoltaic and wind energy generation systems. *J. Clean. Prod.* **2023**, *415*, 137806. [CrossRef]
30. Brillas, E. Solar photoelectro-Fenton: A very effective and cost-efficient electrochemical advanced oxidation process for the removal of organic pollutants from synthetic and real wastewaters. *Chemosphere* **2023**, *327*, 138532. [CrossRef]
31. Mamaghani, A.H.; Haghighat, F.; Lee, C.-S. Photocatalytic oxidation technology for indoor environment air purification: The state-of-the-art. *Appl. Catal. B Environ.* **2017**, *203*, 247–269. [CrossRef]

32. Hammad, M.; Angel, S.; Al-Kamal, A.K.; Asghar, A.; Amin, A.S.; Kräenbring, M.-A.; Wiedemann, H.T.A.; Vinayakumar, V.; Ali, M.Y.; Fortugno, P.; et al. Synthesis of novel LaCoO$_3$/graphene catalysts as highly efficient peroxymonosulfate activator for the degradation of organic pollutants. *Chem. Eng. J.* **2023**, *454*, 139900. [CrossRef]
33. Liu, G.-Y.; Li, K.-K.; Jia, J.; Zhang, Y.-T. Coal-based graphene as a promoter of TiO$_2$ catalytic activity for the photocatalytic degradation of organic dyes. *Carbon* **2023**, *203*, 899. [CrossRef]
34. Min, D.H.; Han, X.; Li, N.; Jung, M.G.; Lee, S.J.; Park, H.W.; Lee, J.Y.; Park, H.S. Catalytic active interfacial B–C bonds of boron nanosheet/reduced graphene oxide heterostructures for efficient oxygen reduction reaction. *Compos. Part B Eng.* **2023**, *252*, 110496. [CrossRef]
35. Lu, K.-Q.; Yuan, L.; Xin, X.; Xu, Y.-J. Hybridization of graphene oxide with commercial graphene for constructing 3D metal-free aerogel with enhanced photocatalysis. *Appl. Catal. B Environ.* **2018**, *226*, 16–22. [CrossRef]
36. Bilal, M.; Rashid, E.U.; Zdarta, J.; Jesionowski, T. Graphene-based nanoarchitectures as ideal supporting materials to develop multifunctional nanobiocatalytic systems for strengthening the biotechnology industry. *Chem. Eng. J.* **2023**, *452*, 139509. [CrossRef]
37. Yu, W.; Liu, L.; Yang, Y.; Li, N.; Chen, Y.; Yin, X.; Niu, J.; Wang, J.; Ding, S. N, O-diatomic dopants activate catalytic activity of 3D self-standing graphene carbon aerogel for long-cycle and high-efficiency Li-CO$_2$ batteries. *Chem. Eng. J.* **2023**, *465*, 142787. [CrossRef]
38. Esfandiari, M.; Lalbakhsh, A.; Shehni, P.N.; Jarchi, S.; Ghaffari-Miab, M.; Mahtaj, H.N.; Reisenfeld, S.; Alibakhshikenari, M.; Koziel, S.; Szczepanski, S. Recent and emerging applications of Graphene-based metamaterials in electromagnetics. *Mater. Des.* **2022**, *221*, 110920. [CrossRef]
39. Esfandiyari, M.; Lalbakhsh, A.; Jarchi, S.; Ghaffari-Miab, M.; Mahtaj, H.N.; Simorangkir, R.B.V.B. Tunable terahertz filter/antenna-sensor using graphene-based metamaterials. *Mater. Des.* **2022**, *220*, 110855. [CrossRef]
40. Chen, Y.; Zhang, A.; Ding, L.; Liu, Y.; Lu, H. A three-dimensional absorber hybrid with polar oxygen functional groups of MWNTs/graphene with enhanced microwave absorbing properties. *Compos. Part B Eng.* **2017**, *108*, 386–392. [CrossRef]
41. Lin, K.T.; Lin, H.; Yang, T.; Jia, B. Structured graphene metamaterial selective absorbers for high efficiency and omnidirectional solar thermal energy conversion. *Nat. Commun.* **2020**, *11*, 1389. [CrossRef] [PubMed]
42. Xu, Y.-L.; Li, E.-P.; Wei, X.-C.; Yi, D. A Novel Tunable Absorber Based on Vertical Graphene Strips. *IEEE Microw. Wirel. Compon. Lett.* **2016**, *26*, 10–12. [CrossRef]
43. Zhang, J.; Tian, J.; Li, L. A Dual-Band Tunable Metamaterial Near-Unity Absorber Composed of Periodic Cross and Disk Graphene Arrays. *IEEE Photonics J.* **2018**, *10*, 1–12. [CrossRef]
44. Borah, D.; Bhattacharyya, N.S. Design and Development of Expanded Graphite-Based Non-metallic and Flexible Metamaterial Absorber for X-band Applications. *J. Electron. Mater.* **2017**, *46*, 226–232. [CrossRef]
45. Chen, X.; Jia, X.; Wu, Z.; Tang, Z.; Zeng, Y.; Wang, X.; Fu, X.; Zou, Y. A Graphite-Based Metamaterial Microwave Absorber. *IEEE Antennas Wirel. Propag. Lett.* **2019**, *18*, 1016–1020. [CrossRef]
46. Guo, T.; Sun, Y.; Evans, J.; Wang, N.; Fu, Y.; He, S. Thermal management with a highly emissive and thermally conductive graphite absorber. *Aip Adv.* **2019**, *9*, 025224. [CrossRef]
47. Liu, H.; Wang, Z.-H.; Li, L.; Fan, Y.-X.; Tao, Z.-Y. Vanadium dioxide-assisted broadband tunable terahertz metamaterial absorber. *Sci. Rep.* **2019**, *9*, 5751. [CrossRef]
48. Singhal, S. Wide angle insensitive and polarization independent graphite based superwideband absorber. *Opt. Quantum Electron.* **2022**, *54*, 671. [CrossRef]
49. Norouzi, M.; Jarchi, S.; Ghaffari-Miab, M.; Esfandiari, M.; Lalbakhsh, A.; Koziel, S.; Reisenfeld, S.; Moloudian, G. 3D metamaterial ultra-wideband absorber for curved surface. *Sci. Rep.* **2023**, *13*, 1043. [CrossRef]
50. Borah, D.; Bhattacharyya, N.S. Design, fabrication and characterization of flexible and ultrathin microwave metamaterial absorber. In Proceedings of the International Conference on Innovations in Electronics, Signal Processing and Communication (IESC), Shillong, India, 6–7 April 2017; pp. 190–193.
51. Rani, N.; Saha, S. Graphite based metal-free and polarization-insensitive multiband THz absorber with wide incident angle. *Optik* **2022**, *266*, 169601. [CrossRef]
52. Soni, A.K.; Giri, P.; Varshney, G. Metal-free super-wideband THz absorber for electromagnetic shielding. *Phys. Scr.* **2021**, *96*, 125866. [CrossRef]
53. Soni, A.K.; Varshney, G. Multiband Generation and Absorption Enhancement in a Graphite-Based Metal-Free Absorber. *Plasmonics* **2021**, *16*, 241–252. [CrossRef]
54. Varshney, G. Wideband THz Absorber: By Merging the Resonance of Dielectric Cavity and Graphite Disk Resonator. *IEEE Sens. J.* **2021**, *21*, 1635–1643. [CrossRef]
55. Varshney, G.; Rani, N.; Pandey, V.S.; Yaduvanshi, R.S.; Singh, D. Graphite/graphene disk stack-based metal-free wideband terahertz absorber. *J. Opt. Soc. Am. B-Opt. Phys.* **2021**, *38*, 530–538. [CrossRef]
56. Korkmaz, S.; Kariper, I.A. Graphene and graphene oxide based aerogels: Synthesis, characteristics and supercapacitor applications. *J. Energy Storage* **2020**, *27*, 101038. [CrossRef]
57. Liu, X.; Liang, B.; Long, J. Research progress on graphene aerogel in catalysis. *N. Chem. Mater.* **2021**, *49*, 60–63.
58. Shaikh, J.S.; Shaikh, N.S.; Mishra, Y.K.; Pawar, S.S.; Parveen, N.; Shewale, P.M.; Sabale, S.; Kanjanaboos, P.; Prasertdham, S.; Lokhande, C.D. The implementation of graphene-based aerogel in the field of supercapacitor. *Nanotechnology* **2021**, *32*, 362001. [CrossRef]

59. Zhong, K.; Zhang, C.; Zhong, Y.; Cui, S.; Shen, X. Research progress in the preparation and adsorption capability of graphene aerogel composite materials. *Ind. Water Treat.* **2019**, *39*, 1–6.
60. Wu, F.; Xie, A.; Sun, M.; Wang, Y.; Wang, M. Reduced graphene oxide (RGO) modified spongelike polypyrrole (PPy) aerogel for excellent electromagnetic absorption. *J. Mater. Chem. A* **2015**, *3*, 14358–14369. [CrossRef]
61. Jiang, Y.; Chen, Y.; Liu, Y.-J.; Sui, G.-X. Lightweight spongy bone-like graphene@SiC aerogel composites for high-performance microwave absorption. *Chem. Eng. J.* **2018**, *337*, 522–531. [CrossRef]
62. Gorgolis, G.; Galiotis, C. Graphene aerogels: A review. *2D Mater.* **2017**, *4*, 032001. [CrossRef]
63. Jing, J.; Qian, X.; Si, Y.; Liu, G.; Shi, C. Recent Advances in the Synthesis and Application of Three-Dimensional Graphene-Based Aerogels. *Molecules* **2022**, *27*, 924. [CrossRef] [PubMed]
64. Li, A.; Pei, C.; Zhu, Z.; An, J.; Qin, X.; Bao, X. Progress in graphene aerogels. *Mod. Chem. Ind.* **2013**, *33*, 20–23.
65. Wang, Z.; Liu, L.; Zhang, Y.; Huang, Y.; Liu, J.; Zhang, X.; Liu, X.; Teng, H.; Zhang, X.; Zhang, J.; et al. A Review of Graphene-Based Materials/Polymer Composite Aerogels. *Polymers* **2023**, *15*, 1888. [CrossRef] [PubMed]
66. Wang, Z.; Wei, R.; Gu, J.; Liu, H.; Liu, C.; Luo, C.; Kong, J.; Shao, Q.; Wang, N.; Guo, Z.; et al. Ultralight, highly compressible and fire-retardant graphene aerogel with self-adjustable electromagnetic wave absorption. *Carbon* **2018**, *139*, 1126–1135. [CrossRef]
67. Li, J.; Li, J.; Meng, H.; Xie, S.; Zhang, B.; Li, L.; Ma, H.; Zhang, J.; Yu, M. Compressible and fire-resistant graphene aerogel as a highly efficient and recyclable absorbent for organic liquids. *J. Mater. Chem. A* **2014**, *2*, 2934–2941. [CrossRef]
68. Hong, J.-Y.; Sohn, E.-H.; Park, S.; Park, H.S. Highly-efficient and recyclable oil absorbing performance of functionalized graphene aerogel. *Chem. Eng. J.* **2015**, *269*, 229–235. [CrossRef]
69. Yao, Y.; Zhao, Y. Three-Dimensional Porous Graphene Networks and Hybrids for Lithium-Ion Batteries and Supercapacitors. *Chem* **2017**, *2*, 171–200. [CrossRef]
70. Cao, L.; Wang, C.; Huang, Y. Structure optimization of graphene aerogel-based composites and applications in batteries and supercapacitors. *Chem. Eng. J.* **2023**, *454*, 140094. [CrossRef]
71. Xie, W.; Yao, F.; Gu, H.; Du, A.; Lei, Q.; Naik, N.; Guo, Z. Magnetoresistive and piezoresistive polyaniline nanoarrays in-situ polymerized surrounding magnetic graphene aerogel. *Adv. Compos. Hybrid Mater.* **2022**, *5*, 1003–1016. [CrossRef]
72. Hao, P.; Zhao, Z.; Leng, Y.; Tian, J.; Sang, Y.; Boughton, R.I.; Wong, C.P.; Liu, H.; Yang, B. Graphene-based nitrogen self-doped hierarchical porous carbon aerogels derived from chitosan for high performance supercapacitors. *Nano Energy* **2015**, *15*, 9–23. [CrossRef]
73. Tayebi, M.; Tayyebi, A.; Masoumi, Z.; Lee, B.-K. Photocorrosion suppression and photoelectrochemical (PEC) enhancement of ZnO via hybridization with graphene nanosheets. *Appl. Surf. Sci.* **2020**, *502*, 144189. [CrossRef]
74. Wang, Z.; Huang, J.; Mao, J.; Guo, Q.; Chen, Z.; Lai, Y. Metal-organic frameworks and their derivatives with graphene composites: Preparation and applications in electrocatalysis and photocatalysis. *J. Mater. Chem. A* **2020**, *8*, 2934–2961. [CrossRef]
75. Li, Y.-H.; Tang, Z.-R.; Xu, Y.-J. Multifunctional graphene-based composite photocatalysts oriented by multifaced roles of graphene in photocatalysis. *Chin. J. Catal.* **2022**, *43*, 708–730. [CrossRef]
76. Subramanyam, P.; Vinodkumar, T.; Nepak, D.; Deepa, M.; Subrahmanyam, C. Mo-doped BiVO4@reduced graphene oxide composite as an efficient photoanode for photoelectrochemical water splitting. *Catal. Today* **2019**, *325*, 73–80. [CrossRef]
77. Ranjan, R.; Kumar, M.; Sinha, A.S.K. Development and characterization of rGO supported CdS MoS$_2$ photoelectrochemical catalyst for splitting water by visible light. *Int. J. Hydrogen Energy* **2019**, *44*, 16176–16189. [CrossRef]
78. Qiu, H.J.; Du, P.; Hu, K.; Gao, J.; Li, H.; Liu, P.; Ina, T.; Ohara, K.; Ito, Y.; Chen, M. Metal and Nonmetal Codoped 3D Nanoporous Graphene for Efficient Bifunctional Electrocatalysis and Rechargeable Zn-Air Batteries. *Adv. Mater.* **2019**, *31*, e1900843. [CrossRef]
79. Liu, J.; Kong, X.; Zheng, L.; Guo, X.; Liu, X.; Shui, J. Rare Earth Single-Atom Catalysts for Nitrogen and Carbon Dioxide Reduction. *ACS Nano* **2020**, *14*, 1093–1101. [CrossRef]
80. Bie, C.; Yu, H.; Cheng, B.; Ho, W.; Fan, J.; Yu, J. Design, Fabrication, and Mechanism of Nitrogen-Doped Graphene-Based Photocatalyst. *Adv. Mater.* **2021**, *33*, e2003521. [CrossRef]
81. Hirscher, M.; Yartys, V.A.; Baricco, M.; von Colbe, J.B.; Blanchard, D.; Bowman, R.C.; Broom, D.P.; Buckley, C.E.; Chang, F.; Chen, P.; et al. Materials for hydrogen-based energy storage—Past, recent progress and future outlook. *J. Alloys Compd.* **2020**, *827*, 153548. [CrossRef]
82. Li, Y.; Kimura, S. Economic competitiveness and environmental implications of hydrogen energy and fuel cell electric vehicles in ASEAN countries: The current and future scenarios. *Energy Policy* **2021**, *148*, 111980. [CrossRef]
83. Thomas, J.M.; Edwards, P.P.; Dobson, P.J.; Owen, G.P. Decarbonising energy: The developing international activity in hydrogen technologies and fuel cells. *J. Energy Chem.* **2020**, *51*, 405–415. [CrossRef] [PubMed]
84. Ajanovic, A.; Sayer, M.; Haas, R. The economics and the environmental benignity of different colors of hydrogen. *Int. J. Hydrogen Energy* **2022**, *47*, 24136–24154. [CrossRef]
85. Abe, J.O.; Popoola, A.P.I.; Ajenifuja, E.; Popoola, O.M. Hydrogen energy; economy; storage: Review; recommendation. *Int. J. Hydrogen Energy* **2019**, *44*, 15072–15086. [CrossRef]
86. Tarhan, C.; Çil, M.A. A study on hydrogen, the clean energy of the future: Hydrogen storage methods. *J. Energy Storage* **2021**, *40*, 102676. [CrossRef]
87. Fujishima, A.; Honda, K. Electrochemical Photolysis of Water at a Semiconductor Electrode. *Nature* **1972**, *238*, 37–38. [CrossRef]
88. She, H.; Yue, P.; Ma, X.; Huang, J.; Wang, L.; Wang, Q. Fabrication of BiVO$_4$ photoanode cocatalyzed with NiCo-layered double hydroxide for enhanced photoactivity of water oxidation. *Appl. Catal. B Environ.* **2020**, *263*, 118280. [CrossRef]

89. Pan, J.; Wang, P.; Wang, P.; Yu, Q.; Wang, J.; Song, C.; Zheng, Y.; Li, C. The photocatalytic overall water splitting hydrogen production of g-C_3N_4/CdS hollow core–shell heterojunction via the HER/OER matching of Pt/MnO_x. *Chem. Eng. J.* **2021**, *405*, 126622. [CrossRef]
90. Guayaquil-Sosa, J.F.; Serrano-Rosales, B.; Valadés-Pelayo, P.J.; de Lasa, H. Photocatalytic hydrogen production using mesoporous TiO_2 doped with Pt. *Appl. Catal. B Environ.* **2017**, *211*, 337–348. [CrossRef]
91. Mei, F.; Li, Z.; Dai, K.; Zhang, J.; Liang, C. Step-scheme porous g-C_3N_4/$Zn_{0.2}Cd_{0.8}$S-DETA composites for efficient and stable photocatalytic H_2 production. *Chin. J. Catal.* **2020**, *41*, 41–49. [CrossRef]
92. An, X.; Yu, J.C.; Wang, Y.; Hu, Y.; Yu, X.; Zhang, G. WO_3 nanorods/graphene nanocomposites for high-efficiency visible-light-driven photocatalysis and NO_2 gas sensing. *J. Mater. Chem.* **2012**, *22*, 8525–8531. [CrossRef]
93. Marlinda, A.R.; Yusoff, N.; Sagadevan, S.; Johan, M.R. Recent developments in reduced graphene oxide nanocomposites for photoelectrochemical water-splitting applications. *Int. J. Hydrogen Energy* **2020**, *45*, 11976–11994. [CrossRef]
94. Desai, M.A.; Vyas, A.N.; Saratale, G.D.; Sartale, S.D. Zinc oxide superstructures: Recent synthesis approaches and application for hydrogen production via photoelectrochemical water splitting. *Int. J. Hydrogen Energy* **2019**, *44*, 2091–2127. [CrossRef]
95. Kumaravel, V.; Imam, M.; Badreldin, A.; Chava, R.; Do, J.; Kang, M.; Abdel-Wahab, A. Photocatalytic Hydrogen Production: Role of Sacrificial Reagents on the Activity of Oxide, Carbon, and Sulfide Catalysts. *Catalysts* **2019**, *9*, 276. [CrossRef]
96. Guo, Q.; Zhou, C.; Ma, Z.; Yang, X. Fundamentals of TiO(2) Photocatalysis: Concepts, Mechanisms, and Challenges. *Adv. Mater.* **2019**, *31*, e1901997. [CrossRef] [PubMed]
97. Verma, P.; Singh, A.; Rahimi, F.A.; Sarkar, P.; Nath, S.; Pati, S.K.; Maji, T.K. Charge-transfer regulated visible light driven photocatalytic H(2) production and CO(2) reduction in tetrathiafulvalene based coordination polymer gel. *Nat. Commun.* **2021**, *12*, 7313. [CrossRef] [PubMed]
98. Shehzad, N.; Tahir, M.; Johari, K.; Murugesan, T.; Hussain, M. Fabrication of highly efficient and stable indirect Z-scheme assembly of AgBr/TiO_2 via graphene as a solid-state electron mediator for visible light induced enhanced photocatalytic H_2 production. *Appl. Surf. Sci.* **2019**, *463*, 445–455. [CrossRef]
99. Li, S.H.; Qi, M.Y.; Tang, Z.R.; Xu, Y.J. Nanostructured metal phosphides: From controllable synthesis to sustainable catalysis. *Chem. Soc. Rev.* **2021**, *50*, 7539–7586. [CrossRef]
100. Song, Y.-D.; Wang, L.; Wu, L.-M. Theoretical study of the CO, NO, and N_2 adsorptions on Li-decorated graphene and boron-doped graphene. *Can. J. Chem.* **2018**, *96*, 30–39. [CrossRef]
101. Lotfi, E.; Neek-Amal, M. Temperature distribution in graphene doped with nitrogen and graphene with grain boundary. *J. Mol. Graph. Model.* **2017**, *74*, 100–104. [CrossRef]
102. Sturala, J.; Luxa, J.; Plutnar, J.; Janoušek, Z.; Sofer, Z.; Pumera, M. Selenium covalently modified graphene: Towards gas sensing. *2D Mater.* **2019**, *6*, 034006. [CrossRef]
103. Yoo, M.S.; Lee, H.C.; Lee, S.B.; Cho, K. Cu-Phosphorus Eutectic Solid Solution for Growth of Multilayer Graphene with Widely Tunable Doping. *Adv. Funct. Mater.* **2020**, *31*, 2006499. [CrossRef]
104. Chahal, S.; Nair, A.K.; Ray, S.J.; Yi, J.; Vinu, A.; Kumar, P. Microwave flash synthesis of phosphorus and sulphur ultradoped graphene. *Chem. Eng. J.* **2022**, *450*, 138447. [CrossRef]
105. Liu, A.; Li, W.; Jin, H.; Yu, X.; Bu, Y.; He, Y.; Huang, H.; Wang, S.; Wang, J. The enhanced electrocatalytic activity of graphene co-doped with chlorine and fluorine atoms. *Electrochim. Acta* **2015**, *177*, 36–42. [CrossRef]
106. Tongay, S.; Hwang, J.; Tanner, D.B.; Pal, H.K.; Maslov, D.; Hebard, A.F. Supermetallic conductivity in bromine-intercalated graphite. *Phys. Rev. B* **2010**, *81*, 115428. [CrossRef]
107. Albero, J.; Peng, Y.; García, H. Photocatalytic CO_2 Reduction to C2+ Products. *ACS Catal.* **2020**, *10*, 5734–5749. [CrossRef]
108. Guo, Q.; Ma, Z.; Zhou, C.; Ren, Z.; Yang, X. Single Molecule Photocatalysis on TiO(2) Surfaces. *Chem. Rev.* **2019**, *119*, 11020–11041. [CrossRef]
109. He, F.; Meng, A.; Cheng, B.; Ho, W.; Yu, J. Enhanced photocatalytic H2-production activity of WO_3/TiO_2 step-scheme heterojunction by graphene modification. *Chin. J. Catal.* **2020**, *41*, 9–20. [CrossRef]
110. Kiranakumar, H.V.; Thejas, R.; Naveen, C.S.; Khan, M.I.; Prasanna, G.D.; Reddy, S.; Oreijah, M.; Guedri, K.; Bafakeeh, O.T.; Jameel, M. A review on electrical and gas-sensing properties of reduced graphene oxide-metal oxide nanocomposites. *Biomass Convers. Biorefinery* **2022**. [CrossRef]
111. Padmanabhan, N.T.; Thomas, N.; Louis, J.; Mathew, D.T.; Ganguly, P.; John, H.; Pillai, S.C. Graphene coupled TiO_2 photocatalysts for environmental applications: A review. *Chemosphere* **2021**, *271*, 129506. [CrossRef]
112. Purabgola, A.; Mayilswamy, N.; Kandasubramanian, B. Graphene-based TiO_2 composites for photocatalysis & environmental remediation: Synthesis and progress. *Environ. Sci. Pollut. Res.* **2022**, *29*, 32305–32325.
113. Qian, R.; Zong, H.; Schneider, J.; Zhou, G.; Zhao, T.; Li, Y.; Yang, J.; Bahnemann, D.W.; Pan, J.H. Charge carrier trapping, recombination and transfer during TiO_2 photocatalysis: An overview. *Catal. Today* **2019**, *335*, 78–90. [CrossRef]
114. Shi, Z.; Zhang, Y.; Shen, X.; Duoerkun, G.; Zhu, B.; Zhang, L.; Li, M.; Chen, Z. Fabrication of g-C_3N_4/BiOBr heterojunctions on carbon fibers as weaveable photocatalyst for degrading tetracycline hydrochloride under visible light. *Chem. Eng. J.* **2020**, *386*, 124010. [CrossRef]
115. Basavarajappa, P.S.; Patil, S.B.; Ganganagappa, N.; Reddy, K.R.; Raghu, A.V.; Reddy, C.V. Recent progress in metal-doped TiO_2, non-metal doped/codoped TiO_2 and TiO_2 nanostructured hybrids for enhanced photocatalysis. *Int. J. Hydrogen Energy* **2020**, *45*, 7764–7778. [CrossRef]

116. Lee, T.; Bui, H.T.; Yoo, J.; Ra, M.; Han, S.H.; Kim, W.; Kwon, W. Formation of TiO$_2$@Carbon Core/Shell Nanocomposites from a Single Molecular Layer of Aromatic Compounds for Photocatalytic Hydrogen Peroxide Generation. *ACS Appl. Mater. Interfaces* **2019**, *11*, 41196–41203. [CrossRef] [PubMed]
117. Hong, Y.; Cho, Y.; Go, E.M.; Sharma, P.; Cho, H.; Lee, B.; Lee, S.M.; Park, S.O.; Ko, M.; Kwak, S.K.; et al. Unassisted photocatalytic H$_2$O$_2$ production under visible light by fluorinated polymer-TiO$_2$ heterojunction. *Chem. Eng. J.* **2021**, *418*, 129346. [CrossRef]
118. Kamal, K.M.; Narayan, R.; Chandran, N.; Popović, S.; Nazrulla, M.A.; Kovač, J.; Vrtovec, N.; Bele, M.; Hodnik, N.; Kržmanc, M.M.; et al. Synergistic enhancement of photocatalytic CO$_2$ reduction by plasmonic Au nanoparticles on TiO$_2$ decorated N-graphene heterostructure catalyst for high selectivity methane production. *Appl. Catal. B Environ.* **2022**, *307*, 121181. [CrossRef]
119. Baran, T.; Wojtyła, S.; Minguzzi, A.; Rondinini, S.; Vertova, A. Achieving efficient H$_2$O$_2$ production by a visible-light absorbing, highly stable photosensitized TiO$_2$. *Appl. Catal. B Environ.* **2019**, *244*, 303–312. [CrossRef]
120. Wang, X.Y.; Yao, X.; Narita, A.; Mullen, K. Heteroatom-Doped Nanographenes with Structural Precision. *Acc. Chem. Res.* **2019**, *52*, 2491–2505. [CrossRef]
121. Wang, H.; Yuan, X.; Wu, Y.; Zeng, G.; Dong, H.; Chen, X.; Leng, L.; Wu, Z.; Peng, L. In situ synthesis of In$_2$S$_3$@MIL-125(Ti) core–shell microparticle for the removal of tetracycline from wastewater by integrated adsorption and visible-light-driven photocatalysis. *Appl. Catal. B Environ.* **2016**, *186*, 19–29. [CrossRef]
122. Tayebi, M.; Kolaei, M.; Tayyebi, A.; Masoumi, Z.; Belbasi, Z.; Lee, B.-K. Reduced graphene oxide (RGO) on TiO$_2$ for an improved photoelectrochemical (PEC) and photocatalytic activity. *Sol. Energy* **2019**, *190*, 185–194. [CrossRef]
123. Shu, R.; Wan, Z.; Zhang, J.; Wu, Y.; Liu, Y.; Shi, J.; Zheng, M. Facile Design of Three-Dimensional Nitrogen-Doped Reduced Graphene Oxide/Multi-Walled Carbon Nanotube Composite Foams as Lightweight and Highly Efficient Microwave Absorbers. *ACS Appl. Mater. Interfaces* **2020**, *12*, 4689–4698. [CrossRef] [PubMed]
124. Zhao, Q.; Sun, J.; Li, S.; Huang, C.; Yao, W.; Chen, W.; Zeng, T.; Wu, Q.; Xu, Q. Single Nickel Atoms Anchored on Nitrogen-Doped Graphene as a Highly Active Cocatalyst for Photocatalytic H$_2$ Evolution. *ACS Catal.* **2018**, *8*, 11863–11874. [CrossRef]
125. Liu, Y.; Xu, X.; Zheng, S.; Lv, S.; Li, H.; Si, Z.; Wu, X.; Ran, R.; Weng, D.; Kang, F. Ni single atoms anchored on nitrogen-doped graphene as H2-Evolution cocatalyst of SrTiO3(Al)/CoO for photocatalytic overall water splitting. *Carbon* **2021**, *183*, 763–773. [CrossRef]
126. Li, X.; Wang, J.; Duan, X.; Li, Y.; Fan, X.; Zhang, G.; Zhang, F.; Peng, W. Fine-Tuning Radical/Nonradical Pathways on Graphene by Porous Engineering and Doping Strategies. *ACS Catal.* **2021**, *11*, 4848–4861. [CrossRef]
127. Yang, Y.; Zhu, B.; Wang, L.; Cheng, B.; Zhang, L.; Yu, J. In-situ grown N, S co-doped graphene on TiO$_2$ fiber for artificial photosynthesis of H$_2$O$_2$ and mechanism study. *Appl. Catal. B Environ.* **2022**, *317*, 121788. [CrossRef]
128. Kurra, N.; Jiang, Q.; Nayak, P.; Alshareef, H.N. Laser-derived graphene: A three-dimensional printed graphene electrode and its emerging applications. *Nano Today* **2019**, *24*, 81–102. [CrossRef]
129. Yang, L.; Peng, Y.; Luo, X.; Dan, Y.; Ye, J.; Zhou, Y.; Zou, Z. Beyond C$_3$N$_4$ pi-conjugated metal-free polymeric semiconductors for photocatalytic chemical transformations. *Chem. Soc. Rev.* **2021**, *50*, 2147–2172. [CrossRef]
130. Lu, C.; Meng, J.; Zhang, J.; Chen, X.; Du, M.; Chen, Y.; Hou, C.; Wang, J.; Ju, A.; Wang, X.; et al. Three-Dimensional Hierarchically Porous Graphene Fiber-Shaped Supercapacitors with High Specific Capacitance and Rate Capability. *ACS Appl. Mater. Interfaces* **2019**, *11*, 25205–25217. [CrossRef]
131. Xiong, C.; Li, B.; Lin, X.; Liu, H.; Xu, Y.; Mao, J.; Duan, C.; Li, T.; Ni, Y. The recent progress on three-dimensional porous graphene-based hybrid structure for supercapacitor. *Compos. Part B Eng.* **2019**, *165*, 10–46. [CrossRef]
132. Zhang, J.; Xin, B.; Shan, C.; Zhang, W.; Dionysiou, D.D.; Pan, B. Roles of oxygen-containing functional groups of O-doped g-C$_3$N$_4$ in catalytic ozonation: Quantitative relationship and first-principles investigation. *Appl. Catal. B Environ.* **2021**, *292*, 120155. [CrossRef]
133. Han, C.; Su, P.; Tan, B.; Ma, X.; Lv, H.; Huang, C.; Wang, P.; Tong, Z.; Li, G.; Huang, Y.; et al. Defective ultra-thin two-dimensional g-C$_3$N$_4$ photocatalyst for enhanced photocatalytic H$_2$ evolution activity. *J. Colloid Interface Sci.* **2021**, *581 Pt A*, 159–166. [CrossRef] [PubMed]
134. Liu, X.; Liu, H.; Wang, Y.; Yang, W.; Yu, Y. Nitrogen-rich g-C$_3$N$_4$@AgPd Mott-Schottky heterojunction boosts photocatalytic hydrogen production from water and tandem reduction of NO$_3^-$ and NO$_2^-$. *J. Colloid Interface Sci.* **2021**, *581 Pt B*, 619–626. [CrossRef] [PubMed]
135. Wang, S.; Xu, M.; Peng, T.; Zhang, C.; Li, T.; Hussain, I.; Wang, J.; Tan, B. Porous hypercrosslinked polymer-TiO$_2$-graphene composite photocatalysts for visible-light-driven CO$_2$ conversion. *Nat. Commun.* **2019**, *10*, 676. [CrossRef]
136. Lee, K.; Yoon, H.; Ahn, C.; Park, J.; Jeon, S. Strategies to improve the photocatalytic activity of TiO$_2$: 3D nanostructuring and heterostructuring with graphitic carbon nanomaterials. *Nanoscale* **2019**, *11*, 7025–7040. [CrossRef] [PubMed]
137. Xu, J.; Chen, X.; Xu, Y.; Du, Y.; Yan, C. Ultrathin 2D Rare-Earth Nanomaterials: Compositions, Syntheses, and Applications. *Adv. Mater.* **2020**, *32*, e1806461. [CrossRef]
138. Zhang, L.; Long, R.; Zhang, Y.; Duan, D.; Xiong, Y.; Zhang, Y.; Bi, Y. Direct Observation of Dynamic Bond Evolution in Single-Atom Pt/C$_3$N$_4$ Catalysts. *Angew. Chem. Int. Ed. Engl.* **2020**, *59*, 6224–6229. [CrossRef]
139. Jiang, J.; Yu, J.; Cao, S. Au/PtO nanoparticle-modified g-C$_3$N$_4$ for plasmon-enhanced photocatalytic hydrogen evolution under visible light. *J. Colloid Interface Sci.* **2016**, *461*, 56–63. [CrossRef]

140. Wang, Y.; Wang, H.; Li, J.; Zhao, X. Facile synthesis of metal free perylene imide-carbon nitride membranes for efficient photocatalytic degradation of organic pollutants in the presence of peroxymonosulfate. *Appl. Catal. B Environ.* **2020**, *278*, 118981. [CrossRef]
141. Xiao, J.; Xie, Y.; Rabeah, J.; Bruckner, A.; Cao, H. Visible-Light Photocatalytic Ozonation Using Graphitic C_3N_4 Catalysts: A Hydroxyl Radical Manufacturer for Wastewater Treatment. *Acc. Chem. Res.* **2020**, *53*, 1024–1033. [CrossRef]
142. Wen, P.; Sun, Y.; Li, H.; Liang, Z.; Wu, H.; Zhang, J.; Zeng, H.; Geyer, S.M.; Jiang, L. A highly active three-dimensional Z-scheme ZnO/Au/g-C_3N_4 photocathode for efficient photoelectrochemical water splitting. *Appl. Catal. B Environ.* **2020**, *263*, 118180. [CrossRef]
143. Li, W.; Chu, X.-S.; Wang, F.; Dang, Y.-Y.; Liu, X.-Y.; Wang, X.-C.; Wang, C.-Y. Enhanced cocatalyst-support interaction and promoted electron transfer of 3D porous g-C_3N_4/GO-M (Au, Pd, Pt) composite catalysts for hydrogen evolution. *Appl. Catal. B Environ.* **2021**, *288*, 120034. [CrossRef]
144. Wang, J.W.; Qiao, L.Z.; Nie, H.D.; Huang, H.H.; Li, Y.; Yao, S.; Liu, M.; Zhang, Z.M.; Kang, Z.H.; Lu, T.B. Facile electron delivery from graphene template to ultrathin metal-organic layers for boosting CO_2 photoreduction. *Nat. Commun.* **2021**, *12*, 813. [CrossRef]
145. Yang, J.; Miao, H.; Jing, J.; Zhu, Y.; Choi, W. Photocatalytic activity enhancement of PDI supermolecular via π-π action and energy level adjusting with graphene quantum dots. *Appl. Catal. B Environ.* **2021**, *281*, 119547. [CrossRef]
146. Huang, X.; Qi, X.; Boey, F.; Zhang, H. Graphene-based composites. *Chem. Soc. Rev.* **2012**, *41*, 666–686. [CrossRef]
147. He, J.; Hartmann, G.; Lee, M.; Hwang, G.S.; Chen, Y.; Manthiram, A. Freestanding 1T MoS_2/graphene heterostructures as a highly efficient electrocatalyst for lithium polysulfides in Li-S batteries. *Energy Environ. Sci.* **2019**, *12*, 344–350. [CrossRef]
148. Huang, H.; Shi, H.; Das, P.; Qin, J.; Li, Y.; Wang, X.; Su, F.; Wen, P.; Li, S.; Lu, P.; et al. The Chemistry and Promising Applications of Graphene and Porous Graphene Materials. *Adv. Funct. Mater.* **2020**, *30*, 1909035. [CrossRef]
149. Liu, W.; Li, M.; Jiang, G.; Li, G.; Zhu, J.; Xiao, M.; Zhu, Y.; Gao, R.; Yu, A.; Feng, M.; et al. Graphene Quantum Dots-Based Advanced Electrode Materials: Design, Synthesis and Their Applications in Electrochemical Energy Storage and Electrocatalysis. *Adv. Energy Mater.* **2020**, *10*, 2001275. [CrossRef]
150. Zhang, Z.; Zhang, Q.; Chen, Y.; Bao, J.; Zhou, X.; Xie, Z.; Wei, J.; Zhou, Z. The First Introduction of Graphene to Rechargeable Li-CO_2 Batteries. *Angew. Chem.-Int. Ed.* **2015**, *54*, 6550–6553. [CrossRef]
151. Ding, W.; Wei, Z.; Chen, S.; Qi, X.; Yang, T.; Hu, J.; Wang, D.; Wan, L.-J.; Alvi, S.F.; Li, L. Space-Confinement-Induced Synthesis of Pyridinic- and Pyrrolic-Nitrogen-Doped Graphene for the Catalysis of Oxygen Reduction. *Angew. Chem.-Int. Ed.* **2013**, *52*, 11755–11759. [CrossRef]
152. Dou, S.; Tao, L.; Huo, J.; Wang, S.; Dai, L. Etched and doped Co_9S_8/graphene hybrid for oxygen electrocatalysis. *Energy Environ. Sci.* **2016**, *9*, 1320–1326. [CrossRef]
153. Ma, Z.; Dou, S.; Shen, A.; Tao, L.; Dai, L.; Wang, S. Sulfur-Doped Graphene Derived from Cycled Lithium-Sulfur Batteries as a Metal-Free Electrocatalyst for the Oxygen Reduction Reaction. *Angew. Chem.-Int. Ed.* **2015**, *54*, 1888–1892. [CrossRef] [PubMed]
154. Yang, J.; Voiry, D.; Ahn, S.J.; Kang, D.; Kim, A.Y.; Chhowalla, M.; Shin, H.S. Two-Dimensional Hybrid Nanosheets of Tungsten Disulfide and Reduced Graphene Oxide as Catalysts for Enhanced Hydrogen Evolution. *Angew. Chem.-Int. Ed.* **2013**, *52*, 13751–13754. [CrossRef]
155. Zhang, Y.; Zhou, Q.; Zhu, J.; Yan, Q.; Dou, S.X.; Sun, W. Nanostructured Metal Chalcogenides for Energy Storage and Electrocatalysis. *Adv. Funct. Mater.* **2017**, *27*, 1702317. [CrossRef]
156. Zeng, M.; Liu, Y.; Zhao, F.; Nie, K.; Han, N.; Wang, X.; Huang, W.; Song, X.; Zhong, J.; Li, Y. Metallic Cobalt Nanoparticles Encapsulated in Nitrogen-Enriched Graphene Shells: Its Bifunctional Electrocatalysis and Application in Zinc-Air Batteries. *Adv. Funct. Mater.* **2016**, *26*, 4397–4404. [CrossRef]
157. Chia, X.; Pumera, M. Characteristics and performance of two-dimensional materials for electrocatalysis. *Nat. Catal.* **2018**, *1*, 909–921. [CrossRef]
158. Huang, H.; Yan, M.; Yang, C.; He, H.; Jiang, Q.; Yang, L.; Lu, Z.; Sun, Z.; Xu, X.; Bando, Y.; et al. Graphene Nanoarchitectonics: Recent Advances in Graphene-Based Electrocatalysts for Hydrogen Evolution Reaction. *Adv. Mater.* **2019**, *31*, e1903415. [CrossRef]
159. Jahan, M.; Liu, Z.; Loh, K.P. A Graphene Oxide and Copper-Centered Metal Organic Framework Composite as a Tri-Functional Catalyst for HER, OER, and ORR. *Adv. Funct. Mater.* **2013**, *23*, 5363–5372. [CrossRef]
160. Prabhu, P.; Jose, V.; Lee, J.M. Heterostructured Catalysts for Electrocatalytic and Photocatalytic Carbon Dioxide Reduction. *Adv. Funct. Mater.* **2020**, *30*, 1910768. [CrossRef]
161. Chen, J.; Li, H.; Fan, C.; Meng, Q.; Tang, Y.; Qiu, X.; Fu, G.; Ma, T. Dual Single-Atomic Ni-N_4 and Fe-N_4 Sites Constructing Janus Hollow Graphene for Selective Oxygen Electrocatalysis. *Adv. Mater.* **2020**, *32*, 2003134. [CrossRef]
162. Li, M.F.; Liu, Y.G.; Zeng, G.M.; Liu, N.; Liu, S.B. Graphene and graphene-based nanocomposites used for antibiotics removal in water treatment: A review. *Chemosphere* **2019**, *226*, 360–380. [CrossRef] [PubMed]
163. Bano, Z.; Mazari, S.A.; Saeed, R.M.Y.; Majeed, M.A.; Xia, M.; Memon, A.Q.; Abro, R.; Wang, F. Water decontamination by 3D graphene based materials: A review. *J. Water Process Eng.* **2020**, *36*, 101404. [CrossRef]
164. Wang, X.; Yin, R.; Zeng, L.; Zhu, M. A review of graphene-based nanomaterials for removal of antibiotics from aqueous environments. *Environ. Pollut.* **2019**, *253*, 100–110. [CrossRef]
165. Vaizoğullar, A.I. ZnO/ZrO_2 composites: Synthesis characterization and photocatalytic performance in the degradation of oxytetracycline antibiotic. *Mater. Technol.* **2019**, *34*, 433–443. [CrossRef]

166. Cerro-Lopez, M.; Méndez-Rojas, M.A. Application of Nanomaterials for Treatment of Wastewater Containing Pharmaceuticals. In *Ecopharmacovigilance*; Springer: Cham, Switzerland, 2017; pp. 201–219.
167. Ibrahim, F.A.; Al-Ghobashy, M.A.; El-Rahman, M.K.A.; Abo-Elmagd, I.F. Optimization and in line potentiometric monitoring of enhanced photocatalytic degradation kinetics of gemifloxacin using TiO_2 nanoparticles/H_2O_2. *Environ. Sci. Pollut. Res. Int.* **2017**, *24*, 23880–23892. [CrossRef]
168. Yang, Y.; Song, W.; Lin, H.; Wang, W.; Du, L.; Xing, W. Antibiotics and antibiotic resistance genes in global lakes: A review and meta-analysis. *Environ. Int.* **2018**, *116*, 60–73. [CrossRef] [PubMed]
169. Liu, Y.; He, X.; Duan, X.; Fu, Y.; Dionysiou, D.D. Photochemical degradation of oxytetracycline: Influence of pH and role of carbonate radical. *Chem. Eng. J.* **2015**, *276*, 113–121. [CrossRef]
170. Liu, L.; Wu, W.; Zhang, J.; Lv, P.; Xu, L.; Yan, Y. Progress of research on the toxicology of antibiotic pollution in aquatic organisms. *Acta Ecol. Sin.* **2018**, *38*, 36–41. [CrossRef]
171. Guo, H.; Li, Z.; Lin, S.; Li, D.; Jiang, N.; Wang, H.; Han, J.; Li, J. Multi-catalysis induced by pulsed discharge plasma coupled with graphene-Fe_3O_4 nanocomposites for efficient removal of ofloxacin in water: Mechanism, degradation pathway and potential toxicity. *Chemosphere* **2021**, *265*, 129089. [CrossRef]
172. Dong, S.; Zhao, Y.; Yang, J.; Liu, X.; Li, W.; Zhang, L.; Wu, Y.; Sun, J.; Feng, J.; Zhu, Y. Visible-light responsive PDI/rGO composite film for the photothermal catalytic degradation of antibiotic wastewater and interfacial water evaporation. *Appl. Catal. B Environ.* **2021**, *291*, 120127. [CrossRef]
173. Juengchareonpoon, K.; Wanichpongpan, P.; Boonamnuayvitaya, V. Graphene oxide and carboxymethylcellulose film modified by citric acid for antibiotic removal. *J. Environ. Chem. Eng.* **2020**, *9*, 104637. [CrossRef]
174. Guo, H.; Wang, Y.; Yao, X.; Zhang, Y.; Li, Z.; Pan, S.; Han, J.; Xu, L.; Qiao, W.; Li, J.; et al. A comprehensive insight into plasma-catalytic removal of antibiotic oxytetracycline based on graphene-$TiO2$-Fe_3O_4 nanocomposites. *Chem. Eng. J.* **2021**, *425*, 130614. [CrossRef]
175. Romao, J.; Barata, D.; Ribeiro, N.; Habibovic, P.; Fernandes, H.; Mul, G. High throughput screening of photocatalytic conversion of pharmaceutical contaminants in water. *Environ. Pollut.* **2017**, *220 Pt B*, 1199–1207. [CrossRef]
176. Shaniba, C.; Akbar, M.; Ramseena, K.; Raveendran, P.; Narayanan, B.N.; Ramakrishnan, R.M. Sunlight-assisted oxidative degradation of cefixime antibiotic from aqueous medium using TiO_2/nitrogen doped holey graphene nanocomposite as a high performance photocatalyst. *J. Environ. Chem. Eng.* **2020**, *8*, 102204. [CrossRef]
177. Qiao, D.; Li, Z.; Duan, J.; He, X. Adsorption and photocatalytic degradation mechanism of magnetic graphene oxide/ZnO nanocomposites for tetracycline contaminants. *Chem. Eng. J.* **2020**, *400*, 125952. [CrossRef]
178. Wu, Z.; Liang, Y.; Yuan, X.; Zou, D.; Fang, J.; Jiang, L.; Zhang, J.; Yang, H.; Xiao, Z. MXene Ti_3C_2 derived Z–scheme photocatalyst of graphene layers anchored TiO_2/g–C_3N_4 for visible light photocatalytic degradation of refractory organic pollutants. *Chem. Eng. J.* **2020**, *394*, 124921. [CrossRef]
179. Guo, H.; Li, Z.; Xiang, L.; Jiang, N.; Zhang, Y.; Wang, H.; Li, J. Efficient removal of antibiotic thiamphenicol by pulsed discharge plasma coupled with complex catalysis using graphene-WO_3-Fe_3O_4 nanocomposites. *J. Hazard. Mater.* **2021**, *403*, 123673. [CrossRef]
180. Yang, W.; Wang, Y. Enhanced electron and mass transfer flow-through cell with C_3N_4-MoS_2 supported on three-dimensional graphene photoanode for the removal of antibiotic and antibacterial potencies in ampicillin wastewater. *Appl. Catal. B Environ.* **2021**, *282*, 119574. [CrossRef]
181. Zhu, L.; Ji, J.; Liu, J.; Mine, S.; Matsuoka, M.; Zhang, J.; Xing, M. Designing 3D-MoS_2 Sponge as Excellent Cocatalysts in Advanced Oxidation Processes for Pollutant Control. *Angew. Chem. Int. Ed. Engl.* **2020**, *59*, 13968–13976. [CrossRef]
182. Zhu, T.-T.; Su, Z.-X.; Lai, W.-X.; Zhang, Y.-B.; Liu, Y.-W. Insights into the fate and removal of antibiotics and antibiotic resistance genes using biological wastewater treatment technology. *Sci. Total Environ.* **2021**, *776*, 145906. [CrossRef]
183. Ullah, S.; Hashmi, M.; Hussain, N.; Ullah, A.; Sarwar, M.N.; Saito, Y.; Kim, S.H.; Kim, I.S. Stabilized nanofibers of polyvinyl alcohol (PVA) crosslinked by unique method for efficient removal of heavy metal ions. *J. Water Process Eng.* **2020**, *33*, 101111. [CrossRef]
184. Citulski, J.; Farahbakhsh, K.; Kent, F. Optimization of phosphorus removal in secondary effluent using immersed ultrafiltration membranes with in-line coagulant pretreatment—Implications for advanced water treatment and reuse applications. *J. Environ. Eng. Sci.* **2013**, *8*, 359–370. [CrossRef]
185. Zhang, P.; Hou, D.; Li, X.; Pehkonen, S.; Varma, R.S.; Wang, X. Greener and size-specific synthesis of stable Fe-Cu oxides as earth-abundant adsorbents for malachite green. *J. Mater. Chem. A Mater.* **2018**, *6*, 9229–9236.
186. Dai, Y.; Zhang, N.; Xing, C.; Cui, Q.; Sun, Q. The adsorption, regeneration and engineering applications of biochar for removal organic pollutants: A review. *Chemosphere* **2019**, *223*, 12–27. [CrossRef]
187. Verma, A.; Thakur, S.; Mamba, G.; Prateek; Gupta, R.K.; Thakur, P.; Thakur, V.K. Graphite modified sodium alginate hydrogel composite for efficient removal of malachite green dye. *Int. J. Biol. Macromol.* **2020**, *148*, 1130–1139. [CrossRef] [PubMed]
188. Arabkhani, P.; Asfaram, A. Development of a novel three-dimensional magnetic polymer aerogel as an efficient adsorbent for malachite green removal. *J. Hazard. Mater.* **2020**, *384*, 121394. [CrossRef] [PubMed]
189. Nawaz, H.; Umar, M.; Ullah, A.; Razzaq, H.; Zia, K.M.; Liu, X. Polyvinylidene fluoride nanocomposite super hydrophilic membrane integrated with Polyaniline-Graphene oxide nano fillers for treatment of textile effluents. *J. Hazard. Mater.* **2021**, *403*, 123587. [CrossRef] [PubMed]

190. Mittal, H.; Al Alili, A.; Morajkar, P.P.; Alhassan, S.M. GO crosslinked hydrogel nanocomposites of chitosan/carboxymethyl cellulose—A versatile adsorbent for the treatment of dyes contaminated wastewater. *Int. J. Biol. Macromol.* **2021**, *167*, 1248–1261. [CrossRef]
191. Jiang, L.; Wen, Y.; Zhu, Z.; Liu, X.; Shao, W. A Double cross-linked strategy to construct graphene aerogels with highly efficient methylene blue adsorption performance. *Chemosphere* **2021**, *265*, 129169. [CrossRef]
192. Pashaei-Fakhri, S.; Peighambardoust, S.J.; Foroutan, R.; Arsalani, N.; Ramavandi, B. Crystal violet dye sorption over acrylamide/graphene oxide bonded sodium alginate nanocomposite hydrogel. *Chemosphere* **2021**, *270*, 129419. [CrossRef]
193. Cao, M.; Shen, Y.; Yan, Z.; Wei, Q.; Jiao, T.; Shen, Y.; Han, Y.; Wang, Y.; Wang, S.; Xia, Y.; et al. Extraction-like removal of organic dyes from polluted water by the graphene oxide/PNIPAM composite system. *Chem. Eng. J.* **2021**, *405*, 126647. [CrossRef]
194. Hao, J.; Wang, Z.; Xiao, C.; Zhao, J.; Chen, L. In situ reduced graphene oxide-based polyurethane sponge hollow tube for continuous oil removal from water surface. *Environ. Sci. Pollut. Res. Int.* **2018**, *25*, 4837–4845. [CrossRef] [PubMed]
195. Cao, N.; Yang, B.; Barras, A.; Szunerits, S.; Boukherroub, R. Polyurethane sponge functionalized with superhydrophobic nanodiamond particles for efficient oil/water separation. *Chem. Eng. J.* **2017**, *307*, 319–325. [CrossRef]
196. Ma, W.; Li, Y.; Zhang, M.; Gao, S.; Cui, J.; Huang, C.; Fu, G. Biomimetic Durable Multifunctional Self-Cleaning Nanofibrous Membrane with Outstanding Oil/Water Separation, Photodegradation of Organic Contaminants, and Antibacterial Performances. *ACS Appl. Mater. Interfaces* **2020**, *12*, 34999–35010. [CrossRef] [PubMed]
197. Khalilifard, M.; Javadian, S. Magnetic superhydrophobic polyurethane sponge loaded with Fe_3O_4@oleic acid@graphene oxide as high performance adsorbent oil from water. *Chem. Eng. J.* **2021**, *408*, 127369. [CrossRef]
198. Alammar, A.; Park, S.-H.; Williams, C.J.; Derby, B.; Szekely, G. Oil-in-water separation with graphene-based nanocomposite membranes for produced water treatment. *J. Membr. Sci.* **2020**, *603*, 118007. [CrossRef]
199. Cao, J.; Chen, Y.; Zhang, J.; Wang, X.; Wang, J.; Shi, C.; Ning, Y.; Wang, X. Preparation and application of nanofluid flooding based on polyoxyethylated graphene oxide nanosheets for enhanced oil recovery. *Chem. Eng. Sci.* **2022**, *247*, 117023. [CrossRef]
200. Abdalla, O.; Wahab, M.A.; Abdala, A. Mixed matrix membranes containing aspartic acid functionalized graphene oxide for enhanced oil-water emulsion separation. *J. Environ. Chem. Eng.* **2020**, *8*, 104269. [CrossRef]
201. Liang, Y.; Li, C.; Li, S.; Su, B.; Hu, M.Z.; Gao, X.; Gao, C. Graphene quantum dots (GQDs)-polyethyleneimine as interlayer for the fabrication of high performance organic solvent nanofiltration (OSN) membranes. *Chem. Eng. J.* **2020**, *380*, 122462. [CrossRef]
202. Zhou, Q.; Yang, N.; Li, Y.; Ren, B.; Ding, X.; Bian, H.; Yao, X. Total concentrations and sources of heavy metal pollution in global river and lake water bodies from 1972 to 2017. *Glob. Ecol. Conserv.* **2020**, *22*, e00925. [CrossRef]
203. Sankaran, R.; Show, P.L.; Ooi, C.-W.; Ling, T.C.; Shu-Jen; Chen, S.-Y.; Chang, Y.-K. Feasibility assessment of removal of heavy metals and soluble microbial products from aqueous solutions using eggshell wastes. *Clean Technol. Environ. Policy* **2020**, *22*, 773–786. [CrossRef]
204. Cheng, S.Y.; Show, P.L.; Lau, B.F.; Chang, J.S.; Ling, T.C. New Prospects for Modified Algae in Heavy Metal Adsorption. *Trends Biotechnol.* **2019**, *37*, 1255–1268. [CrossRef] [PubMed]
205. Burakov, A.E.; Galunin, E.V.; Burakova, I.V.; Kucherova, A.E.; Agarwal, S.; Tkachev, A.G.; Gupta, V.K. Adsorption of heavy metals on conventional and nanostructured materials for wastewater treatment purposes: A review. *Ecotoxicol. Environ. Saf.* **2018**, *148*, 702–712. [CrossRef] [PubMed]
206. O'connell, D.W.; Birkinshaw, C.; O'dwyer, T.F. Heavy metal adsorbents prepared from the modification of cellulose: A review. *Bioresour. Technol.* **2008**, *99*, 6709–6724. [CrossRef]
207. Kong, Q.; Shi, X.; Ma, W.; Zhang, F.; Yu, T.; Zhao, F.; Zhao, D.; Wei, C. Strategies to improve the adsorption properties of graphene-based adsorbent towards heavy metal ions and their compound pollutants: A review. *J. Hazard. Mater.* **2021**, *415*, 125690. [CrossRef] [PubMed]
208. Liang, Q.; Ploychompoo, S.; Chen, J.; Zhou, T.; Luo, H. Simultaneous Cr(VI) reduction and bisphenol A degradation by a 3D Z-scheme Bi_2S_3-$BiVO_4$ graphene aerogel under visible light. *Chem. Eng. J.* **2020**, *384*, 123256. [CrossRef]
209. Chen, R.; Cheng, Y.; Wang, P.; Wang, Q.; Wan, S.; Huang, S.; Su, R.; Song, Y.; Wang, Y. Enhanced removal of Co(II) and Ni(II) from high-salinity aqueous solution using reductive self-assembly of three-dimensional magnetic fungal hyphal/graphene oxide nanofibers. *Sci. Total Environ.* **2021**, *756*, 143871. [CrossRef]
210. Bao, S.; Yang, W.; Wang, Y.; Yu, Y.; Sun, Y. One-pot synthesis of magnetic graphene oxide composites as an efficient and recoverable adsorbent for Cd(II) and Pb(II) removal from aqueous solution. *J. Hazard. Mater.* **2020**, *381*, 120914. [CrossRef]
211. Pakulski, D.; Gorczynski, A.; Marcinkowski, D.; Czepa, W.; Chudziak, T.; Witomska, S.; Nishina, Y.; Patroniak, V.; Ciesielski, A.; Samori, P. High-sorption terpyridine-graphene oxide hybrid for the efficient removal of heavy metal ions from wastewater. *Nanoscale* **2021**, *13*, 10490–10499. [CrossRef]
212. Tobaldi, D.M.; Dvoranova, D.; Lajaunie, L.; Rozman, N.; Figueiredo, B.; Seabra, M.P.; Skapin, A.S.; Calvino, J.J.; Brezova, V.; Labrincha, J.A. Graphene-TiO_2 hybrids for photocatalytic aided removal of VOCs and nitrogen oxides from outdoor environment. *Chem. Eng. J.* **2021**, *405*, 126651. [CrossRef]
213. Shajari, S.; Kowsari, E.; Seifvand, N.; Ajdari, F.B.; Chinnappan, A.; Ramakrishna, S.; Saianand, G.; Najafi, M.D.; Haddadi-Asl, V.; Abdpour, S. Efficient Photocatalytic Degradation of Gaseous Benzene and Toluene over Novel Hybrid PIL@TiO_2/m-GO Composites. *Catalysts* **2021**, *11*, 126. [CrossRef]
214. Li, J.; Li, X.; Zhang, X.; Zhang, J.; Duan, Y.; Li, X.; Jiang, D.; Kozawa, T.; Naito, M. Development of graphene aerogels with high strength and ultrahigh adsorption capacity for gas purification. *Mater. Des.* **2021**, *208*, 109903. [CrossRef]

215. Xiang, Q.; Cheng, B.; Yu, J. Photokatalysatoren auf Graphenbasis für die Produktion von Solarbrennstoffen. *Angew. Chem.* **2015**, *127*, 11508–11524. [CrossRef]
216. Inoue, T.; Fujishima, A.; Konishi, S.; Honda, K. Photoelectrocatalytic reduction of carbon dioxide in aqueous suspensions of semiconductor powders. *Nature* **1979**, *277*, 637–638. [CrossRef]
217. Ali, S.; Razzaq, A.; In, S.-I. Development of graphene based photocatalysts for CO_2 reduction to C1 chemicals: A brief overview. *Catal. Today* **2019**, *335*, 39–54. [CrossRef]
218. Thoai, D.B.; Hu, Y.Z.; Koch, S.W. Influence of the confinement potential on the electron-hole-pair states in semiconductor microcrystallites. *Phys. Rev. B Condens. Matter* **1990**, *42*, 11261–11266. [CrossRef] [PubMed]
219. Khan, M.F.; Iqbal, M.Z.; Iqbal, M.W.; Eom, J. Improving the electrical properties of graphene layers by chemical doping. *Sci. Technol. Adv. Mater.* **2014**, *15*, 055004. [CrossRef]
220. Tian, P.; Zang, J.; Jia, S.; Zhang, Y.; Gao, H.; Zhou, S.; Wang, W.; Xu, H.; Wang, Y. Preparation of S/N co-doped graphene through a self-generated high gas pressure for high rate supercapacitor. *Appl. Surf. Sci.* **2018**, *456*, 781–788. [CrossRef]
221. Rao, C.N.R.; Gopalakrishnan, K.; Govindaraj, A. Synthesis, properties and applications of graphene doped with boron, nitrogen and other elements. *Nano Today* **2014**, *9*, 324–343. [CrossRef]
222. He, Y.; Chen, Q.; Yang, S.; Lu, C.; Feng, M.; Jiang, Y.; Cao, G.; Zhang, J.; Liu, C. Micro-crack behavior of carbon fiber reinforced Fe_3O_4/graphene oxide modified epoxy composites for cryogenic application. *Compos. Part A Appl. Sci. Manuf.* **2018**, *108*, 12–22. [CrossRef]
223. Singh, D.P.; Herrera, C.E.; Singh, B.; Singh, S.; Singh, R.K.; Kumar, R. Graphene oxide: An efficient material and recent approach for biotechnological and biomedical applications. *Mater. Sci. Eng. C Mater. Biol. Appl.* **2018**, *86*, 173–197. [CrossRef] [PubMed]
224. Singh, R.K.; Kumar, R.; Singh, D.P.; Savu, R.; Moshkalev, S.A. Progress in microwave-assisted synthesis of quantum dots (graphene/carbon/semiconducting) for bioapplications: A review. *Mater. Today Chem.* **2019**, *12*, 282–314. [CrossRef]
225. Awasthi, K.; Kumar, R.; Tiwari, R.S.; Srivastava, O.N. Large scale synthesis of bundles of aligned carbon nanotubes using a natural precursor: Turpentine oil. *J. Exp. Nanosci.* **2010**, *5*, 498–508. [CrossRef]
226. Xue, Y.; Wu, B.; Bao, Q.; Liu, Y. Controllable synthesis of doped graphene and its applications. *Small* **2014**, *10*, 2975–2991. [CrossRef] [PubMed]
227. Wang, Y.; Chen, Y.; Lacey, S.D.; Xu, L.; Xie, H.; Li, T.; Danner, V.A.; Hu, L. Reduced graphene oxide film with record-high conductivity and mobility. *Mater. Today* **2018**, *21*, 186–192. [CrossRef]
228. Du, J.; Duan, J.; Yang, X.; Duan, Y.; Zhou, Q.; Tang, Q. p-Type Charge Transfer Doping of Graphene Oxide with $(NiCo)_{1-y}Fe_yO_x$ for Air-Stable, All-Inorganic $CsPbIBr_2$ Perovskite Solar Cells. *Angew. Chem. Int. Ed. Engl.* **2021**, *60*, 10608–10613. [CrossRef]
229. Meng, F.; Li, J.; Cushing, S.K.; Zhi, M.; Wu, N. Solar hydrogen generation by nanoscale p-n junction of p-type molybdenum disulfide/n-type nitrogen-doped reduced graphene oxide. *J. Am. Chem. Soc.* **2013**, *135*, 10286–10289. [CrossRef]
230. Nag, A.; Mitra, A.; Mukhopadhyay, S.C. Graphene and its sensor-based applications: A review. *Sens. Actuators A Phys.* **2018**, *270*, 177–194. [CrossRef]
231. Razaq, A.; Bibi, F.; Zheng, X.; Papadakis, R.; Jafri, S.H.M.; Li, H. Review on Graphene-, Graphene Oxide-, Reduced Graphene Oxide-Based Flexible Composites: From Fabrication to Applications. *Materials* **2022**, *15*, 1012. [CrossRef]
232. Chen, X.; Chen, B. Macroscopic and spectroscopic investigations of the adsorption of nitroaromatic compounds on graphene oxide, reduced graphene oxide, and graphene nanosheets. *Environ. Sci. Technol.* **2015**, *49*, 6181–6189. [CrossRef]
233. Bi, W.; Li, X.; You, R.; Chen, M.; Yuan, R.; Huang, W.; Wu, X.; Chu, W.; Wu, C.; Xie, Y. Surface Immobilization of Transition Metal Ions on Nitrogen-Doped Graphene Realizing High-Efficient and Selective CO_2 Reduction. *Adv. Mater.* **2018**, *30*, e1706617. [CrossRef] [PubMed]
234. Khan, M.S.; Jhankal, D.; Shakya, P.; Sharma, A.K.; Banerjee, M.K.; Sachdev, K. Ultraslim and highly flexible supercapacitor based on chemical vapor deposited nitrogen-doped bernal graphene for wearable electronics. *Carbon* **2023**, *208*, 227–237. [CrossRef]
235. Sinha, S.; Barman, P.B.; Hazra, S.K. Probing the electronic properties of chemically synthesised doped and undoped graphene derivative. *Mater. Sci. Eng. B* **2023**, *287*, 116145. [CrossRef]
236. Backes, C.; Abdelkader, A.M.; Alonso, C.; Andrieux-Ledier, A.; Arenal, R.; Azpeitia, J.; Balakrishnan, N.; Banszerus, L.; Barjon, J.; Bartali, R.; et al. Production and processing of graphene and related materials. *2D Mater.* **2020**, *7*, 022001. [CrossRef]
237. Khudair, S.A.M.; Jappor, H.R. Adsorption of Gas Molecules on Graphene Doped with Mono and Dual Boron as Highly Sensitive Sensors and Catalysts. *J. Nanostruct.* **2020**, *10*, 217–229.
238. Chen, X.; Chen, S.; Wang, J. Screening of catalytic oxygen reduction reaction activity of metal-doped graphene by density functional theory. *Appl. Surf. Sci.* **2016**, *379*, 291–295. [CrossRef]
239. Dai, X.; Shen, T.; Feng, Y.; Yang, B.; Liu, H. DFT investigations on photoelectric properties of graphene modified by metal atoms. *Ferroelectrics* **2020**, *568*, 143–154. [CrossRef]
240. Jiang, M.; Zhang, W.; Zhao, K.; Guan, F.; Wang, Y. Investigations on the electronic structure and optical properties of (Ga, N, Ga-N) doped graphene by first-principle calculations. *Int. J. Mod. Phys. B* **2021**, *35*, 2150067. [CrossRef]
241. Kumar, P.; Mungse, H.P.; Khatri, O.P.; Jain, S.L. Nitrogen-doped graphene-supported copper complex: A novel photocatalyst for CO2 reduction under visible light irradiation. *RSC Adv.* **2015**, *5*, 54929–54935. [CrossRef]
242. Tang, Z.; He, W.; Wang, Y.; Wei, Y.; Yu, X.; Xiong, J.; Wang, X.; Zhang, X.; Zhao, Z.; Liu, J. Ternary heterojunction in rGO-coated Ag/Cu_2O catalysts for boosting selective photocatalytic CO_2 reduction into CH_4. *Appl. Catal. B Environ.* **2022**, *311*, 121371. [CrossRef]

243. Yang, A.; Luo, J.; Xie, Z. First-principles study of Graphene/ZnV$_2$O$_6$(001) heterostructure photocatalyst. *J. Mater. Res. Technol.* **2021**, *15*, 1479–1486. [CrossRef]
244. Li, F.; Zhang, L.; Tong, J.; Liu, Y.; Xu, S.; Cao, Y.; Cao, S. Photocatalytic CO$_2$ conversion to methanol by Cu$_2$O/graphene/TNA heterostructure catalyst in a visible-light-driven dual-chamber reactor. *Nano Energy* **2016**, *27*, 320–329. [CrossRef]
245. Sun, R.; Yin, H.; Zhang, Z.; Wang, Y.; Liang, T.; Zhang, S.; Jing, L. Graphene-Modulated PDI/g-C$_3$N$_4$ All-Organic S-Scheme Heterojunction Photocatalysts for Efficient CO$_2$ Reduction under Full-Spectrum Irradiation. *J. Phys. Chem. C* **2021**, *125*, 23830–23839. [CrossRef]
246. Radovic, L.R.; Salgado-Casanova, A.J.A.; Mora-Vilches, C.V. On the active sites for the oxygen reduction reaction catalyzed by graphene-based materials. *Carbon* **2020**, *156*, 389–398. [CrossRef]
247. Ren, S.; Cui, M.; Liu, C.; Wang, L. A comprehensive review on ultrathin, multi-functionalized, and smart graphene and graphene-based composite protective coatings. *Corros. Sci.* **2023**, *212*, 110939. [CrossRef]
248. Zhang, T.; Gao, X.; Li, J.; Xiao, L.; Gao, H.; Zhao, F.; Ma, H. Progress on the application of graphene-based composites toward energetic materials: A review. *Def. Technol.* **2023**. [CrossRef]
249. Choi, M.G.; Park, S.; Lee, H.; Kim, S. Correlating surface structures and nanoscale friction of CVD Multi-Layered graphene. *Appl. Surf. Sci.* **2022**, *584*, 152572. [CrossRef]
250. Muschi, M.; Devautour-Vinot, S.; Aureau, D.; Heymans, N.; Sene, S.; Emmerich, R.; Ploumistos, A.; Geneste, A.; Steunou, N.; Patriarche, G.; et al. Metal–organic framework/graphene oxide composites for CO$_2$ capture by microwave swing adsorption. *J. Mater. Chem. A* **2021**, *9*, 13135–13142. [CrossRef]
251. Sun, Z.; Hu, Y.H. 3D Graphene Materials from the Reduction of CO$_2$. *Acc. Mater. Res.* **2021**, *2*, 48–58. [CrossRef]
252. Szczęśniak, B.; Phuriragpitikhon, J.; Choma, J.; Jaroniec, M. Mechanochemical synthesis of three-component graphene oxide/ordered mesoporous carbon/metal-organic framework composites. *J. Colloid Interface Sci.* **2020**, *577*, 163–172. [CrossRef]
253. Chebrolu, N.R.; Chittari, B.L.; Jung, J. Flat bands in twisted double bilayer graphene. *Phys. Rev. B* **2019**, *99*, 235417. [CrossRef]
254. Li, C.; Zhang, C.; Xu, S.; Huo, Y.; Jiang, S.; Yang, C.; Li, Z.; Zhao, X.; Zhang, S.; Man, B. Experimental and theoretical investigation for a hierarchical SERS activated platform with 3D dense hot spots. *Sens. Actuators B Chem.* **2018**, *263*, 408–416. [CrossRef]
255. Shen, Y.; Yang, S.; Zhou, P.; Sun, Q.; Wang, P.; Wan, L.; Li, J.; Chen, L.; Wang, X.; Ding, S.; et al. Evolution of the band-gap and optical properties of graphene oxide with controllable reduction level. *Carbon* **2013**, *62*, 157–164. [CrossRef]
256. Hunt, A.; Kurmaev, E.Z.; Moewes, A. Band gap engineering of graphene oxide by chemical modification. *Carbon* **2014**, *75*, 366–371. [CrossRef]
257. Hunt, A.; Dikin, D.A.; Kurmaev, E.Z.; Lee, Y.H.; Luan, N.V.; Chang, G.S.; Moewes, A. Modulation of the band gap of graphene oxide: The role of AA-stacking. *Carbon* **2014**, *66*, 539–546. [CrossRef]
258. Sugawara, K.; Suzuki, K.; Sato, M.; Sato, T.; Takahashi, T. Enhancement of band gap and evolution of in-gap states in hydrogen-adsorbed monolayer graphene on SiC(0001). *Carbon* **2017**, *124*, 584–587. [CrossRef]
259. Wang, X.; Yang, X.; Wang, B.; Wang, G.; Wan, J. Significant band gap induced by uniaxial strain in graphene/blue phosphorene bilayer. *Carbon* **2018**, *130*, 120–126. [CrossRef]
260. Méndez-Romero, U.A.; Pérez-García, S.A.; Xu, X.; Wang, E.; Licea-Jiménez, L. Functionalized reduced graphene oxide with tunable band gap and good solubility in organic solvents. *Carbon* **2019**, *146*, 491–502. [CrossRef]
261. Pradhan, J.; Srivastava, S.K.; Palanivelu, M.; Kothalamuthu, S.; Balakrishnan, S.; Sarkar, S.; Mathew, S.; Venkatesan, T. Band gap opening and surface morphology of monolayer graphene induced by single ion impacts of argon monomer and dimer ions. *Carbon* **2021**, *184*, 322–330. [CrossRef]
262. Liu, Z.; Shi, W.; Yang, T.; Zhang, Z. Magic angles and flat Chern bands in alternating-twist multilayer graphene system. *J. Mater. Sci. Technol.* **2022**, *111*, 28–34. [CrossRef]
263. Dompreh, K.A.; Sekyi-Arthur, D.; Mensah, S.Y.; Adu, K.W.; Edziah, R.; Amekpewu, M. Effect of band-gap tuning on absorption of phonons and acoustoelectric current in graphene nanoribbon. *Phys. E Low-Dimens. Syst. Nanostructures* **2023**, *147*, 115516. [CrossRef]
264. Nandee, R.; Chowdhury, M.A.; Shahid, A.; Hossain, N.; Rana, M. Band gap formation of 2D materialin graphene: Future prospect and challenges. *Results Eng.* **2022**, *15*, 100474. [CrossRef]
265. Chen, D.-M.; Shenai, P.M.; Zhao, Y. Tight binding description on the band gap opening of pyrene-dispersed graphene. *Phys. Chem. Chem. Phys.* **2011**, *13*, 1515–1520. [CrossRef]
266. Gong, E.; Shahzad, A.; Hiragond, C.B.; Kim, H.S.; Powar, N.S.; Kim, D.; Kim, H.; In, S.-I. Solar fuels: Research and development strategies to accelerate photocatalytic CO$_2$ conversion into hydrocarbon fuels. *Energy Environ. Sci.* **2022**, *15*, 880–937. [CrossRef]
267. Wu, J.-T.; Shi, X.-H.; Wei, Y.-J. Tunable band structures of polycrystalline graphene by external and mismatch strains. *Acta Mech. Sin.* **2012**, *28*, 1539–1544. [CrossRef]
268. Yuan, W.; Zhou, Y.; Li, Y.; Li, C.; Peng, H.; Zhang, J.; Liu, Z.; Dai, L.; Shi, G. The edge-and basal-plane-specific electrochemistry of a single-layer graphene sheet. *Sci. Rep.* **2013**, *3*, 2248. [CrossRef] [PubMed]

Disclaimer/Publisher's Note: The statements, opinions and data contained in all publications are solely those of the individual author(s) and contributor(s) and not of MDPI and/or the editor(s). MDPI and/or the editor(s) disclaim responsibility for any injury to people or property resulting from any ideas, methods, instructions or products referred to in the content.

Review

Application of Polypyrrole-Based Electrochemical Biosensor for the Early Diagnosis of Colorectal Cancer

Xindan Zhang, Xiao Tan, Ping Wang * and Jieling Qin *

Tongji University Cancer Center, Shanghai Tenth People's Hospital, School of Medicine, Tongji University, Shanghai 200092, China
* Correspondence: pwangecnu@163.com (P.W.); qinjieling770@hotmail.com (J.Q.)

Abstract: Although colorectal cancer (CRC) is easy to treat surgically and can be combined with postoperative chemotherapy, its five-year survival rate is still not optimistic. Therefore, developing sensitive, efficient, and compliant detection technology is essential to diagnose CRC at an early stage, providing more opportunities for effective treatment and intervention. Currently, the widely used clinical CRC detection methods include endoscopy, stool examination, imaging modalities, and tumor biomarker detection; among them, blood biomarkers, a noninvasive strategy for CRC screening, have shown significant potential for early diagnosis, prediction, prognosis, and staging of cancer. As shown by recent studies, electrochemical biosensors have attracted extensive attention for the detection of blood biomarkers because of their advantages of being cost-effective and having sound sensitivity, good versatility, high selectivity, and a fast response. Among these, nano-conductive polymer materials, especially the conductive polymer polypyrrole (PPy), have been broadly applied to improve sensing performance due to their excellent electrical properties and the flexibility of their surface properties, as well as their easy preparation and functionalization and good biocompatibility. This review mainly discusses the characteristics of PPy-based biosensors, their synthetic methods, and their application for the detection of CRC biomarkers. Finally, the opportunities and challenges related to the use of PPy-based sensors for diagnosing CRC are also discussed.

Keywords: polypyrrole; electrochemical biosensor; biomarkers; early diagnosis; colorectal cancer

1. Introduction

As a result of the influences of individual lifestyle, food safety, and the ecological environment on health and the accelerating processes of industrialization, urbanization, and global population aging, the morbidity and mortality of chronic diseases are increasing with a constant upward trend, seriously endangering population health. Colorectal cancer (CRC) has one of the highest incidences among malignant tumors all over the world, which is related to genetics, gender, age, race, living environment, lifestyle, eating habits, and medicine use [1]. CRC develops from genetic and epigenetic variation and progresses to adenoma and malignancy through subsequent changes, such as transcription, translation, and abnormal protein expression, in a multi-stage process. There were over 1.9 million new CRC cases and over 900,000 deaths worldwide in 2020 [1]. Early diagnosis of CRC can enhance the survival rate with successful treatment and improve CRC outcomes by offering care at the earliest possible stage.

Recently, many techniques have been developed to diagnose CRC clinically, such as colonoscopy, sigmoidoscopy, colon capsule endoscopy (CCE), computed tomography colonography (CTC), the antibody-based fecal immunochemical test (FIT), the immune-based fecal occult blood test (FOBT), and biomarker determination [2]. Colonoscopy and sigmoidoscopy (endoscopic examination of the distal colon) are practical tools for CRC screening with higher sensitivity that can clearly show lesions, and they can be utilized as auxiliary options during CRC surgery. However, these methods are invasive and costly and require special facilities and sedatives with low patient compliance [3,4]. CCE is a potential noninvasive screening tool for CRC, but it is difficult to control and lacks accuracy in the determination of the type and size of lesions [5]. CTC is an auxiliary imaging method that can be used to identify colonic lesions in three-dimensional (3D) images with high sensitivity, and it is less invasive and suitable for visualization of the entire colon [6,7]. Nevertheless, CTC is also limited due to the uncomfortable bowel preparation, radiological safety concerns, and poor specificity in the detection of small tumors [8]. Standard non-invasive assays, such as FIT and FOBT, enable the quantitative measurement of hemoglobin content in human fecal samples [9,10]. However, they lack specificity in the detection of precancerous lesions and are related to false positives in clinical tests [11]. In contrast, the detection of CRC biomarkers, including Kirsten rat sarcoma viral oncogene homolog (KRAS), V-RAF murine sarcoma viral oncogene homolog B1 (BRAF), tumor protein 53 (TP53), microRNA-21 (miRNA-21), carcinoembryonic antigen (CEA), carbohydrate antigens (CA19-9, CA72-4, CA125), interleukin-6 (IL-6), and vascular endothelial growth factor (VEGF), is more sensitive and has important practical significance for the early screening of CRC [12,13].

Biomarkers mainly refer to biological molecules—i.e., DNA, RNA, miRNA, and proteins—that are markers of normal or abnormal states [14]. The conventional detection methods include the polymerase chain reaction (PCR) method [15], the enzyme-linked immunosorbent assay (ELISA) [16], DNA microarrays [17], Northern blot techniques [18], electrophoresis [19], radioimmunoassays (RIAs) [20], immunohistochemistry [21], chromatography-based technologies [22], the fluorescence method [23,24], and the chemiluminescence method [25]. Although these methods have been used to obtain relatively accurate results, detecting trace biomarkers at the early stage of CRC is far from effective. Moreover, the limitations include complex operation, a time-consuming detection process, the need for high-level technology and expensive equipment, and low sensitivity. Thus, there is still a strong demand for specific tools for the early detection of CRC that are efficient, sensitive, accurate, convenient, and fast.

Electrochemical-based detection methods, including electrochemistry, electrochemiluminescence (ECL), and photoelectrochemistry (PEC) methods, can convert analytical signals generated by target molecules into readable electrical signals, and they have the inherent advantages of excellent sensitivity and selectivity, simple operation, rapid detection, cost-effectiveness, simultaneous detection of multiple biomarkers, and potential for miniaturization [26]. Electrochemical techniques can be divided into five categories in accordance with the measurable signals: amperometry, potentiometry, resistance methods, voltammetry, and conductometry (Figure 1). Among them, voltammetry and electrochemical impedance spectroscopy (EIS) are often used for the construction of electrochemical biosensors, sensor characterization, and quantitative analysis [27,28]. Voltammetry has the advantages of simple operation, intuitive atlas analysis, widespread effectiveness, and high sensitivity. EIS has no significant influence on the system during the measurement process. It provides more interface structure and electrode dynamics information than other electrochemical methods.

Electrochemical sensors are sensitive sensing devices that use electrochemical technology to detect analytes; they are generally composed of sensitive elements, transducers, and transformation circuits. Electrochemical sensors can be categorized according to their different target molecules as biosensors, gas sensors, ion sensors, etc. The biometric element is a sensitive element with a molecular recognition ability that has a significant place in the construction of biosensors. In accordance with the different biometric elements, biosensors can be divided into enzyme sensors, immune sensors, aptamer sensors, and so on. Moreover, the electrode reaction in the sensor occurs at the interface between the solution and the electrode, and the properties of the interface have a significant influence on the reaction. Therefore, reasonable modifications of the interface play a decisive role in improving sensing efficiency.

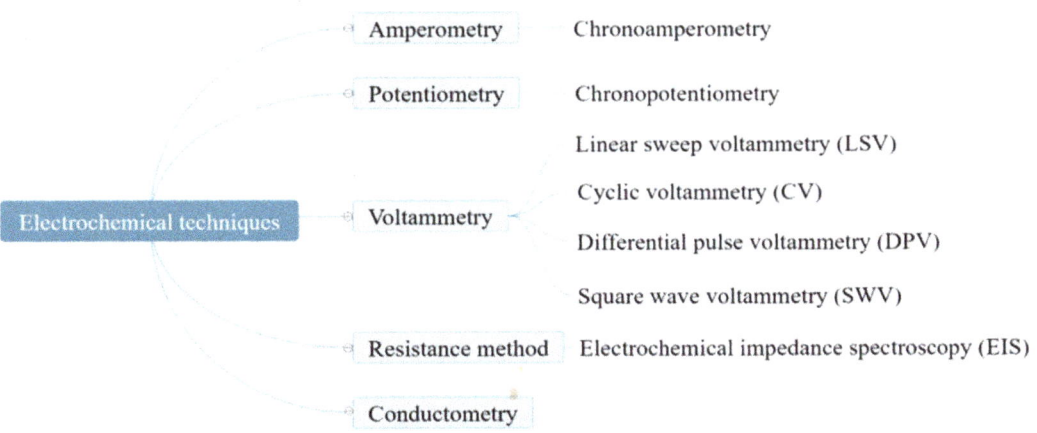

Figure 1. Electrochemical technologies commonly used in electrochemical detection.

Nanomaterials and nanocomposites with different structures [29,30], including conductive polymers (CPs), carbon-based materials, metal nanomaterials, metal oxide nanomaterials, and silicon nanomaterials, are often utilized in biosensors to increase the surface area, fix the biometric elements, catalyze the electrochemical reactions, enhance the electrical conductivity of the electrode surface, and label the biomolecules, improving the detection performance of the sensors.

Among the various nanomaterials, increasing attention has been focused on CPs, which are known as "synthetic metals" because of their outstanding electrical, optical, and magnetic characteristics [31]. Furthermore, several CPs, such as PPy, polyaniline (PANI), polythiophene (PTh), and poly(3,4-ethylene dioxythiophene) (PEDOT), have gained extensive attention for practical applications relating to biomedicine, electronics, energy equipment, etc. because of their biocompatibility, high surface area, good environmental stability, inherent electrical conductivity, and other physical properties [32,33]. Specifically, these CPs can be used for a wide range of applications in electrochemical biosensors, as shown in Table 1. Wang et al. utilized polypyrrole nanowires (PPyNWs) and polyamidoamine dendrimer (PAMAM) to design an miRNA biosensor with a high surface area and high electrical conductivity that showed significantly improved sensitivity in the determination of miRNA [34]. Compared to other CPs, PPy has been widely studied and applied, especially for the development of implantable, flexible, and wearable electronic equipment, due to its electrical versatility, which ranges from that of an insulator to near that of a metal; outstanding optical, thermoelectric, and electrical characteristics; easy synthesis and functionalization; low electropolymerization potential; stability under environmental conditions; and biocompatibility [35–37].

Table 1. Conductive polymers commonly used in electrochemical biosensors.

Conductive Polymers	Ref.
Polypyrrole (PPy)	[38–41]
Polyaniline (PANI)	[42–44]
Polythiophene (PTh)	[45–47]
Poly(3,4-ethylene dioxythiophene) (PEDOT)	[48–50]

2. Polypyrrole Biosensors

2.1. Physical and Chemical Characteristics of PPy

Pyrrole monomer is a five-membered heterocyclic molecule composed of C and N that appears as a colorless oil-like liquid at room temperature. PPy, a heterocyclic conjugated conductive polymer that usually appears as an amorphous black solid with good electroconductivity, processability, and chemical stability, is easy to form through polymerization of pyrrole in various organic electrolytes. Although conventional PPy has high rigidity, poor mechanical ductility, and poor solubility in common organic solvents, as well as deficiencies in its optical, electrical, and biological properties, nanostructured PPy has improved electrochemical activity, better electrical conductivity and biocompatibility, enhanced optical properties, good mechanical properties, and is easy to process because of the nanostructure and larger surface area, making it widely utilized in biomedical applications [51,52].

2.2. Synthesis and Modification of PPy

The polymerization of pyrrole can be carried out using chemical, electrochemical, ultrasonic, electrospinning, and even biotechnological methods with different morphologies (Figure 2), among which chemical oxidation polymerization and electropolymerization are commonly used [53–59]. During the synthesis of PPy, the CP is able to carry biomolecules or functional groups for specific biometric functions through physicochemical means [60,61]; i.e., physical adsorption, embedding, affinity, covalent immobilization, etc.

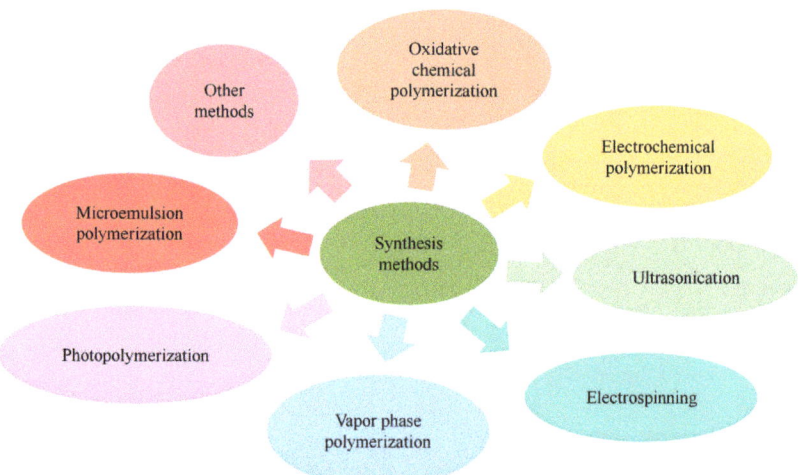

Figure 2. Synthesis methods for PPy.

Oxidative chemical polymerization of PPy is inexpensive and suitable for large-scale production. Andriukonis et al. proved that $[Fe(CN)_6]^{3-}$ can induce the synthesis of PPy [62]. Mao and Zhang pointed out that $FeCl_3$, H_2O_2, and other oxidants can be utilized to oxidize Py and polymerize PPy [63]. Furthermore, pyrrole monomer can also be oxidized using oxidoreductases (e.g., peroxidase, glucose oxidase, etc.) through enzymatic reactions in an environmentally friendly fashion with a suitable pH and room temperature conditions [64,65], and the embedded enzyme can maintain its catalytic activity when encapsulated in the polymer particles or layers formed during the preparation of enzyme-based biosensors. In addition, the formation of PPy can be implemented in cells, where the PPy induced by microorganisms is mainly deposited in the cytoderm and between the cell membrane and the cytoderm [66,67].

Electrochemical polymerization can be used to form PPy in situ with or without embedding materials (nucleic acids, enzymes, receptor proteins, and antibodies) to improve the mechanical properties and solubility [34,68–71]. In the electrochemical preparation of PPy, selection of the appropriate electrodeposition methods and electrochemical parameters, including the applied voltage, current, potential window, potential scanning rate, and duration, can ensure that the morphology, thickness, conductivity, doping, and dedoping of the PPy are regulated well [72]. Moreover, the electrochemical properties of PPy, such as the conductivity, morphology, thickness, structure, and porosity, can also be modulated by controlling the type and concentration of dopants, electrolytes, and solvent, as well as the pH value, temperature, and monomer concentration of the bulk solution [54,73,74]. In addition to normal PPy preparation, electropolymerization can be utilized for the synthesis of molecularly imprinted polymer (MIP) films on the electrode surface with high stability and low cost [72,75,76]. During the preparation of the MIP, additional electrochemical operations can be performed for overoxidation after the formation of the initial electrodeposited PPy layer [77,78]. Although peroxidation will destroy the π–π conjugate system of CP and inhibit the polymerization process, during the construction of the MIP, oxygen-containing groups, such as hydroxyl (-OH), carbonyl (-CH = O), and carboxyl (-COOH), can be generated adjacent to the embedded molecules, forming a specific environment conducive to the attachment of imprinted template molecules. Moreover, excessive oxidation is able to promote template removal and regeneration based on the MIP layer.

In addition to the direct synthesis of PPy via chemical oxidation and electrochemical oxidation, ultrasound can also be used to promote the polymerization of Py. The ultrasonic cavitation effect produced in the ultrasound process can heat the solvent and atomize it locally, making it possible to prepare polypyrrole with a small size and uniform shape in a short time [79]. Electrospinning is another convenient method that can be used to produce ultra-fine polymer nanofibers with a porous structure and high specific surface area [80]. Vapor phase polymerization does not require a solution environment and can be used to produce high-purity PPy nanomaterials on different types of substrates. However, the formed PPy has insufficient adhesion with the substrate surface [81]. Photopolymerization is a technology with which Py monomers can be polymerized under visible light or ultraviolet light, using laser-generating free radicals for the preparation of porous PPy with a high specific surface area. This approach allows for excellent control of the size of PPy and facilitates micro-machining and direct polymerization on substrates with solubility or temperature limitations [82].

2.3. Applications of PPy

As shown in Figure 3, PPy—with different morphologies including nanoparticles, nanotubes, nanowires, nanorods, nanocapsules, thin films, nanofibers, and hydrogels—can be utilized as a conductive material, electrical display material, electrochromic material, or photoluminescent quencher [35,83,84] in the construction of chemical sensors, biosensors, optical sensors, actuators, flexible electronics, transistors, electrochemical batteries, biofuel cells, photovoltaic cells, electrochromic displays, wearable and implantable/connectable biomedical tools, and other sensors [31,85–94]. Yang et al. prepared a high-performance biosensor utilizing PPy nanowires (PPyNWs) with outstanding conductivity and a large surface area to detect hydrogen peroxide and miRNA [95]. Rong et al. reported a nanocomposite film containing PPy for the selective adsorption and removal of Pb [96]. Mohamed et al. designed a nanocomposite using PPy nanofibers (PPyNFs) with photocatalytic activity and the capacity for dye adsorption to remove dye from raw water samples [97]. Han et al. fabricated a conductive hydrogel combining PPy and silk hydrogel for use in the preparation of flexible and wearable sensors [91]. Fan and his colleagues utilized a PPy sponge with micron-sized pores, good mechanical capacities, and the capacity for light absorption for use in a functional solar steam generator [98].

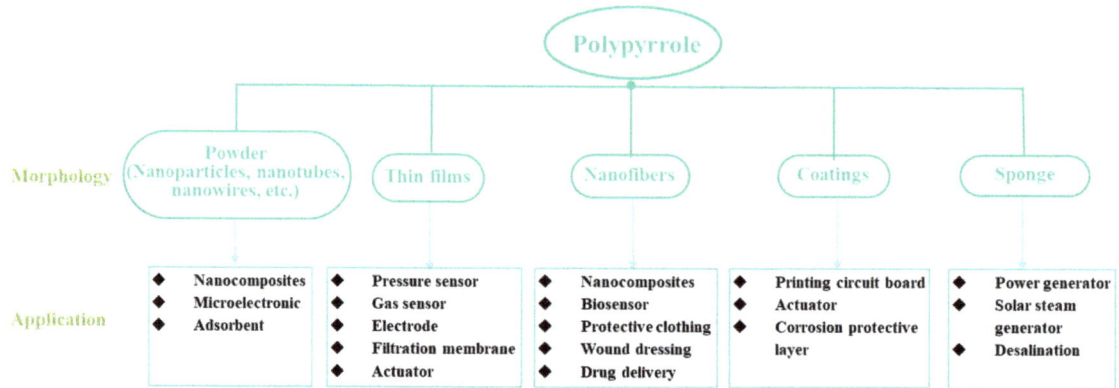

Figure 3. Synthesized PPy morphologies and their applications.

2.4. Application of PPy-Based Biosensors

PPy and its derivatives are some of the most effective nanomaterials for improving the sensing performance of different biosensors (see Table 2). PPy nanomaterials and their composites, which have unique optical, electrochemical, and other physical and chemical properties, have great potential for the enhancement of sensing performance, including the response and recovery time, stability, selectivity, and sensitivity [99].

Table 2. Applications of PPy-based biosensors with different functions.

Type of Biosensor	Function of PPy	Ref.
Enzyme-based biosensors	♦ Entrap and immobilize the enzyme ♦ Reduce the oxidation potential of the substrate ♦ Improve the biocompatibility of the biosensor ♦ Promote electron transfer	[100–102]
Immunobiosensors	♦ Offer a proper environment for the immobilization of biomolecules ♦ Increase surface area ♦ Improve the biocompatibility and conductivity of the biosensor	[103–105]
Aptamer-based biosensors	♦ Immobilize aptamers ♦ Enlarge the specific surface area ♦ Enhance the biocompatibility and electroconductivity of the biosensor	[106,107]
MIP-based biosensors	♦ Manually create specific molecular recognition sites ♦ Enhance the specificity, biocompatibility, and electroconductivity of the biosensor	[108–111]
Nanocatalytic biosensors	♦ Serve as a suitable immobilization matrix and disperse metal nanoparticles effectively ♦ Improve the catalytic activity of enzyme mimics ♦ Provide good conductivity	[112–114]

2.4.1. Enzyme-Based Biosensors

PPy is widely used as a substrate for the preparation of enzyme-based biosensors. During this process, PPy is used for enzyme fixation and facilitates electron transfer between the active center of the immobilized enzyme and the electrode. The permeability of the PPy layer for enzyme substrates and reaction products is relatively low, contributing to the increase in the apparent Michaelis constant of the enzyme and thereby expanding the detection range for analytes [115]. Apetrei et al. described a PPy-based enzymatic sensor in which the polyphenol oxidase extract could be utilized as a catalyst for both the synthesis of PPy and the self-encapsulation of the enzyme (Figure 4A) [100]. Dutta et al. synthesized an amperometric biosensor in which PPy was used to entrap and immobilize acetylcholinesterase, facilitate electron transfer, and reduce the oxidation potential of the reaction substrate [101]. Shi et al. synthesized overoxidized PPy to modify an electrode, and it had a catalytic oxidation ability that made it possible to reduce the oxidation potential of ascorbic acid and then increase the sensitivity of the biosensor [102].

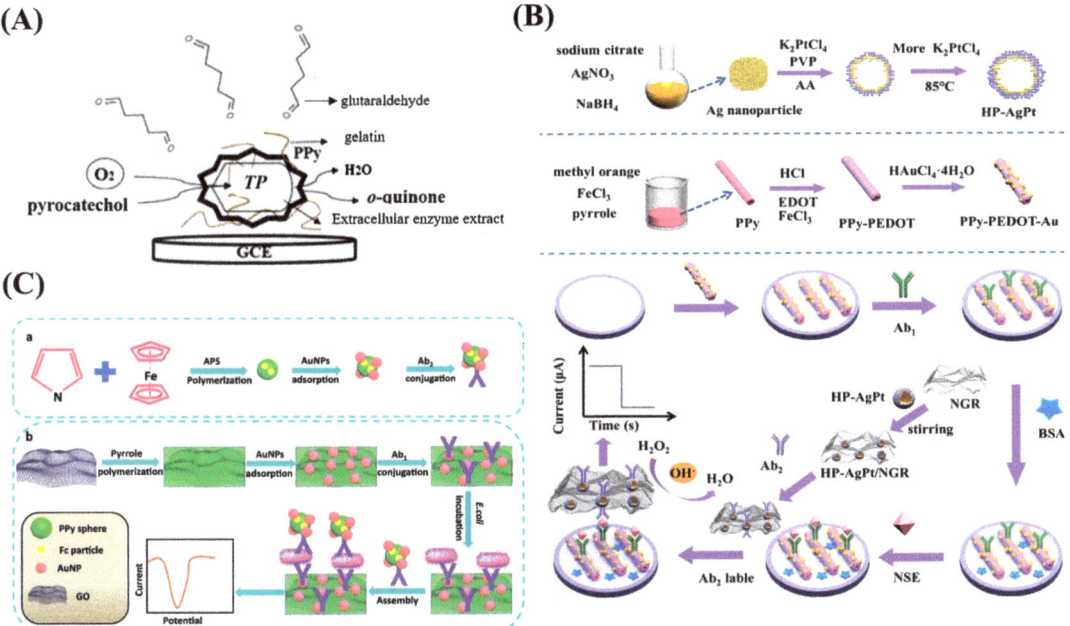

Figure 4. (**A**) Schematic diagram of PPy-based enzymatic sensor. Reproduced from [100] with permission from Elsevier. (**B**) Schematic diagram of the electrochemical platform for polypyrrole–poly(3,4-ethylenedioxythiophene)–gold (PPy-PEDOT-Au). Reproduced from [104] with permission from Elsevier. (**C**) Schematic diagram of the fabrication process for polypyrrole-reduced graphene oxide/gold nanoparticles (PPy-rGO/AuNPs) for use in biosensors. Reproduced from [105] with permission from Elsevier.

2.4.2. Immunobiosensors

Antibody/antigen-based biosensors are sensing devices based on the affinity interaction between antibodies and antigens [116]. PPy is commonly applied in these sensors because of its low oxidation potential and good biocompatibility. During biosensor preparation, PPy can serve as both an immobilizing substrate for biometric components and as a signal transduction system [99]. Tang et al. fabricated PPy-PEDOT-Au to fix the antibody in an electrochemical immunosensor due to the excellent electron transfer efficiency and environmental stability of PPy (Figure 4B) [104]. Zou et al. utilized PPy with good biocompatibility and a high surface area to improve the dispersion of AuNPs, constructing a biocompatible platform and immobilizing more antibodies to detect trace amounts of *E. coli* K12 (Figure 4C) [105].

2.4.3. Aptamer-Based Biosensors

Aptamers, which are artificial single-stranded oligonucleotide fragments, can precisely capture ligands, such as DNA, RNA, or proteins, with high affinity and selectivity. Aptamers are easily synthesized and more stable than antibodies and antigens, and they have been widely utilized in electrochemical biosensors to detect low concentrations of target molecules in the blood. Duan et al. designed an aptasensor to detect lipopolysaccharide in which PPyNWs with -COOH ensured the immobilization of aptamers and enlarged the specific surface area (Figure 5A) [106]. In addition, a nitrogen-doped graphene (NG)/PPy nanocomposite was synthesized in order to design an aptasensor that, with the help of PPy, would exhibit a large surface area and enhanced electroconductivity [107].

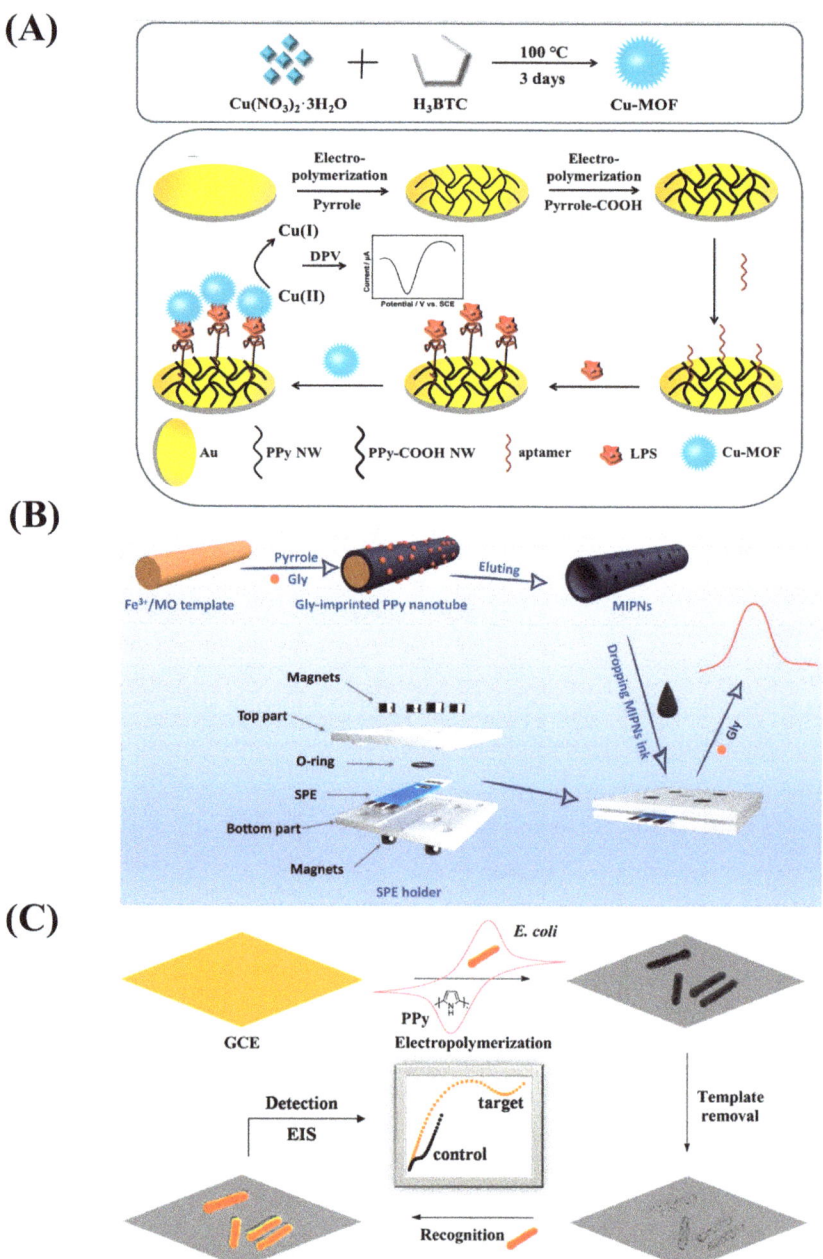

Figure 5. (**A**) Schematic diagram of the preparation procedure for the polypyrrole nanowire (PPyNW)-based aptasensor. Reproduced from [106] with permission from Elsevier. (**B**) Schematic diagram of

the preparation of molecularly imprinted polypyrrole nanotubes (MIPNs) and the MIPN-based glyphosate platform. Reproduced from [108] with permission from Elsevier. (**C**) Schematic diagram of the fabrication process for the biosensor based on imprinted polypyrrole film from bacteria. Reproduced from [109] with permission from the American Chemical Society.

2.4.4. MIP-Based Biosensors

MIP is synthesized by polymerizing one or more monomers in the presence of the template molecules and then removing the template analytes, forming a complementary 3D imprint in the polymer matrix. It is a kind of polymer with specific recognition sites that are created artificially and complement the imprinted analyte, and it is also known as a "bionic receptor" [117]. Considering its biocompatibility, excellent electron transmission rate, and good environmental stability [108,110], molecularly imprinted polypyrrole (MIPPy) can be polymerized and impressed simultaneously using amperometric, potentiometric, or potential scanning methods or polymerized on a template-modified electrode [75]. It can be used for biomedical and environmental monitoring applications, and it is cost-effective and exhibits excellent selectivity, sensitivity, and chemical/thermal stability [75,111]. Ding et al. described an electrochemical method that can be used to detect glyphosate with a detection limit of 1.94 ng/mL based on the construction of molecularly imprinted polypyrrole nanotubes (MIPNs) that exhibited excellent conductivity and specificity (Figure 5B) [108]. Wu et al. designed an electrochemical platform utilizing imprinted polypyrrole from bacteria that could sensitively detect *E. coli* O157. Owning to the presence of MIPPy, the biosensor had high selectivity and specificity and a rapid detection ability and was easy to prepare (Figure 5C) [109].

2.4.5. Nanocatalytic Biosensors

As natural enzymes are limited by their instability, demanding environmental conditions, and complex preparation processes, nanocatalytic biosensors have been developed that provide improved electrocatalytic behavior, stability, and selectivity because of their satisfactory surface area, morphology, and high conductivity [118–120]. After Fe_3O_4 nanoparticles were reported to exhibit catalytic activities similar to enzymes in 2007 [121], a growing number of catalytically active nanomaterials have been discovered that can be used as alternatives to enzymes. Metals and metal oxides have outstanding electronic properties, and their nanoenzymes have received significant attention for sensing and electrocatalysis applications [118]. Li et al. prepared nanocomposites capable of catalyzing glucose oxidation by depositing gold nanoparticles on the surface of PPyNFs to non-enzymatically detect glucose, using PPyNFs as carriers to avoid the need for harsh oxidation pretreatment and disperse the AuNPs well, thereby improving their catalytic activity and reducing Au consumption (Figure 6A) [112]. Jeong and his colleagues proposed an electrochemical method based on a chitosan–PPy/TiO_2 sensor in which TiO_2 NPs, as the electrochemical catalyst, are deposited using the plasma process, demonstrating outstanding catalytic activity, reactivity, sensitivity, and selectivity in the detection of glucose (Figure 6B) [113]. Meng et al. fabricated a non-enzymatic biosensor utilizing Cu_xO-modified PPyNWs to sensitively detect glucose, and the PPyNWs showed excellent electrical performance and produced a good immobilization matrix for nanoparticles (Figure 6C) [114].

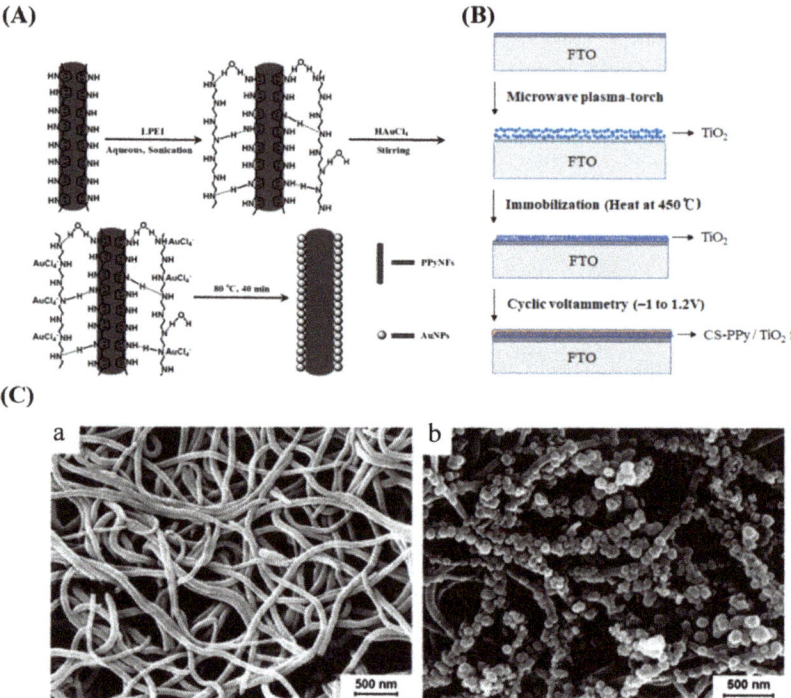

Figure 6. (**A**) Schematic diagram of the synthesis of polypyrrole nanofiber-supporting Au nanoparticles (Au/PPyNFs). Reproduced from [112] with permission from Elsevier. (**B**) Schematic diagram of the preparation of chitosan–polypyrrole/titanium oxide (CS-PPy/TiO$_2$) nanocomposite films on a fluorine-doped tin oxide-coated glass slide (FTO). Reproduced from [113] with permission from MDPI. (**C**) Scanning electron microscope images of PPyNWs (**a**) and copper oxide (Cu$_x$O) nanoparticle-modified PPyNWs (**b**). Reproduced from [114] with permission from Elsevier.

3. PPy-Based Biosensors for CRC Biomarker Detection

CRC is currently the second-leading cause of death worldwide [122]. The electrochemical detection of biomarkers in the blood (such as DNA, RNA, miRNA, proteins, and other molecules), a low-cost, simple, rapid, specific, sensitive, and noninvasive strategy, exhibits tremendous potential for the early diagnosis of CRC [27].

3.1. Circulating Cell-Free DNA (ccf-DNA)

Ccf-DNA is actively released by tumor cells through apoptosis, necrosis, or exosomes. It is present in significantly higher amounts than in healthy people, highlighting its crucial practical significance for the early diagnosis of malignant tumors. Ccf-DNA is attractive and easily accessible, providing a new, non-invasive method for CRC detection and characterization. Many studies have shown the diagnostic, predictive, and prognostic significance of abnormal ccf-DNA with genetic and epigenetic variations in the plasma/serum of CRC patients [123].

3.1.1. DNA Mutation

Abnormal genetic mutations in the blood have been evaluated as one of the most promising diagnostic tools for CRC. The protooncogene KRAS is an early candidate in this context, as its mutations involve some of the most frequently mutated oncogenes and have a certain practical significance for the clinical diagnosis of CRC [122]. BRAF, belonging to the RAF gene family, is a direct downstream effector of KRAS and is connected to CRC development [123]. The adenomatous polyposis coli (APC) gene, a tumor-suppressor gene with the encoded protein in the Wnt signaling pathway, was one of the early genetic factors used in the screening of CRC [124]. It is claimed that TP53 is the guardian of the genome, and its mutation can disable the functional activity of wild-type p53 (wtp53) and produce oncogenic properties [125]. As the TP53 mutation occurs in 50–70% of CRC patients, monitoring the changes in the TP53 gene and/or its encoded protein in CRC patients may contribute to the early diagnosis and detection of clinical conditions.

As shown in Figure 7A, Wang and colleagues coated PPy-covered MWNT-Ru(bpy)$_3^{2+}$ composite materials on the surface of a Au electrode to prepare a DNA sensor for wtp53 sequence detection using ECL [126]. ECL, which has the advantages of low background signals, high sensitivity, good versatility and controllability, and a wide detection range, is generated using electrochemical reactions that trigger light signals. Considering its large specific surface area and prominent conductivity, PPy was utilized as a stable modification layer for ssDNA attachment to improve the wtp53 sensing performance, resulting in a detection range of 0.2 pM–200 pM and an LOD of 0.1 pM. The authors also fabricated another DNA-based biosensor modified by electrospinning composite MWNT-PA6-PPy nanofibers (Figure 7B) [127]. Compared with traditional planar materials, PPy nanofibers have better mechanical strength, uniformity, porosity, and reusability, as well as satisfactory biocompatibility and a high surface area, making it possible to immobilize more ssDNA and increase the hybridization sensitivity for determination of trace amounts of wtp53.

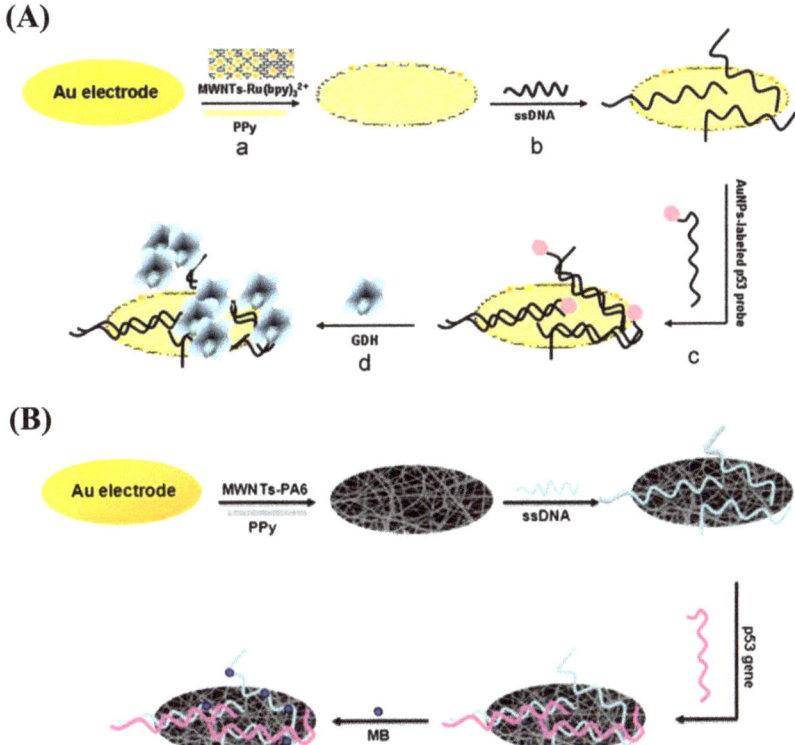

Figure 7. (**A**) Schematic illustration of the fabrication of a PPy-covered multiwalled carbon nanotube and ruthenium (II) tris-(bipyridine) (MWNT-Ru(bpy)$_3^{2+}$-PPy) biosensor for the wild-type p53 sequence (wtp53) assay based on electrochemical luminescence (ECL). Reproduced from [126] with permission from Elsevier. (**B**) Schematic illustration of the preparation of a multi-walled carbon nanotube–nylon 6–polypyrrole (MWNT-PA6-PPy) biosensor for wtp53 detection based on electrospinning technology. Reproduced from [127] with permission from Elsevier.

3.1.2. DNA Methylation

DNA methylation is another ccf-DNA-based technique for the early diagnosis of CRC that employs genes such as SEPT9, SCTR, SDC2, SFRP2, TMEFF2, NGFR, and CG10673833 [128–131]. Sun et al. demonstrated that aberrant methylation of the SEPT9 gene (^{m}SEPT9) in the blood can be used as a marker in the early diagnosis and screening of CRC [128]. Li et al. showed that hypermethylation of the SCTR gene resulted in good accuracy in the diagnosis of CRC and its precursor lesions [129].

3.2. MiRNA-Based Biomarkers

MicroRNAs are abundant and endogenous noncoding RNAs with a small size and hairpin structure. MiRNAs can regulate and control physiological processes, such as the proliferation and differentiation of cells, by regulating the expression of various genes. The development of numerous diseases (cancer, neurodegenerative diseases, cardiovascular diseases, etc.) is related to abnormal microRNA expression [132]. It has been shown that miRNAs play significant roles in the progress of CRC, and evaluation of the strange expression of miRNAs, such as miR-21, miR-92a, miR-451a, miR-29a, miR-23a, miR-141, let-7a, miR-1229, miR-223, miR-1246, miR-150, and miR-378, has shown great clinical value in CRC screening, prognosis, prediction, and treatment [133–136]. Among these miRNAs, miR-21 is one of the most widely researched for the diagnosis, prediction, and treatment of CRC [134,136–138]. Pothipor et al. synthesized a gold nanoparticle/polypyrrole/graphene (AuNP/PPy/GP) nanocomposite for the selective detection of miR-21 (Figure 8A) [139]. The use of PPy in this work improved the dispersion of AuNPs on the electrode surface, facilitating the fixation of an miR-21 probe with a corresponding detection range of 1.0 fM–1.0 nM and LOD of 0.020 fM that can be used to detect miR-21 in clinical trials. Tian et al. designed a PPy-AuNP superlattice (AuNS) biosensor for the detection of miR-21 (Figure 8B) [140]. Compared to randomly arranged nanoparticles, the presence of the conductive polymer PPy can induce AuNPs to assemble into AuNS structures with a larger surface area, better electron transfer performance, and more active sites. Furthermore, using PPy ligand can facilitate quantitative and accurate control of the distance between adjacent particles, enabling miR-21 determination with a 100 aM–1 nM detection range and limit of detection (LOD) of 78 aM.

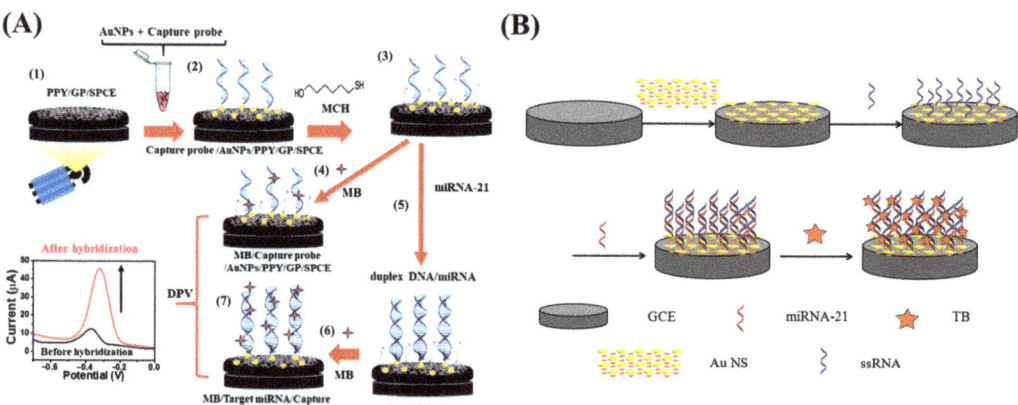

Figure 8. (**A**) Schematic representation of the strategy based on a gold nanoparticle/polypyrrole/graphene (AuNP/PPy/GP) nanocomposite. Reproduced from [139] with permission from the Royal Society of Chemistry. (**B**) Schematic representation of the approach using a PPy-AuNP superlattice (PPy-AuNS). Reproduced from [140] with permission from Elsevier.

Kaplan et al. sensitively detected miRNA-21 using electropolymerized PPy on the surface of a pencil graphite electrode (PGE) and achieved maximum doping of the anti-miR-21 probe in PPy (Figure 9A) [141]. PPy has high conductivity and a porous structure, making it possible to improve charge transfer, increase the number of fixed probe molecules, and reduce the non-specific binding of MDB and other molecules, and the designed biosensor demonstrated improved selectivity and an LOD of 0.17 nM. Yang et al. prepared PDA-PPy-NS with π-electron coupling and an ultra-narrow band-gap by polymerizing PPy onto hybrid polydopamine nanosheets (PDA-NSs) and used nucleic acid dye (Cy5)-labeled ssDNA as probes to detect miRNA-21 (Figure 9B) [83]. The mixed PDA-PPy-NS nanoquencher showed a better fluorescence quenching ability than PDA-NSs owing to the presence of the narrow band-gap PPy and its excellent π-electron delocalization ability, promoting intermolecular electron coupling and realizing fluorescence quenching. The nanoquencher/probe was demonstrated to have remarkable specificity, stability, and sensitivity and an LOD of 23.1 pM, indicating the tremendous potential of nanoquencher-based sensors for detection with real samples.

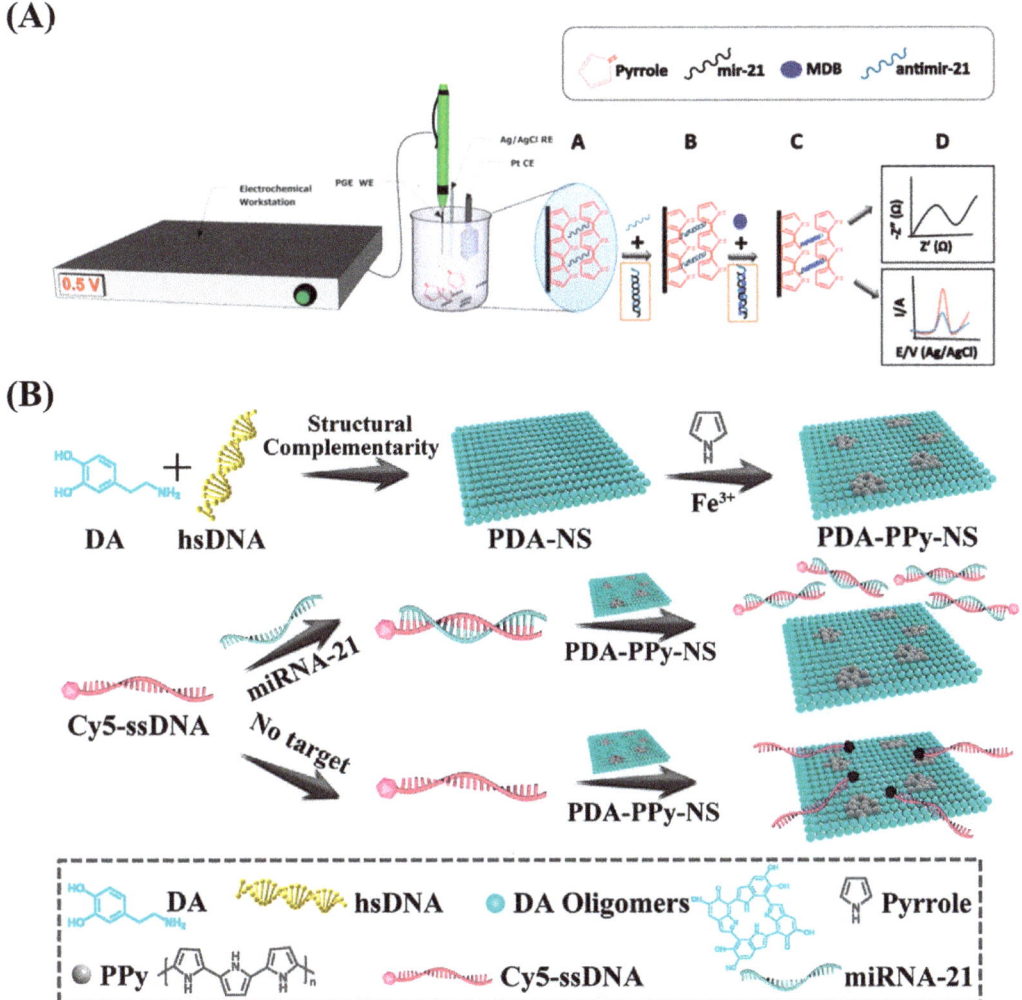

Figure 9. (**A**) Schematic diagram of the biosensor fabrication process utilizing PPy electrodeposition. Reproduced from [141] with permission from Elsevier. (**B**) Schematic diagram of the preparation

procedure for the hybrid polydopamine/polypyrrole nanosheet (PDA-PPy-NS) biosensor for miRNA-21 determination. Reproduced from [83] with permission from the American Chemical Society.

3.3. Specific Protein Biomarkers

In addition to DNA and RNA, various proteins secreted from tumor cells can also facilitate the early diagnosis of CRC, including CEA, CA19-9, CA72-4, CA125, IL-6, IL-8, MUC1, and p53, as shown in Table 3. As the table demonstrates, conductive PPy is commonly applied to improve electrochemical sensing performance, including the detection range and LOD, in the detection of specific CRC protein biomarkers.

Table 3. PPy-based biosensors for the determination of CRC protein biomarkers.

Protein Biomarkers	Biosensor Components	Detection Method	Detection Range	LOD [I]	Ref.
CEA [II]	2-NS-PPy [III]/PEE [IV]-PPy/2-NS-PPy/AuNP/Apt/CEA	EIS	10^{-1}–10^3 ng/mL	0.033 ng/mL	[142]
	PPy foam/Cu/ITO [V]/PET [VI]/Kapton/PDMS [VII]/cAb [VIII]/CEA/PtNP-labeled dAb [IX]	Resistance determination	0.2–60 ng/mL	0.13 ng/mL	[143]
	GCE [X]/PPy@AuNP-luminol-anti-CEA/CEA	ECL	10^{-5}–10 ng/mL	3 fg/mL	[144]
	ITO/PANI/PPy-Ag/Ab$_1$/CEA/ZnO@AgNC [XI]-Ab$_2$	ECL	10^{-3}–100 ng/mL	0.4 pg/mL	[145]
	AuNP/NH$_2$-GS [XII]/Ab$_1$/CEA/Au@PdND [XIII]/Fe^{2+}-CS/PPy NT/Ab$_2$	i-t/SWV	5×10^{-5}–50 ng/mL	17 fg/mL	[146]
	GCE/PPy hydrogel/AuNP/anti-CEA/CEA	DPV	10^{-6}–200 ng/mL	0.16 fg/mL	[147]
CA72-4 [XIV]	GCE/PPy-NH$_2$GO-Ag$_2$Se@CdSe/Ab/CEA	ECL	10^{-4}–20 U/mL	2.1×10^{-5} U/mL	[148]
CA125 [XV]	ITO/MB-mAb$_1$/CEA/PPy-Ag-pAb$_2$	LSV	0.001–300 U/mL	7.6 mU/mL	[149]
	Au-SPE/MIPPy [XVI]/CA125	SWV/SPR [XVII]	0.01–500 U/mL	0.01 U/mL	[150]
IL-6 [XVIII]	SPGE [XIX]/PPy/AuPts/Apt [XX]/IL-6	EIS	10^{-6}–15 μg/mL	0.33 pg/mL	[151]
	ITO/PPCE [XXI]/IL-6 receptor/IL-6	EIS/CV	0.02–16 pg/mL	6.0 fg/mL	[152]
	ITO/AB [XXII]/EpxS-PPyr [XXIII]/IL-6 receptor/IL-6	EIS/CV	0.01–50 pg/mL	3.2 fg/mL	[153]
	PEEK [XXIV]/PETE [XXV]/PPyNW/mAb/IL-6	EIS	1–50 pg/mL	0.36 pg/mL	[154]
VEGF [XXVI]	Glass substrate/CPNT [XXVII]/Apt/VEGF	FET [XXVIII]	-	400 fM	[155]
	Flexible substrate/PPy-NDFLG [XXIX]/Apt/VEGF	FET	-	100 fM	[156]

Notes: I. LOD: limitation of detection; II. CEA: carcinoembryonic antigen; III. 2-NS-PPy: PPy doped with 2-naphthalene sulfonate; IV. PEE: pentaerythritol ethoxylate; V. ITO: indium tin oxide; VI. PET: poly(ethylene terephthalate); VII. PDMS: poly(dimethylsiloxane); VIII. cAb: capture antibody; IX. dAb: detection antibody; X. GCE: glassy carbon electrode; XI. AgNC: silver nanocluster. XII. NH$_2$-GS: amino-functionalized graphene sheet; XIII. PdND: palladium nanodendrite; XIV. CA72-4: carbohydrate antigen 72-4; XV. CA125: carbohydrate antigen 125; XVI. MIPPy: molecularly imprinted polypyrrole; XVII. SPR: surface plasmon resonance; XVIII. IL-6: interleukin-6; XIX. SPGE: screen-printed graphite electrodes; XX. Apt: aptasensor; XXI. PPCE: polypyrrole polymer-containing epoxy side group; XXII. AB: acetylene black; XXIII. EpxS-PPyr: epoxy-substituted poly(pyrrole) polymer; XXIV. PEEK: polyether ether ketone; XXV. PETE: poly(ethylene terephthalate); XXVI. VEGF: vascular endothelial growth factor; XXVII. CPNT: carboxylated polypyrrole nanotube; XXVIII. FET: field-effect transistor; XXIX. PPy-NDFLG: PPy-transformed N-doped few-layer graphene.

3.3.1. Carcinoembryonic Antigen (CEA)

The polymeric glycoprotein CEA is a tumor-associated antigen that is overexpressed in CRC, gastric cancer, breast cancer, lung cancer, and other cancers and has a certain value for the evaluation of tumor status and therapeutic effect [157]. CEA is abnormally expressed in more than 90% of CRC patients. The monitoring of CEA concentration in serum/plasma is an effective strategy for CRC diagnosis and measurement of disease progression.

Tavares et al. designed a self-powered and self-signaled biosensing platform for the detection of CEA using MIPPy and the DSSC method (Figure 10A) [158]. Specifically, they assembled the MIPPy as a biorecognition element on the PEDOT layer of an FTO-conductive glass substrate and used it as the counter electrode of the DSSC. The DSSC/biosensor device was connected to an electrochromic cell to produce a color gradient for the CEA. The concentration of CEA ranged from 0.1 ng/mL to 100 µg/mL. This strategy can be used for clinical point-of-care (POC) analysis with high independence.

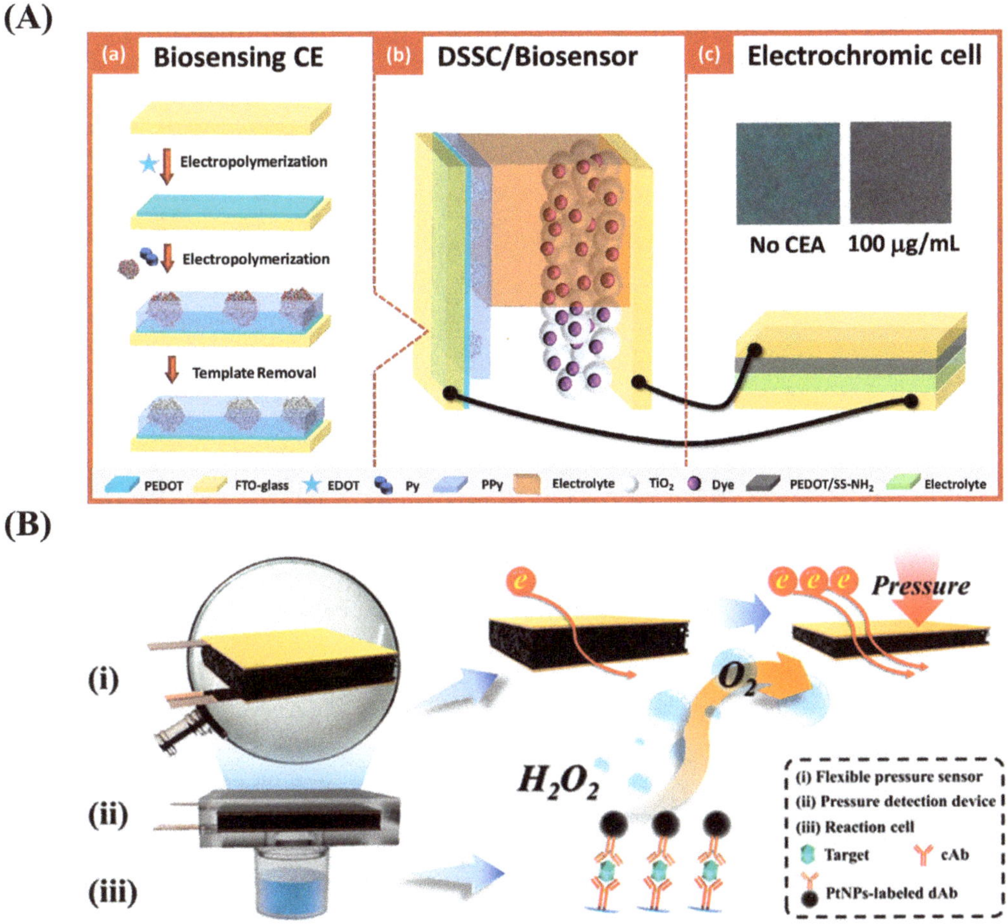

Figure 10. (**A**) Schematic diagram of the design of a self-powered and self-signaled biosensing platform based on biosensing, molecular imprinting technology, a dye-sensitized solar cell (DSSC), and electrochromic technology. Reproduced from [158] with permission from Elsevier. (**B**) Schematic diagram of a flexible pressure sensor based on the elastic three-dimensional (3D) structure of PPy foam. Reproduced from [143] with permission from the American Chemical Society.

PPy can also be used to prepare flexible pressure sensors that convert external force information into electrical signals in real time. Yu et al. prepared a pressure-based immune sensor based on 3D PPy foam (Figure 10B) [143]. The PtNPs attached to dAb catalyze the decomposition of H_2O_2 and produce oxygen in the sealed device. Consequently, CEA concentration can be detected using pressure changes in the range of 0.2–60 ng/mL with an LOD of 0.13 ng/mL.

Zhu et al. constructed a novel ECL immune-based biosensor using anti-CEA–luminol–AuNP@PPy (Figure 11A) [144]. The PPy nanostructure made it possible to enhance the conductivity of the biosensor and provided a high specific surface area for the combination with the AuNPs, thus enabling the attachment of abundant amounts of the ECL reagent (luminol) and promoting the immobilization of antibodies. This strategy showed a detection range of 0.01 pg/mL–10 ng/mL and an LOD of 3 fg/mL, making it an effective tool for the clinical detection of CEA.

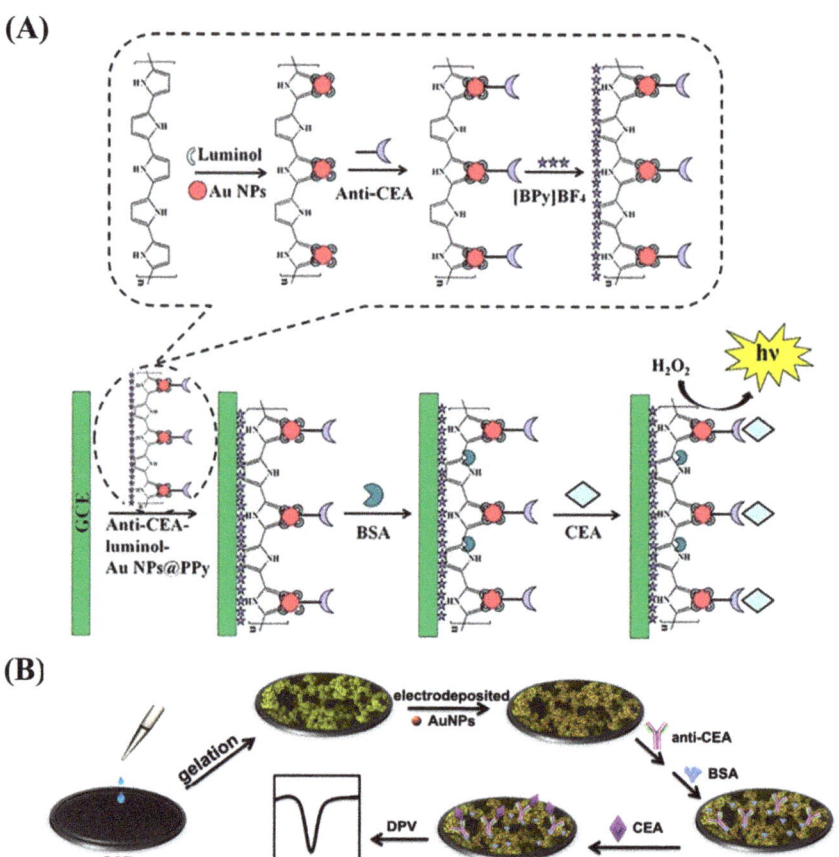

Figure 11. (**A**) Schematic diagram of the preparation of an anti-CEA–luminol–AuNP@PPy-based immunosensor for CEA detection. Reproduced from [144] with permission from Nature Publishing Group. (**B**) Schematic diagram of the electrochemical method for CEA determination based on 3D continuous conducting network nanocomposites composed of PPy hydrogel loaded with AuNPs. Reproduced from [147] with permission from Nature Publishing Group.

CP hydrogels, such as PPy hydrogel, have been applied to construct biosensors; in particular, those with 3D nanostructures. Compared to other materials, PPy hydrogel has the advantages of prominent conductivity, good biocompatibility, a large specific surface area, and easy processing [147]. Rong et al. designed an electrochemical immunosensing platform to measure CEA based on 3D continuous conducting network nanocomposites composed of PPy hydrogel loaded with AuNPs (Figure 11B) [147]. The hydrogel had good biocompatibility and electronic properties and offered a larger space that made it possible to immobilize more biomolecules. The nanostructure-based sensor had a highly porous 3D network, high specificity, and good stability. It showed a broad linear range (1 fg/mL–200 ng/mL) and an LOD of 0.16 fg/mL.

Abnormal changes in CEA concentration in the blood are generally associated with cancer progression, and the sensitivity of this biomarker increases with tumor stage. Therefore, CEA is the preferred marker for monitoring CRC progression and prognosis. However, abnormal elevation of CEA in the blood occurs not only in CRC but also in various other diseases. In addition, CEA has low sensitivity in the early stages of CRC. Hence, this biomarker is ineffective for screening and detecting CRC early [123].

3.3.2. Carbohydrate Antigens

Abnormal expression of CA19-9, CA72-4, CA125, CA242, and other carbohydrate antigens also has a certain correlation with CRC [157,159–161]. CA19-9 was first discovered in 1981, and elevated CA19-9 levels can be used both as an aid in the diagnosis of CRC and as a reference index to assess the development of CRC.

CA72-4 is a cancer biomarker with a certain diagnostic value that can provide diagnostic information regarding recurrent CRC. Lv et al. proposed an ECL immunosensor using a novel Ag2Se@CdSe nanomaterial nanoneedle modified with polypyrrole-intercalated aminated graphene (PPy-NH$_2$GO) for CA72-4 detection (Figure 12A) [148]. The PPy-functionalized NH$_2$GO had a high surface area that made it possible to immobilize significant amounts of Ag$_2$Se@CdSe, and it demonstrated a low LOD of 2.1×10^{-5} U/mL and a detection range of 10^{-4}–20 U/mL.

An electrochemical magnetic immunoassay platform was described by Huang et al. that used anti-CA125 antibody (mAb$_1$)-conjugated magnetic beads as the capture probe and an anti-CA125 antibody (pAb$_2$)-labeled Ag-PPy nanostructure as the detection probe (Figure 12B) [149]. Compared with AgNPs, use of Ag-PPy as the electroactivity indicator can further improve sensors' analytical performance.

However, due to the limited applicability of markers such as CEA and CA19-9, several other proteins have been highlighted as potential biomarkers associated with CRC.

3.3.3. Interleukin-6 (IL-6)

IL-6 is an inflammatory cytokine with hematopoietic and immunomodulatory functions; although not a specific biomarker of CRC, IL-6 is closely associated with CRC occurrence, development, staging, invasion, and metastasis.

Tertis et al. constructed an electrochemical aptasensor to sensitively detect IL-6 in human serum by depositing a nanocomposite consisting of PPy nanoparticles (PPyNPs) and AuNPs onto SPCE (Figure 13A) [151]. Both the PPyNPs and AuNPs coexisted on the electrode surface, providing an appropriate environment for immobilization of IL-6 aptamers. Moreover, conjugated polypyrrole polymer-containing epoxy side group (PPCE) is a novel conjugated polymer polymerized by the Py monomer that contains an epoxy active side group. As shown in Figure 13B, Elif Burcu Aydın designed an immunosensor using PPCE-modified ITO electrodes [152]. The PPCE polymer formed a large specific surface area on the ITO electrode, making it possible to fix the biomolecular IL-6 receptor, and the sensor had good conductivity, stability, and biocompatibility, which enhanced its sensitivity.

PPyNWs exhibit good electrical characteristics and can enhance the sensitivity of biosensors because of their high specific surface area. Cruz et al. prepared PPyNWs on PEEK and PETE flexible thermoplastics using nanocontact printing technology and controlled chemical technology and then functionalized them with diazo chemistry and a crosslinking agent to immobilize the IL-6 antibody, enabling IL-6 detection with a wide linear range from 1 pg/mL to 50 pg/mL and an LOD of 0.36 pg/mL [154].

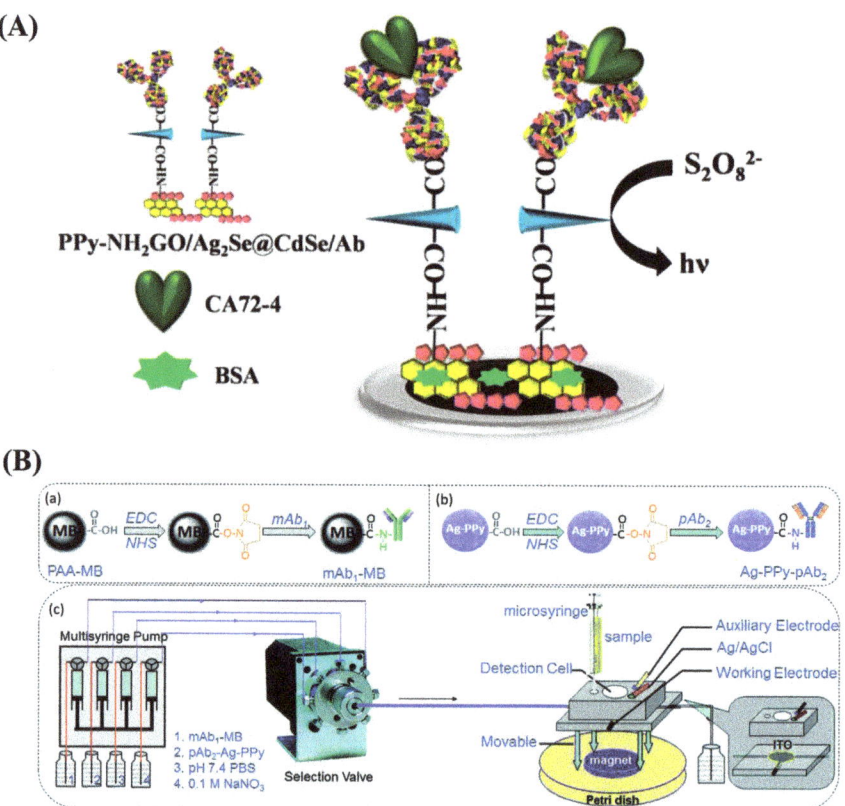

Figure 12. (**A**) Schematic diagram of the polypyrrole-intercalated aminated graphene/Ag$_2$Se@CdSe (PPy-NH$_2$GO/Ag$_2$Se@CdSe)-based immunosensor for the CA72-4 assay. Reproduced from [148] with permission from the American Chemical Society. (**B**) Schematic diagram of the platform for CA125 measurement: (**a**) the preparation of the capture probe using mAb$_1$-conjugated magnetic beads (mAb$_1$-MB). (**b**) the preparation of the detection probe using pAb$_2$-labeled Ag-PPy nanostructure (Ag-PPy- pAb$_2$). (**c**) magneto-controlled microfluidic device with an electrochemical detection cell. Reproduced from [149] with permission from the Royal Society of Chemistry.

3.3.4. Vascular Endothelial Growth Factor (VEGF)

Angiogenesis is closely connected to the growth of solid tumors and the metastasis of cancer cells, and VEGF participates in the regulation of angiogenesis by stimulating the corresponding receptors and is important for the development of blood vessels. Nogués et al. pointed out that high-level expression of VEGF in the serum of CRC patients seems to be a promising tumor biomarker [162].

By immobilizing anti-VEGF RNA aptamers onto a field-effect transistor (FET) modified with carboxylated polypyrrole nanotubes (CPNTs) with excellent conductivity, Kwon et al. developed a biosensor capable of recognizing VEGF, as shown in Figure 14A [155]. The FET platform is capable of detecting VEGF at concentrations as low as 400 fM. In a related study, the author led a team in combining PPy-transformed N-doped few-layer graphene (PPy-NDFLG) with an RNA aptamer for the fabrication of a high-performance and flexible FET biosensor (Figure 14B) [156]. The sensor demonstrated good mechanical flexibility, high sensitivity, high selectivity, a rapid response, reusability, durability, and a low LOD of 100 fM.

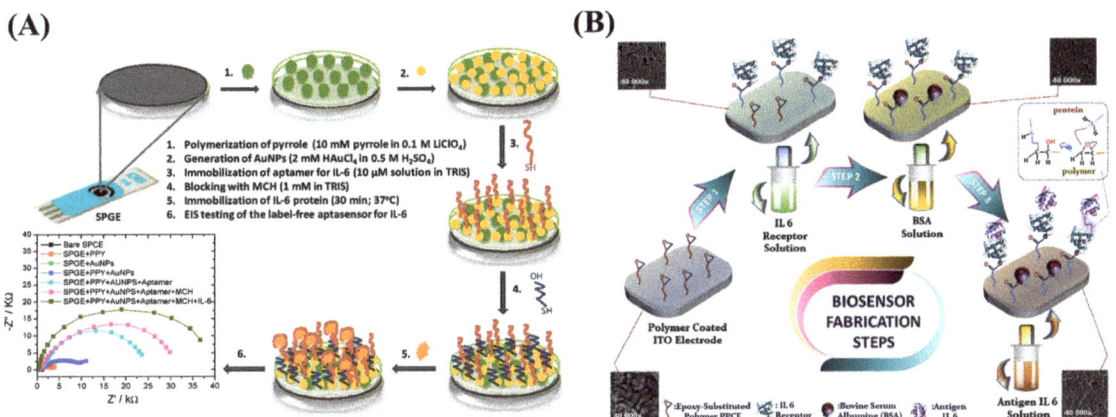

Figure 13. (**A**) Schematic diagram of the AuNP/PPyNP-based aptasensor for the IL-6 assay. Reproduced from [151] with permission from Elsevier. (**B**) Schematic diagram of the fabrication process for the PPCE-modified biosensor for IL-6 measurement. Reproduced from [152] with permission from Elsevier.

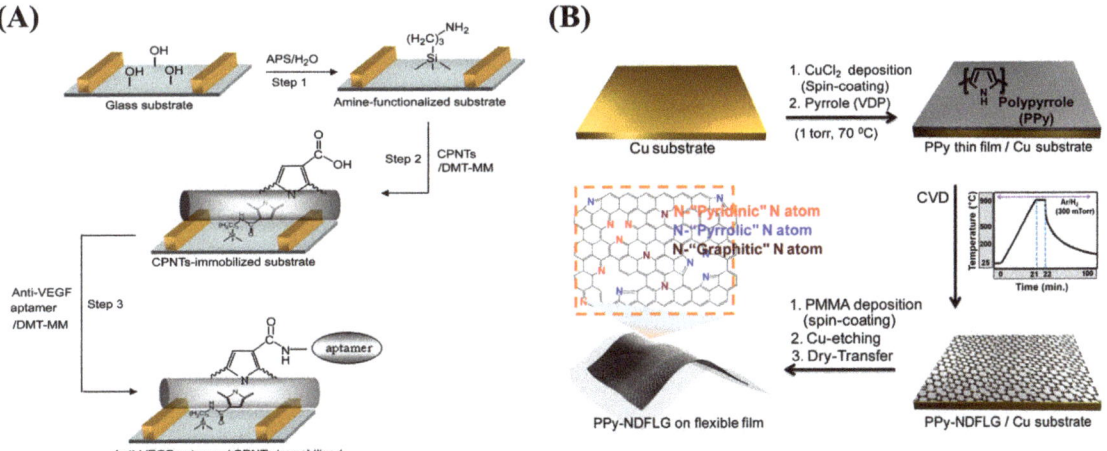

Figure 14. (**A**) Schematic diagram of the fabrication of a CPNT-modified aptasensor for the VEGF assay. Reproduced from [155] with permission from Elsevier. (**B**) Schematic diagram of the process of synthesizing flexible PPy-NDFLG. Reproduced from [156] with permission from the American Chemical Society.

3.3.5. Other CRC-Related Protein Biomarkers

Mucin 1 (MUC1) is a membrane-associated macromolecule glycoprotein that is overexpressed in most adenocarcinomas [163,164]. Detection of abnormal increases in MUC1 in the blood can offer new opportunities for CRC early diagnosis, tumor staging, and clinical treatment.

Huang et al. designed a microfluidic aptasensor by combining the use of an MUC1 aptamer as a detection probe with PPyNWs. The sensor can be used for sensitive, rapid, label-free, and real-time detection of MUC1 [165]. The PPyNW-modified biosensor showed significantly enhanced sensitivity, conductivity, and biocompatibility. In addition, serum angiopoietin, MST1/STK4, S100A9 TIMP1, ITGB4, Cyr61, and CXCL-8 in the blood have also been proven to have potential uses as biomarkers in the diagnosis and detection of colorectal cancer, and they could be employed in electrochemical sensor detection after further verification [166–170].

3.4. Opportunities and Challenges Related to PPy-Based Sensors in CRC Diagnosis

Currently, most discovered CRC biomarkers are not specific to CRC. Combined detection of multiple biomarkers in the blood could sufficiently increase the accuracy of CRC diagnosis, resulting in an effective detection strategy. It has been shown that, compared with the detection of single biomarkers, simultaneous determination of the serum biomarkers MMP-7, TIMP-1, and CEA increased the sensitivity and specificity of CRC diagnosis [171]. However, studies are still required to further determine the clinical significance of single CRC marker analysis and combined detection of groups of biomarkers as early detection tools for CRC and to develop additional analytical means that can enhance the accuracy and specificity of CRC diagnosis.

As a heterocyclic conjugated polymer with good electroconductivity, processability, and chemical stability, PPy is rarely used for the detection of multiple CRC biomarkers. The electrochemical deposition of PPy could enable sensing of coating designs with different physical characteristics and the development of arrays of electrochemical biosensors in which a single sensor would respond differently to similar mixtures of the analyte, opening applications for the detection of CRC biomarkers. Moreover, the biocompatibility of PPy provides it with potential for use in the design of implantable biomedical devices. However, PPy lacks selectivity for target molecules, and the modification of suitable biometric molecules is essential. In addition, while PPy may not always show the best results alone in an electrochemical biosensor, copolymers blended with other CPs could also be applied for suitable trace sensing. Moreover, PPy may degrade over time during detection, and efforts should be devoted to improving the stability of sensor response. In addition, in past decades, researchers mainly focused on the interfacial design, employing different PPy polymerization methods to improve the stability of the enzymes. However, research on the interface interaction between PPy and biorecognition elements, which has the potential to improve the sensitivity of biosensors, is lacking. Therefore, the interface interaction between PPy and biorecognition elements should be focused on and discussed in future studies with the aim of increasing sensing performance.

4. Conclusions

CRC is a type of cancer with high morbidity and mortality worldwide. Analyzing the concentration changes in CRC-associated biomolecules through electrochemical biosensing technology is significant for the early diagnosis, prognosis, and prediction of CRC. The fundamental purpose of this review paper was to introduce the design of PPy biosensors and their applications for the electrochemical measurement of CRC biomarkers. The conductive polymer PPy is a nanomaterial that has attracted much attention because of its specific characteristics, such as its excellent electrical properties, the flexibility of its surface properties, its easy preparation and functionalization, and its good biocompatibility, and it is often used in the design and improvement of biosensors. The different forms of PPy, synthesized via oxidation chemistry or electrochemical synthesis, can be used for various purposes. Moreover, various properties of PPy can be significantly improved by embedding, doping, or dedoping specific materials during or after its formation. PPy and its derivatives can be applied in enzyme-based biosensors, immunobiosensors, aptamer-based biosensors, MIP-based biosensors, and nanocatalytic biosensors and improve their sensing performance. Herein, it was shown that electrochemical-based PPy-based sensors, which have the inherent advantages of excellent sensitivity and selectivity, rapid detection, cost-effectiveness, and being capable of simultaneous detection of multiple CRC biomarkers, can be used to develop highly important detection strategies for CRC. However, the potential of such biosensors for simultaneous detection of multiple biomarkers needs further research and development. It is necessary to skillfully combine PPy with other nanomaterials to effectively improve the detection performance of biosensors. The stability of PPy during detection also needs to be further enhanced. In addition, more attention should be focused on improving the interfacial synergy between PPy and biorecognition elements, thereby improving the sensing performance of assays.

Author Contributions: X.Z. wrote the original draft; X.T. visualized the data and reviewed the original draft; P.W. and J.Q. conceived the presented idea, conducted the investigations, and reviewed and edited the manuscript. All authors have read and agreed to the published version of the manuscript.

Funding: This work was financially supported by the National Natural Science Foundation of China (82003150), the Shanghai Sailing Program (20YF1453400), the Shanghai Medical Innovation Project (21Y11905800), and the "Chenguang Program" supported by the Shanghai Education Development Foundation and the Shanghai Municipal Education Commission (20CG25).

Data Availability Statement: Data available in a public (institutional, general or subject specific) repository that issues datasets with DOIs.

Conflicts of Interest: The authors declare no conflict of interest.

References

1. Xi, Y.; Xu, P. Global colorectal cancer burden in 2020 and projections to 2040. *Transl. Oncol.* **2021**, *14*, 101174. [CrossRef] [PubMed]
2. Kanth, P.; Inadomi, J.M. Screening and prevention of colorectal cancer. *BMJ* **2021**, *374*, n1855. [CrossRef]
3. Millien, V.O.; Mansour, N.M. Bowel Preparation for Colonoscopy in 2020: A Look at the Past, Present, and Future. *Curr. Gastroenterol. Rep.* **2020**, *22*, 28. [CrossRef]
4. Kim, S.Y.; Kim, H.-S.; Park, H.J. Adverse events related to colonoscopy: Global trends and future challenges. *World J. Gastroenterol.* **2019**, *25*, 190–204. [CrossRef] [PubMed]
5. Vuik, F.E.R.; Nieuwenburg, S.A.V.; Moen, S.; Spada, C.; Senore, C.; Hassan, C.; Pennazio, M.; Rondonotti, E.; Pecere, S.; Kuipers, E.J.; et al. Colon capsule endoscopy in colorectal cancer screening: A systematic review. *Endoscopy* **2021**, *53*, 815–824. [CrossRef]
6. Bai, W.; Yu, D.; Zhu, B.; Yu, X.; Duan, R.; Li, Y.; Yu, W.; Hua, W.; Kou, C. Diagnostic accuracy of computed tomography colonography in patients at high risk for colorectal cancer: A meta-analysis. *Color. Dis.* **2020**, *22*, 1528–1537. [CrossRef] [PubMed]
7. Dreyer, P.; Duncan, A. CT colonography: For screening and monitoring disease. *Radiol. Technol.* **2021**, *92*, 595CT–608CT.
8. Wilkins, T.; McMechan, D.; Talukder, A. Colorectal cancer screening and prevention. *Am. Fam. Physician* **2018**, *97*, 658–665.
9. Ladabaum, U.; Dominitz, J.A.; Kahi, C.; Schoen, R.E. Strategies for Colorectal Cancer Screening. *Gastroenterology* **2020**, *158*, 418–432. [CrossRef]
10. Li, J.N.; Yuan, S.Y. Fecal occult blood test in colorectal cancer screening. *J. Dig. Dis.* **2019**, *20*, 62–64. [CrossRef]
11. Zhang, W.; Xiao, G.; Chen, J.; Wang, L.; Hu, Q.; Wu, J.; Zhang, W.; Song, M.; Qiao, J.; Xu, C. Electrochemical biosensors for measurement of colorectal cancer biomarkers. *Anal. Bioanal. Chem.* **2021**, *413*, 2407–2428. [CrossRef] [PubMed]
12. Nikolouzakis, T.K.; Vassilopoulou, L.; Fragkiadaki, P.; Sapsakos, T.M.; Papadakis, G.Z.; Spandidos, D.; Tsatsakis, A.; Tsiaoussis, J. Improving diagnosis, prognosis and prediction by using biomarkers in CRC patients (Review). *Oncol. Rep.* **2018**, *39*, 2455–2472. [CrossRef] [PubMed]
13. Nikolaou, S.; Qiu, S.; Fiorentino, F.; Rasheed, S.; Tekkis, P.; Kontovounisios, C. Systematic review of blood diagnostic markers in colorectal cancer. *Tech. Coloproctol.* **2018**, *22*, 481–498. [CrossRef] [PubMed]
14. Zygulska, A.L.; Pierzchalski, P. Novel Diagnostic Biomarkers in Colorectal Cancer. *Int. J. Mol. Sci.* **2022**, *23*, 852. [CrossRef]
15. Monsalve-Lancheros, A.; Ibáñez-Pinilla, M.; Ramírez-Clavijo, S. Detection of mammagloblin by RT-PCR as a biomarker for lymph node metastasis in breast cancer patients: A systematic review and meta-analysis. *PLoS ONE* **2019**, *14*, e0216989. [CrossRef]
16. Arya, S.K.; Estrela, P. Recent Advances in Enhancement Strategies for Electrochemical ELISA-Based Immunoassays for Cancer Biomarker Detection. *Sensors* **2018**, *18*, 2010. [CrossRef]
17. Li, S.; Wu, H.; Mai, H.; Zhen, J.; Li, G.; Chen, S. Microarray-based analysis of whole-genome DNA methylation profiling in early detection of breast cancer. *J. Cell. Biochem.* **2019**, *120*, 658–670. [CrossRef]
18. Oltedal, S.; Skaland, I.; Maple-Grødem, J.; Tjensvoll, K.; Janssen, E.A.M.; Gilje, B.; Smaaland, R.; Heikkilä, R.; Nordgård, O. Expression profiling and intracellular localization studies of the novel Proline-, Histidine-, and Glycine-rich protein 1 suggest an essential role in gastro-intestinal epithelium and a potential clinical application in colorectal cancer diagnostics. *BMC Gastroenterol.* **2018**, *18*, 26. [CrossRef]
19. Kondo, T. Cancer biomarker development and two-dimensional difference gel electrophoresis (2D-DIGE). *Biochim. Biophys. Acta (BBA)-Proteins Proteom.* **2019**, *1867*, 2–8. [CrossRef]
20. El-Bayoumy, A.S.A.-B.; Keshta, A.T.H.; Sallam, K.M.; Ebeid, N.H.; Elsheikh, H.M.; Bayoumy, B.E.-S. Extraction, purification of prostate-specific antigen (PSA), and establishment of radioimmunoassay system as a diagnostic tool for prostate disorders. *J. Immunoass. Immunochem.* **2018**, *39*, 12–29. [CrossRef]

21. Sukswai, N.; Khoury, J.D. Immunohistochemistry Innovations for Diagnosis and Tissue-Based Biomarker Detection. *Curr. Hematol. Malign. Rep.* **2019**, *14*, 368.–375. [CrossRef] [PubMed]
22. Neubert, H.; Shuford, C.M.; Olah, T.V.; Garofolo, F.; Schultz, G.A.; Jones, B.R.; Amaravadi, L.; Laterza, O.F.; Xu, K.; Ackermann, B.L. Protein Biomarker Quantification by Immunoaffinity Liquid Chromatography–Tandem Mass Spectrometry: Current State and Future Vision. *Clin. Chem.* **2020**, *66*, 282–301. [CrossRef]
23. Yin, S.; Kabashima, T.; Zhu, Q.; Shibata, T.; Kai, M. Fluorescence assay of dihydroorotate dehydrogenase that may become a cancer biomarker. *Sci. Rep.* **2017**, *7*, 40670. [CrossRef] [PubMed]
24. Han, C.; Chen, R.; Wu, X.; Shi, N.; Duan, T.; Xu, K.; Huang, T. Fluorescence turn-on immunosensing of HE4 biomarker and ovarian cancer cells based on target-triggered metal-enhanced fluorescence of carbon dots. *Anal. Chim. Acta* **2021**, *1187*, 339160. [CrossRef] [PubMed]
25. Liu, J.; Zhao, J.; Li, S.; Zhang, L.; Huang, Y.; Zhao, S. A novel microchip electrophoresis-based chemiluminescence immunoassay for the detection of alpha-fetoprotein in human serum. *Talanta* **2017**, *165*, 107–111. [CrossRef] [PubMed]
26. Zhou, Y.; Yin, H.; Zhao, W.-W.; Ai, S. Electrochemical, electrochemiluminescent and photoelectrochemical bioanalysis of epigenetic modifiers: A comprehensive review. *Coord. Chem. Rev.* **2020**, *424*, 213519. [CrossRef]
27. Dzulkurnain, N.A.; Mokhtar, M.; Rashid, J.I.A.; Knight, V.F.; Yunus, W.M.Z.W.; Ong, K.K.; Kasim, N.A.M.; Noor, S.A.M. A Review on Impedimetric and Voltammetric Analysis Based on Polypyrrole Conducting Polymers for Electrochemical Sensing Applications. *Polymers* **2021**, *13*, 2728. [CrossRef]
28. Luong, J.H.T.; Narayan, T.; Solanki, S.; Malhotra, B.D. Recent Advances of Conducting Polymers and Their Composites for Electrochemical Biosensing Applications. *J. Funct. Biomater.* **2020**, *11*, 71. [CrossRef]
29. Cho, I.-H.; Kim, D.H.; Park, S. Electrochemical biosensors: Perspective on functional nanomaterials for on-site analysis. *Biomater. Res.* **2020**, *24*, 6. [CrossRef]
30. Lara, S.; Perez-Potti, A. Applications of Nanomaterials for Immunosensing. *Biosensors* **2018**, *8*, 104. [CrossRef]
31. Sharma, S.; Sudhakara, P.; Omran, A.A.B.; Singh, J.; Ilyas, R.A. Recent Trends and Developments in Conducting Polymer Nanocomposites for Multifunctional Applications. *Polymers* **2021**, *13*, 2898. [CrossRef] [PubMed]
32. Terán-Alcocer, Á.; Bravo-Plascencia, F.; Cevallos-Morillo, C.; Palma-Cando, A. Electrochemical Sensors Based on Conducting Polymers for the Aqueous Detection of Biologically Relevant Molecules. *Nanomaterials* **2021**, *11*, 252. [CrossRef] [PubMed]
33. Nezakati, T.; Seifalian, A.; Tan, A.; Seifalian, A.M. Conductive Polymers: Opportunities and Challenges in Biomedical Applications. *Chem. Rev.* **2018**, *118*, 6766–6843. [CrossRef]
34. Wang, D.; Wang, J. A sensitive and label-free electrochemical microRNA biosensor based on Polyamidoamine Dendrimer functionalized Polypyrrole nanowires hybrid. *Microchim. Acta* **2021**, *188*, 173. [CrossRef] [PubMed]
35. Puerres, J.; Ortiz, P.; Cortés, M. Effect of Electrosynthesis Potential on Nucleation, Growth, Adhesion, and Electronic Properties of Polypyrrole Thin Films on Fluorine-Doped Tin Oxide (FTO). *Polymers* **2021**, *13*, 2419. [CrossRef] [PubMed]
36. Mao, J.; Rouabhia, M.; Zhang, Z. Surface modification by assembling: A modular approach based on the match in nanostructures. *J. Mater. Chem. B* **2019**, *7*, 755–762. [CrossRef]
37. Huang, J.; Wang, S.; Xing, Y.; Zhou, W.; Cai, K. Interface-Hybridization-Enhanced Photothermal Performance of Polypyrrole/Polydopamine Heterojunctions on Porous Nanoparticles. *Macromol. Rapid Commun.* **2019**, *40*, e1900263. [CrossRef] [PubMed]
38. Li, J.; Yuan, W.; Luo, S.-X.L.; Bezdek, M.J.; Peraire-Bueno, A.; Swager, T.M. Wireless Lateral Flow Device for Biosensing. *J. Am. Chem. Soc.* **2022**, *144*, 15786–15792. [CrossRef]
39. Wang, J.; Li, B.; Lu, Q.; Li, X.; Weng, C.; Yan, X.; Hong, J.; Zhou, X. A versatile fluorometric aptasensing scheme based on the use of a hybrid material composed of polypyrrole nanoparticles and DNA-silver nanoclusters: Application to the determination of adenosine, thrombin, or interferon-gamma. *Microchim. Acta* **2019**, *186*, 356. [CrossRef]
40. Liu, Y.; Zhu, Z.-Z.; Zheng, G.; Zhang, H.; Zhou, M.; Li, S.; Kong, Y. Dual-template molecularly imprinted electrochemical biosensor for IgG-IgM combined assay based on a dual-signal strategy. *Bioelectrochemistry* **2022**, *148*, 108267. [CrossRef]
41. Nguyet, N.T.; Yen, L.T.H.; Doan, V.Y.; Hoang, N.L.; Van Thu, V.; Lan, H.; Trung, T.; Pham, V.-H.; Tam, P.D. A label-free and highly sensitive DNA biosensor based on the core-shell structured CeO$_2$-NR@Ppy nanocomposite for Salmonella detection. *Mater. Sci. Eng. C* **2019**, *96*, 790–797. [CrossRef] [PubMed]
42. Zhu, C.; Xue, H.; Zhao, H.; Fei, T.; Liu, S.; Chen, Q.; Gao, B.; Zhang, T. A dual-functional polyaniline film-based flexible electrochemical sensor for the detection of pH and lactate in sweat of the human body. *Talanta* **2022**, *242*, 123289. [CrossRef] [PubMed]
43. Uzunçar, S.; Meng, L.; Turner, A.P.; Mak, W.C. Processable and nanofibrous polyaniline:polystyrene-sulphonate (nano-PANI:PSS) for the fabrication of catalyst-free ammonium sensors and enzyme-coupled urea biosensors. *Biosens. Bioelectron.* **2021**, *171*, 112725. [CrossRef]
44. Malmir, M.; Arjomandi, J.; Khosroshahi, A.G.; Moradi, M.; Shi, H. Label-free E-DNA biosensor based on PANi-RGO-G*NPs for detection of cell-free fetal DNA in maternal blood and fetal gender determination in early pregnancy. *Biosens. Bioelectron.* **2021**, *189*, 113356. [CrossRef] [PubMed]
45. Shoja, Y.; Rafati, A.A.; Ghodsi, J. Polythiophene supported MnO$_2$ nanoparticles as nano-stabilizer for simultaneously electrostatically immobilization of d-amino acid oxidase and hemoglobin as efficient bio-nanocomposite in fabrication of dopamine bi-enzyme biosensor. *Mater. Sci. Eng. C Mater. Biol. Appl.* **2017**, *76*, 637–645. [CrossRef] [PubMed]

46. Dervisevic, M.; Senel, M.; Sagir, T.; Isik, S. Highly sensitive detection of cancer cells with an electrochemical cytosensor based on boronic acid functional polythiophene. *Biosens. Bioelectron.* **2017**, *90*, 6–12. [CrossRef] [PubMed]
47. Dervisevic, M.; Senel, M.; Cevik, E. Novel impedimetric dopamine biosensor based on boronic acid functional polythiophene modified electrodes. *Mater. Sci. Eng. C* **2017**, *72*, 641–649. [CrossRef]
48. Zahed, A.; Barman, S.C.; Das, P.S.; Sharifuzzaman; Yoon, H.S.; Yoon, S.H.; Park, J.Y. Highly flexible and conductive poly (3, 4-ethylene dioxythiophene)-poly (styrene sulfonate) anchored 3-dimensional porous graphene network-based electrochemical biosensor for glucose and pH detection in human perspiration. *Biosens. Bioelectron.* **2020**, *160*, 112220. [CrossRef]
49. Qin, J.; Cho, M.; Lee, Y. Ultrasensitive Detection of Amyloid-β Using Cellular Prion Protein on the Highly Conductive Au Nanoparticles–Poly(3,4-ethylene dioxythiophene)–Poly(thiophene-3-acetic acid) Composite Electrode. *Anal. Chem.* **2019**, *91*, 11259–11265. [CrossRef]
50. Madhuvilakku, R.; Alagar, S.; Mariappan, R.; Piraman, S. Glassy carbon electrodes modified with reduced graphene oxide-MoS2-poly (3,4-ethylene dioxythiophene) nanocomposites for the non-enzymatic detection of nitrite in water and milk. *Anal. Chim. Acta* **2020**, *1093*, 93–105. [CrossRef]
51. Baghdadi, N.; Zoromba, M.; Abdel-Aziz, M.; Al-Hossainy, A.; Bassyouni, M.; Salah, N. One-Dimensional Nanocomposites Based on Polypyrrole-Carbon Nanotubes and Their Thermoelectric Performance. *Polymers* **2021**, *13*, 278. [CrossRef] [PubMed]
52. Feng, C.; Huang, J.; Yan, P.; Wan, F.; Zhu, Y.; Cheng, H. Preparation and Properties of Waterborne Polypyrrole/Cement Composites. *Materials* **2021**, *14*, 5166. [CrossRef]
53. Pang, A.L.; Arsad, A.; Ahmadipour, M. Synthesis and factor affecting on the conductivity of polypyrrole: A short review. *Polym. Adv. Technol.* **2021**, *32*, 1428–1454. [CrossRef]
54. Rahaman, M.; Aldalbahi, A.; Almoiqli, M.; Alzahly, S. Chemical and Electrochemical Synthesis of Polypyrrole Using Carrageenan as a Dopant: Polypyrrole/Multi-Walled Carbon Nanotube Nanocomposites. *Polymers* **2018**, *10*, 632. [CrossRef]
55. Manickavasagan, A.; Ramachandran, R.; Chen, S.-M.; Velluchamy, M. Ultrasonic assisted fabrication of silver tungstate encrusted polypyrrole nanocomposite for effective photocatalytic and electrocatalytic applications. *Ultrason. Sonochem.* **2020**, *64*, 104913. [CrossRef]
56. Kang, T.S.; Lee, S.W.; Joo, J.; Lee, J.Y. Electrically conducting polypyrrole fibers spun by electrospinning. *Synth. Met.* **2005**, *153*, 61–64. [CrossRef]
57. Alizadeh, N.; Ataei, A.A.; Pirsa, S. Nanostructured conducting polypyrrole film prepared by chemical vapor deposition on the interdigital electrodes at room temperature under atmospheric condition and its application as gas sensor. *J. Iran. Chem. Soc.* **2015**, *12*, 1585–1594. [CrossRef]
58. Kopecký, D.; Vrňata, M.; Vysloužil, F.; Fitl, P.; Ekrt, O.; Seidl, J.; Myslík, V.; Hofmann, J.; Náhlík, J.; Vlček, J.; et al. Doped polypyrrole for MAPLE deposition: Synthesis and characterization. *Synth. Met.* **2010**, *160*, 1081–1085. [CrossRef]
59. Leonavicius, K.; Ramanaviciene, A.; Ramanavicius, A. Polymerization Model for Hydrogen Peroxide Initiated Synthesis of Polypyrrole Nanoparticles. *Langmuir ACS J. Surf. Colloids* **2011**, *27*, 10970–10976. [CrossRef]
60. Imae, I.; Krukiewicz, K. Self-doped conducting polymers in biomedical engineering: Synthesis, characterization, current applications and perspectives. *Bioelectrochemistry* **2022**, *146*, 108127. [CrossRef]
61. Peng, H.; Zheng, J.; Zhang, B.; Xu, J.; Zhang, M. Fe doped MoS_2/polypyrrole microtubes towards efficient peroxidase mimicking and colorimetric sensing application. *Dalton Trans.* **2021**, *50*, 15380–15388. [CrossRef] [PubMed]
62. Andriukonis, E.; Ramanaviciene, A.; Ramanavicius, A. Synthesis of Polypyrrole Induced by $[Fe(CN)6]^{3-}$ and Redox Cycling of $[Fe(CN)6]^{4-}/[Fe(CN)6]^{3-}$. *Polymers* **2018**, *10*, 749. [CrossRef] [PubMed]
63. Mao, J.; Zhang, Z. Polypyrrole as electrically conductive biomaterials: Synthesis, biofunctionalization, potential applications and challenges. *Adv. Exp. Med. Biol.* **2018**, *1078*, 347–370. [CrossRef] [PubMed]
64. German, N.; Popov, A.; Ramanaviciene, A.; Ramanavicius, A. Enzymatic Formation of Polyaniline, Polypyrrole, and Polythiophene Nanoparticles with Embedded Glucose Oxidase. *Nanomaterials* **2019**, *9*, 806. [CrossRef] [PubMed]
65. German, N.; Ramanaviciene, A.; Ramanavicius, A. Formation of Polyaniline and Polypyrrole Nanocomposites with Embedded Glucose Oxidase and Gold Nanoparticles. *Polymers* **2019**, *11*, 377. [CrossRef]
66. Apetrei, R.-M.; Carac, G.; Ramanaviciene, A.; Bahrim, G.; Tanase, C.; Ramanavicius, A. Cell-assisted synthesis of conducting polymer—Polypyrrole—For the improvement of electric charge transfer through fungal cell wall. *Colloids Surf. B Biointerfaces* **2019**, *175*, 671–679. [CrossRef]
67. Sherman, H.G.; Hicks, J.M.; Jain, A.; Titman, J.J.; Alexander, C.; Stolnik, S.; Rawson, F.J.; Hicks, J. Mammalian-Cell-Driven Polymerisation of Pyrrole. *Chembiochem* **2019**, *20*, 1008–1013. [CrossRef]
68. Fani, M.; Rezayi, M.; Pourianfar, H.R.; Meshkat, Z.; Makvandi, M.; Gholami, M.; Rezaee, S.A. Rapid and label-free electrochemical dna biosensor based on a facile one-step electrochemical synthesis of rgo-ppy-(l-cys)-aunps nanocomposite for the htlv-1 oligonucleotide detection. *Biotechnol. Appl. Biochem.* **2021**, *68*, 626–635. [CrossRef]
69. Ayenimo, J.G.; Adeloju, S.B. Amperometric detection of glucose in fruit juices with polypyrrole-based biosensor with an integrated permselective layer for exclusion of interferences. *Food Chem.* **2017**, *229*, 127–135. [CrossRef]
70. Srinivasan, S.Y.; Gajbhiye, V.; Bodas, D. Development of nano-immunosensor with magnetic separation and electrical detection of Escherichia coli using antibody conjugated Fe_3O_4@Ppy. *Nanotechnology* **2021**, *32*, 085603. [CrossRef]
71. Xu, T.; Chi, B.; Gao, J.; Chu, M.; Fan, W.; Yi, M.; Xu, H.; Mao, C. Novel electrochemical immune sensor based on Hep-PGA-PPy nanoparticles for detection of α-Fetoprotein in whole blood. *Anal. Chim. Acta* **2017**, *977*, 36–43. [CrossRef] [PubMed]

72. Pruna, A.I.; Rosas-Laverde, N.M.; Mataix, D.B. Effect of Deposition Parameters on Electrochemical Properties of Polypyrrole-Graphene Oxide Films. *Materials* **2020**, *13*, 624. [CrossRef] [PubMed]
73. Kesküla, A.; Heinmaa, I.; Tamm, T.; Aydemir, N.; Travas-Sejdic, J.; Peikolainen, A.-L.; Kiefer, R. Improving the Electrochemical Performance and Stability of Polypyrrole by Polymerizing Ionic Liquids. *Polymers* **2020**, *12*, 136. [CrossRef] [PubMed]
74. Mametov, R.; Sagandykova, G.; Monedeiro, F.; Buszewski, B. Development of controlled film of polypyrrole for solid-phase microextraction fiber by electropolymerization. *Talanta* **2021**, *232*, 122394. [CrossRef]
75. Glosz, K.; Stolarczyk, A.; Jarosz, T. Electropolymerised Polypyrroles as Active Layers for Molecularly Imprinted Sensors: Fabrication and Applications. *Materials* **2021**, *14*, 1369. [CrossRef]
76. Ramanavičius, S.; Morkvėnaitė-Vilkončienė, I.; Samukaitė-Bubnienė, U.; Ratautaitė, V.; Plikusienė, I.; Viter, R.; Ramanavičius, A. Electrochemically Deposited Molecularly Imprinted Polymer-Based Sensors. *Sensors* **2022**, *22*, 1282. [CrossRef]
77. Deore, B.; Chen, Z.; Nagaoka, T. Potential-Induced Enantioselective Uptake of Amino Acid into Molecularly Imprinted Overoxidized Polypyrrole. *Anal. Chem.* **2000**, *72*, 3989–3994. [CrossRef]
78. Malitesta, C.; Mazzotta, E.; Picca, R.A.; Poma, A.; Chianella, I.; Piletsky, S.A. MIP sensors—The electrochemical approach. *Anal. Bioanal. Chem.* **2012**, *402*, 1827–1846. [CrossRef]
79. Kowalski, D.; Ueda, M.; Ohtsuka, T. The effect of ultrasonic irradiation during electropolymerization of polypyrrole on corrosion prevention of the coated steel. *Corros. Sci.* **2008**, *50*, 286–291. [CrossRef]
80. Blachowicz, T.; Ehrmann, A. Conductive Electrospun Nanofiber Mats. *Materials* **2019**, *13*, 152. [CrossRef]
81. Lawal, A.T.; Wallace, G.G. Vapour phase polymerisation of conducting and non-conducting polymers: A review. *Talanta* **2014**, *119*, 133–143. [CrossRef] [PubMed]
82. Kasisomayajula, S.; Jadhav, N.; Gelling, V.J. Conductive polypyrrole and acrylate nanocomposite coatings: Mechanistic study on simultaneous photopolymerization. *Prog. Org. Coat.* **2016**, *101*, 440–454. [CrossRef]
83. Yang, M.; Wang, Z.; Ding, T.; Tang, J.; Xie, X.; Xing, Y.; Wang, L.; Zhang, J.; Cai, K. Interfacial Engineering of Hybrid Polydopamine/Polypyrrole Nanosheets with Narrow Band Gaps for Fluorescence Sensing of MicroRNA. *ACS Appl. Mater. Interfaces* **2021**, *13*, 42183–42194. [CrossRef]
84. Okamura, N.; Maeda, T.; Fujiwara, H.; Soman, A.; Unni, K.N.N.; Ajayaghosh, A.; Yagi, S. Photokinetic study on remarkable excimer phosphorescence from heteroleptic cyclometalated platinum(ii) complexes bearing a benzoylated 2-phenylpyridinate ligand. *Phys. Chem. Chem. Phys.* **2017**, *20*, 542–552. [CrossRef] [PubMed]
85. German, N.; Ramanaviciene, A.; Ramanavicius, A. Dispersed Conducting Polymer Nanocomposites with Glucose Oxidase and Gold Nanoparticles for the Design of Enzymatic Glucose Biosensors. *Polymers* **2021**, *13*, 2173. [CrossRef] [PubMed]
86. Zhang, B.G.X.; Spinks, G.M.; Gorkin, R.; Sangian, D.; Di Bella, C.; Quigley, A.F.; Kapsa, R.; Wallace, G.G.; Choong, P.F.M. In vivo biocompatibility of porous and non-porous polypyrrole based trilayered actuators. *J. Mater. Sci. Mater. Med.* **2017**, *28*, 172. [CrossRef]
87. Zhao, P.; Tang, Q.; Zhao, X.; Tong, Y.; Liu, Y. Highly stable and flexible transparent conductive polymer electrode patterns for large-scale organic transistors. *J. Colloid Interface Sci.* **2018**, *520*, 58–63. [CrossRef]
88. Zhang, H.; Huang, L.; Zhai, J.; Dong, S. Water/Oxygen Circulation-Based Biophotoelectrochemical System for Solar Energy Storage and Release. *J. Am. Chem. Soc.* **2019**, *141*, 16416–16421. [CrossRef]
89. Mills, I.N.; Porras, J.A.; Bernhard, S. Judicious Design of Cationic, Cyclometalated Ir(III) Complexes for Photochemical Energy Conversion and Optoelectronics. *Acc. Chem. Res.* **2018**, *51*, 352–364. [CrossRef]
90. Rafique, S.; Rashid, I.; Sharif, R. Cost effective dye sensitized solar cell based on novel Cu polypyrrole multiwall carbon nanotubes nanocomposites counter electrode. *Sci. Rep.* **2021**, *11*, 14830. [CrossRef]
91. Han, Y.; Sun, L.; Wen, C.; Wang, Z.; Dai, J.; Shi, L. Flexible conductive silk-PPy hydrogel toward wearable electronic strain sensors. *Biomed. Mater.* **2022**, *17*, 024107. [CrossRef] [PubMed]
92. Tian, W.; Li, Y.; Zhou, J.; Wang, T.; Zhang, R.; Cao, J.; Luo, M.; Li, N.; Zhang, N.; Gong, H.; et al. Implantable and Biodegradable Micro-Supercapacitor Based on a Superassembled Three-Dimensional Network Zn@PPy Hybrid Electrode. *ACS Appl. Mater. Interfaces* **2021**, *13*, 8285–8293. [CrossRef] [PubMed]
93. Wang, X.; Li, X.; Zhao, X.; Li, M.; Li, Y.; Yang, W.; Ren, J. Polypyrrole-doped conductive self-healing multifunctional composite hydrogels with a dual crosslinked network. *Soft Matter* **2021**, *17*, 8363–8372. [CrossRef] [PubMed]
94. Ramanavicius, S.; Ramanavicius, A. Conducting Polymers in the Design of Biosensors and Biofuel Cells. *Polymers* **2020**, *13*, 49. [CrossRef]
95. Yang, L.; Wang, J.; Lü, H.; Hui, N. Electrochemical sensor based on Prussian blue/multi-walled carbon nanotubes functionalized polypyrrole nanowire arrays for hydrogen peroxide and microRNA detection. *Microchim. Acta* **2021**, *188*, 25. [CrossRef]
96. Rong, Y.; Yan, W.; Wang, Z.; Hao, X.; Guan, G. An electroactive montmorillonite/polypyrrole ion exchange film: Ultrahigh uptake capacity and ion selectivity for rapid removal of lead ions. *J. Hazard. Mater.* **2022**, *437*, 129366. [CrossRef]
97. Mohamed, F.; Abukhadra, M.R.; Shaban, M. Removal of safranin dye from water using polypyrrole nanofiber/Zn-Fe layered double hydroxide nanocomposite (Ppy NF/Zn-Fe LDH) of enhanced adsorption and photocatalytic properties. *Sci. Total Environ.* **2018**, *640–641*, 352–363. [CrossRef]
98. Fan, Y.; Bai, W.; Mu, P.; Su, Y.; Zhu, Z.; Sun, H.; Liang, W.; Li, A. Conductively monolithic polypyrrole 3-D porous architecture with micron-sized channels as superior salt-resistant solar steam generators. *Sol. Energy Mater. Sol. Cells* **2020**, *206*, 110347. [CrossRef]

99. Ramanavicius, S.; Jagminas, A.; Ramanavicius, A. Advances in Molecularly Imprinted Polymers Based Affinity Sensors (Review). *Polymers* **2021**, *13*, 974. [CrossRef]
100. Apetrei, R.-M.; Cârâc, G.; Bahrim, G.-E.; Camurlu, P. Utilization of enzyme extract self-encapsulated within polypyrrole in sensitive detection of catechol. *Enzym. Microb. Technol.* **2019**, *128*, 34–39. [CrossRef]
101. Dutta, R.R.; Puzari, P. Amperometric biosensing of organophosphate and organocarbamate pesticides utilizing polypyrrole entrapped acetylcholinesterase electrode. *Biosens. Bioelectron.* **2014**, *52*, 166–172. [CrossRef] [PubMed]
102. Shi, W.; Liu, C.; Song, Y.; Lin, N.; Zhou, S.; Cai, X. An ascorbic acid amperometric sensor using over-oxidized polypyrrole and palladium nanoparticles composites. *Biosens. Bioelectron.* **2012**, *38*, 100–106. [CrossRef] [PubMed]
103. Sakthivel, R.; Lin, L.-Y.; Lee, T.-H.; Liu, X.; He, J.-H.; Chung, R.-J. Disposable and cost-effective label-free electrochemical immunosensor for prolactin based on bismuth sulfide nanorods with polypyrrole. *Bioelectrochemistry* **2022**, *143*, 107948. [CrossRef]
104. Tang, C.; Wang, P.; Zhou, K.; Ren, J.; Wang, S.; Tang, F.; Li, Y.; Liu, Q.; Xue, L. Electrochemical immunosensor based on hollow porous Pt skin AgPt alloy/NGR as a dual signal amplification strategy for sensitive detection of Neuron-specific enolase. *Biosens. Bioelectron.* **2022**, *197*, 113779. [CrossRef]
105. Zou, Y.; Liang, J.; She, Z.; Kraatz, H.-B. Gold nanoparticles-based multifunctional nanoconjugates for highly sensitive and enzyme-free detection of E.coli K12. *Talanta* **2019**, *193*, 15–22. [CrossRef] [PubMed]
106. Duan, Y.; Wang, N.; Huang, Z.; Dai, H.; Xu, L.; Sun, S.; Ma, H.; Lin, M. Electrochemical endotoxin aptasensor based on a metal-organic framework labeled analytical platform. *Mater. Sci. Eng. C Mater. Biol. Appl.* **2020**, *108*, 110501. [CrossRef]
107. Wang, J.; Zhang, D.; Xu, K.; Hui, N.; Wang, D. Electrochemical assay of acetamiprid in vegetables based on nitrogen-doped graphene/polypyrrole nanocomposites. *Microchim. Acta* **2022**, *189*, 395. [CrossRef]
108. Ding, S.; Lyu, Z.; Li, S.; Ruan, X.; Fei, M.; Zhou, Y.; Niu, X.; Zhu, W.; Du, D.; Lin, Y. Molecularly imprinted polypyrrole nanotubes based electrochemical sensor for glyphosate detection. *Biosens. Bioelectron.* **2021**, *191*, 113434. [CrossRef]
109. Wu, J.; Wang, R.; Lu, Y.; Jia, M.; Yan, J.; Bian, X. Facile Preparation of a Bacteria Imprinted Artificial Receptor for Highly Selective Bacterial Recognition and Label-Free Impedimetric Detection. *Anal. Chem.* **2019**, *91*, 1027–1033. [CrossRef]
110. Ramanavicius, S.; Ramanavicius, A. Development of molecularly imprinted polymer based phase boundaries for sensors design (review). *Adv. Colloid Interface Sci.* **2022**, *305*, 102693. [CrossRef]
111. Moncer, F.; Adhoum, N.; Catak, D.; Monser, L. Electrochemical sensor based on MIP for highly sensitive detection of 5-hydroxyindole-3-acetic acid carcinoid cancer biomarker in human biological fluids. *Anal. Chim. Acta* **2021**, *1181*, 338925. [CrossRef] [PubMed]
112. Li, C.; Su, Y.; Lv, X.; Xia, H.; Shi, H.; Yang, X.; Zhang, J.; Wang, Y. Controllable anchoring of gold nanoparticles to polypyrrole nanofibers by hydrogen bonding and their application in nonenzymatic glucose sensors. *Biosens. Bioelectron.* **2012**, *38*, 402–406. [CrossRef] [PubMed]
113. Jeong, H.; Yoo, J.; Park, S.; Lu, J.; Park, S.; Lee, J. Non-Enzymatic Glucose Biosensor Based on Highly Pure TiO_2 Nanoparticles. *Biosensors* **2021**, *11*, 149. [CrossRef] [PubMed]
114. Meng, F.; Shi, W.; Sun, Y.; Zhu, X.; Wu, G.; Ruan, C.; Liu, X.; Ge, D. Nonenzymatic biosensor based on CuxO nanoparticles deposited on polypyrrole nanowires for improving detectionrange. *Biosens. Bioelectron.* **2013**, *42*, 141–147. [CrossRef] [PubMed]
115. German, N.; Ramanavicius, A.; Voronovic, J.; Ramanaviciene, A. Glucose sensor based on glucose oxidase and gold nanoparticles of different sizes covered by polypyrrole layer. *Colloids Surf. A Physicochem. Eng. Asp.* **2012**, *413*, 224–230. [CrossRef]
116. Meng, L.; Turner, A.P.; Mak, W.C. Soft and flexible material-based affinity sensors. *Biotechnol. Adv.* **2019**, *39*, 107398. [CrossRef] [PubMed]
117. Li, R.; Feng, Y.; Pan, G.; Liu, L. Advances in Molecularly Imprinting Technology for Bioanalytical Applications. *Sensors* **2020**, *19*, 177. [CrossRef]
118. Liu, Q.; Zhang, A.; Wang, R.; Zhang, Q.; Cui, D. A Review on Metal- and Metal Oxide-Based Nanozymes: Properties, Mechanisms, and Applications. *Nano-Micro Lett.* **2021**, *13*, 154. [CrossRef]
119. Bag, S.; Baksi, A.; Nandam, S.H.; Wang, D.; Ye, X.-L.; Ghosh, J.; Pradeep, T.; Hahn, H. Nonenzymatic Glucose Sensing Using $Ni_{60}Nb_{40}$ Nanoglass. *ACS Nano* **2020**, *14*, 5543–5552. [CrossRef]
120. Huang, S.; Zhang, L.; Dai, L.; Wang, Y.; Tian, Y. Nonenzymatic Electrochemical Sensor with Ratiometric Signal Output for Selective Determination of Superoxide Anion in Rat Brain. *Anal. Chem.* **2021**, *93*, 5570–5576. [CrossRef]
121. Gao, L.; Zhuang, J.; Nie, L.; Zhang, J.; Zhang, Y.; Gu, N.; Wang, T.; Feng, J.; Yang, D.; Perrett, S.; et al. Intrinsic peroxidase-like activity of ferromagnetic nanoparticles. *Nat. Nanotechnol.* **2007**, *2*, 577–583. [CrossRef] [PubMed]
122. Zhu, G.; Pei, L.; Xia, H.; Tang, Q.; Bi, F. Role of oncogenic KRAS in the prognosis, diagnosis and treatment of colorectal cancer. *Mol. Cancer* **2021**, *20*, 143. [CrossRef] [PubMed]
123. Bellio, H.; Fumet, J.; Ghiringhelli, F. Targeting BRAF and RAS in Colorectal Cancer. *Cancers* **2021**, *13*, 2201. [CrossRef] [PubMed]
124. Aghabozorgi, A.S.; Bahreyni, A.; Soleimani, A.; Bahrami, A.; Khazaei, M.; Ferns, G.A.; Avan, A.; Hassanian, S.M. Role of adenomatous polyposis coli (APC) gene mutations in the pathogenesis of colorectal cancer; current status and perspectives. *Biochimie* **2018**, *157*, 64–71. [CrossRef]
125. Liebl, M.; Hofmann, T. The Role of p53 Signaling in Colorectal Cancer. *Cancers* **2021**, *13*, 2125. [CrossRef]
126. Wang, X.; Zhang, X.; He, P.; Fang, Y. Sensitive detection of p53 tumor suppressor gene using an enzyme-based solid-state electrochemiluminescence sensing platform. *Biosens. Bioelectron.* **2011**, *26*, 3608–3613. [CrossRef]

127. Wang, X.; Chen, F.; Zhu, K.; Xu, Q.; Tang, M. Novel electrochemical biosensor based on functional composite nanofibers for sensitive detection of p53 tumor suppressor gene. *Anal. Chim. Acta* **2013**, *765*, 63–69. [CrossRef]
128. Sun, J.; Fei, F.; Zhang, M.; Li, Y.; Zhang, X.; Zhu, S.; Zhang, S. The role of mSEPT9 in screening, diagnosis, and recurrence monitoring of colorectal cancer. *BMC Cancer* **2019**, *19*, 450. [CrossRef]
129. Li, D.; Zhang, L.; Fu, J.; Huang, H.; Sun, S.; Zhang, D.; Zhao, L.; Onwuka, J.U.; Zhao, Y.; Cui, B. SCTR hypermethylation is a diagnostic biomarker in colorectal cancer. *Cancer Sci.* **2020**, *111*, 4558–4566. [CrossRef]
130. Zhao, G.; Ma, Y.; Li, H.; Li, S.; Zhu, Y.; Liu, X.; Xiong, S.; Liu, Y.; Miao, J.; Fei, S.; et al. A novel plasma based early colorectal cancer screening assay base on methylated SDC2 and SFRP2. *Clin. Chim. Acta* **2020**, *503*, 84–89. [CrossRef]
131. Luo, H.; Zhao, Q.; Wei, W.; Zheng, L.; Yi, S.; Li, G.; Wang, W.; Sheng, H.; Pu, H.; Mo, H.; et al. Circulating tumor DNA methylation profiles enable early diagnosis, prognosis prediction, and screening for colorectal cancer. *Sci. Transl. Med.* **2020**, *12*, eaax7533. [CrossRef] [PubMed]
132. Mujica, M.L.; Gallay, P.A.; Perrachione, F.; Montemerlo, A.E.; Tamborelli, L.A.; Vaschetti, V.M.; Reartes, D.F.; Bollo, S.; Rodríguez, M.C.; Dalmasso, P.R.; et al. New trends in the development of electrochemical biosensors for the quantification of microRNAs. *J. Pharm. Biomed. Anal.* **2020**, *189*, 113478. [CrossRef] [PubMed]
133. Jung, G.; Hernández-Illán, E.; Moreira, L.; Balaguer, F.; Goel, A. Epigenetics of colorectal cancer: Biomarker and therapeutic potential. *Nat. Rev. Gastroenterol. Hepatol.* **2020**, *17*, 111–130. [CrossRef] [PubMed]
134. Bahreyni, A.; Rezaei, M.; Bahrami, A.; Khazaei, M.; Fiuji, H.; Ryzhikov, M.; Ferns, G.A.; Avan, A.; Hassanian, S.M. Diagnostic, prognostic, and therapeutic potency of microRNA 21 in the pathogenesis of colon cancer, current status and prospective. *J. Cell. Physiol.* **2019**, *234*, 8075–8081. [CrossRef] [PubMed]
135. Zhang, Z.; Zhang, D.; Cui, Y.; Qiu, Y.; Miao, C.; Lu, X. Identification of microRNA-451a as a Novel Circulating Biomarker for Colorectal Cancer Diagnosis. *BioMed Res. Int.* **2020**, *2020*, 5236236. [CrossRef]
136. Fadaka, A.O.; Pretorius, A.; Klein, A. Biomarkers for Stratification in Colorectal Cancer: MicroRNAs. *Cancer Control J. Moffitt Cancer Cent.* **2019**, *26*, 1073274819862784. [CrossRef]
137. Sharif-Askari, N.S.; Sharif-Askari, F.S.; Guraya, S.Y.; Bendardaf, R.; Hamoudi, R. Integrative systematic review meta-analysis and bioinformatics identifies MicroRNA-21 and its target genes as biomarkers for colorectal adenocarcinoma. *Int. J. Surg.* **2019**, *73*, 113–122. [CrossRef]
138. Liu, T.; Liu, D.; Guan, S.; Dong, M. Diagnostic role of circulating MiR-21 in colorectal cancer: A update meta-analysis. *Ann. Med.* **2020**, *53*, 87–102. [CrossRef]
139. Pothipor, C.; Aroonyadet, N.; Bamrungsap, S.; Jakmunee, J.; Ounnunkad, K. A highly sensitive electrochemical microRNA-21 biosensor based on intercalating methylene blue signal amplification and a highly dispersed gold nanoparticles/graphene/polypyrrole composite. *Analyst* **2021**, *146*, 2679–2688. [CrossRef]
140. Tian, L.; Qian, K.; Qi, J.; Liu, Q.; Yao, C.; Song, W.; Wang, Y. Gold nanoparticles superlattices assembly for electrochemical biosensor detection of microRNA-21. *Biosens. Bioelectron.* **2018**, *99*, 564–570. [CrossRef]
141. Kaplan, M.; Kilic, T.; Guler, G.; Mandli, J.; Amine, A.; Ozsoz, M. A novel method for sensitive microRNA detection: Electropolymerization based doping. *Biosens. Bioelectron.* **2017**, *92*, 770–778. [CrossRef] [PubMed]
142. Song, J.; Teng, H.; Xu, Z.; Liu, N.; Xu, L.; Liu, L.; Gao, F.; Luo, X. Free-standing electrochemical biosensor for carcinoembryonic antigen detection based on highly stable and flexible conducting polypyrrole nanocomposite. *Microchim. Acta* **2021**, *188*, 217. [CrossRef] [PubMed]
143. Yu, Z.; Cai, G.; Liu, X.; Tang, D. Platinum Nanozyme-Triggered Pressure-Based Immunoassay Using a Three-Dimensional Polypyrrole Foam-Based Flexible Pressure Sensor. *ACS Appl. Mater. Interfaces* **2020**, *12*, 40133–40140. [CrossRef] [PubMed]
144. Zhu, W.; Wang, Q.; Ma, H.; Lv, X.; Wu, D.; Sun, X.; Du, B.; Wei, Q. Single-step cycle pulse operation of the label-free electrochemiluminescence immunosensor based on branched polypyrrole for carcinoembryonic antigen detection. *Sci. Rep.* **2016**, *6*, 24599. [CrossRef]
145. Zhang, L.; Wang, Y.; Shen, L.; Yu, J.; Ge, S.; Yan, M. Electrochemiluminescence behavior of AgNCs and its application in immunosensors based on PANI/PPy-Ag dendrite-modified electrode. *Analyst* **2017**, *142*, 2587–2594. [CrossRef]
146. Pei, F.; Wang, P.; Ma, E.; Yu, H.; Gao, C.; Yin, H.; Li, Y.; Liu, Q.; Dong, Y. A sandwich-type amperometric immunosensor fabricated by Au@Pd NDs/Fe2+-CS/PPy NTs and Au NPs/NH2-GS to detect CEA sensitively via two detection methods. *Biosens. Bioelectron.* **2018**, *122*, 231–238. [CrossRef]
147. Rong, Q.; Han, H.; Feng, F.; Ma, Z. Network nanostructured polypyrrole hydrogel/Au composites as enhanced electrochemical biosensing platform. *Sci. Rep.* **2015**, *5*, 11440. [CrossRef]
148. Lv, X.; Pang, X.; Li, Y.; Yan, T.; Cao, W.; Du, B.; Wei, Q. Electrochemiluminescent Immune-Modified Electrodes Based on Ag$_2$Se@CdSe Nanoneedles Loaded with Polypyrrole Intercalated Graphene for Detection of CA72-4. *ACS Appl. Mater. Interfaces* **2015**, *7*, 867–872. [CrossRef]
149. Huang, J.; Huang, C.; Zhong, W.; Lin, Y. A magneto-controlled microfluidic device for voltammetric immunoassay of carbohydrate antigen-125 with silver–polypyrrole nanotags. *Anal. Methods* **2020**, *12*, 4211–4219. [CrossRef]
150. Rebelo, T.S.C.R.; Costa, R.; Brandão, A.T.S.C.; Silva, A.F.; Sales, M.G.F.; Pereira, C.M. Molecularly imprinted polymer SPE sensor for analysis of CA-125 on serum. *Anal. Chim. Acta* **2019**, *1082*, 126–135. [CrossRef]
151. Tertiş, M.; Ciui, B.; Suciu, M.; Săndulescu, R.; Cristea, C. Label-free electrochemical aptasensor based on gold and polypyrrole nanoparticles for interleukin 6 detection. *Electrochim. Acta* **2017**, *258*, 1208–1218. [CrossRef]

152. Aydın, E.B. Highly sensitive impedimetric immunosensor for determination of interleukin 6 as a cancer biomarker by using conjugated polymer containing epoxy side groups modified disposable ITO electrode. *Talanta* **2020**, *215*, 120909. [CrossRef] [PubMed]
153. Aydın, E.B.; Aydın, M.; Sezgintürk, M.K. A novel electrochemical immunosensor based on acetylene black/epoxy-substituted-polypyrrole polymer composite for the highly sensitive and selective detection of interleukin 6. *Talanta* **2021**, *222*, 121596. [CrossRef] [PubMed]
154. Garcia-Cruz, A.; Nessark, F.; Lee, M.; Zine, N.; Sigaud, M.; Pruna, R.; Lopez, M.; Marote, P.; Bausells, J.; Jaffrezic-Renault, N.; et al. Efficient fabrication of poly(pyrrole)-nanowires through innovative nanocontact printing, using commercial CD as mold, on flexible thermoplastics substrates: Application for cytokines immunodetection. *Sens. Actuators B Chem.* **2018**, *255*, 2520–2530. [CrossRef]
155. Kwon, O.S.; Park, S.J.; Jang, J. A high-performance VEGF aptamer functionalized polypyrrole nanotube biosensor. *Biomaterials* **2010**, *31*, 4740–4747. [CrossRef] [PubMed]
156. Kwon, O.S.; Park, S.J.; Hong, J.-Y.; Han, A.-R.; Lee, J.S.; Lee, J.S.; Oh, J.H.; Jang, J. Flexible FET-Type VEGF Aptasensor Based on Nitrogen-Doped Graphene Converted from Conducting Polymer. *ACS Nano* **2012**, *6*, 1486–1493. [CrossRef]
157. Lakemeyer, L.; Sander, S.; Wittau, M.; Henne-Bruns, D.; Kornmann, M.; Lemke, J. Diagnostic and Prognostic Value of CEA and CA19-9 in Colorectal Cancer. *Diseases* **2021**, *9*, 21. [CrossRef]
158. Tavares, A.P.; Truta, L.; Moreira, F.; Carneiro, L.P.; Sales, M.G.F. Self-powered and self-signalled autonomous electrochemical biosensor applied to cancinoembryonic antigen determination. *Biosens. Bioelectron.* **2019**, *140*, 111320. [CrossRef]
159. Gao, Y.; Wang, J.; Zhou, Y.; Sheng, S.; Qian, S.Y.; Huo, X. Evaluation of Serum CEA, CA19-9, CA72-4, CA125 and Ferritin as Diagnostic Markers and Factors of Clinical Parameters for Colorectal Cancer. *Sci. Rep.* **2018**, *8*, 2732. [CrossRef]
160. Björkman, K.; Mustonen, H.; Kaprio, T.; Kekki, H.; Pettersson, K.; Haglund, C.; Böckelman, C. CA125: A superior prognostic biomarker for colorectal cancer compared to CEA, CA19-9 or CA242. *Tumor Biol. J. Int. Soc. Oncodev. Biol. Med.* **2021**, *43*, 57–70. [CrossRef]
161. Liu, Y.; Chen, J. Expression Levels and Clinical Significance of Serum miR-497, CEA, CA24-2, and HBsAg in Patients with Colorectal Cancer. *BioMed Res. Int.* **2022**, *2022*, 3541403. [CrossRef] [PubMed]
162. Nogués, A.; Gallardo-Vara, E.; Zafra, M.P.; Mate, P.; Marijuan, J.L.; Alonso, A.; Botella, L.M.; Prieto, M.I. Endoglin (CD105) and VEGF as potential angiogenic and dissemination markers for colorectal cancer. *World J. Surg. Oncol.* **2020**, *18*, 99. [CrossRef] [PubMed]
163. Hazgui, M.; Weslati, M.; Boughriba, R.; Ounissi, D.; Bacha, D.; Bouraoui, S. MUC1 and MUC5AC implication in Tunisian colorectal cancer patients. *Turk. J. Med. Sci.* **2021**, *51*, 309–318. [CrossRef] [PubMed]
164. Niv, Y.; Rokkas, T.; Niv, Y.; Rokkas, T. Mucin Expression in Colorectal Cancer (CRC). *J. Clin. Gastroenterol.* **2019**, *53*, 434–440. [CrossRef]
165. Huang, J.; Luo, X.; Lee, I.; Hu, Y.; Cui, X.T.; Yun, M. Rapid real-time electrical detection of proteins using single conducting polymer nanowire-based microfluidic aptasensor. *Biosens. Bioelectron.* **2011**, *30*, 306–309. [CrossRef]
166. Yu, D.; Sun, J.; Weng, Y.; Luo, L.; Sheng, J.; Xu, Z. Serum angiogenin as a potential biomarker for early detection of colorectal adenomas and colorectal cancer. *Anti-Cancer Drugs* **2021**, *32*, 703–708. [CrossRef]
167. Chen, X.; Sun, J.; Wang, X.; Yuan, Y.; Cai, L.; Xie, Y.; Fan, Z.; Liu, K.; Jiao, X. A Meta-Analysis of Proteomic Blood Markers of Colorectal Cancer. *Curr. Med. Chem.* **2021**, *28*, 1176–1196. [CrossRef]
168. Song, Y.F.; Xu, Z.B.; Zhu, X.J.; Tao, X.; Liu, J.L.; Gao, F.L.; Wu, C.L.; Song, B.; Lin, Q. Serum Cyr61 as a potential biomarker for diagnosis of colorectal cancer. *Clin. Transl. Oncol.* **2017**, *19*, 519–524. [CrossRef]
169. Pączek, S.; Łukaszewicz-Zając, M.; Gryko, M.; Mroczko, P.; Kulczyńska-Przybik, A.; Mroczko, B. CXCL-8 in Preoperative Colorectal Cancer Patients: Significance for Diagnosis and Cancer Progression. *Int. J. Mol. Sci.* **2020**, *21*, 2040. [CrossRef]
170. Jiang, X.; Wang, J.; Wang, M.; Xuan, M.; Han, S.; Li, C.; Li, M.; Sun, X.; Yu, W.; Zhao, Z. ITGB4 as a novel serum diagnosis biomarker and potential therapeutic target for colorectal cancer. *Cancer Med.* **2021**, *10*, 6823–6834. [CrossRef]
171. Huang, X.; Lan, Y.; Li, E.; Li, J.; Deng, Q.; Deng, X. Diagnostic values of MMP-7, MMP-9, MMP-11, TIMP-1, TIMP-2, CEA, and CA19-9 in patients with colorectal cancer. *J. Int. Med. Res.* **2021**, *49*, 03000605211012570. [CrossRef] [PubMed]

Disclaimer/Publisher's Note: The statements, opinions and data contained in all publications are solely those of the individual author(s) and contributor(s) and not of MDPI and/or the editor(s). MDPI and/or the editor(s) disclaim responsibility for any injury to people or property resulting from any ideas, methods, instructions or products referred to in the content.

Article

Electrochemical Performance of a PVDF-HFP-LiClO$_4$-Li$_{6.4}$La$_{3.0}$Zr$_{1.4}$Ta$_{0.6}$O$_{12}$ Composite Solid Electrolyte at Different Temperatures

Xinghua Liang [1], Yujuan Ning [1], Linxiao Lan [1], Guanhua Yang [1,*], Minghua Li [2], Shufang Tang [3] and Jianling Huang [1,*]

[1] Guangxi Key Laboratory of Automobile Components and Vehicle Technology, Guangxi University of Science and Technology, Liuzhou 545006, China
[2] School of Electrical Technology, Guangdong Mechanical & Electrical Polytechnic, Guangzhou 510515, China
[3] College of Automotive Engineering, Liuzhou Institute of Technology, Liuzhou 545616, China
* Correspondence: yghchem@163.com (G.Y.); jlhuang@gxust.edu.cn (J.H.); Tel.: +86-18607736532 (G.Y.); +86-17728100882 (J.H.)

Abstract: The stability and wide temperature performance range of solid electrolytes are the keys to the development of high-energy density all-solid-state lithium-ion batteries. In this work, a PVDF-HFP-LiClO$_4$-Li$_{6.4}$La$_3$Zr$_{1.4}$Ta$_{0.6}$O$_{12}$ (LLZTO) composite solid electrolyte was prepared using the solution pouring method. The PVDF-HFP-LiClO$_4$-LLZTO composite solid electrolyte shows excellent electrochemical performance in the temperature range of 30 to 60 °C. By assembling this electrolyte into the battery, the LiFePO$_4$/PVDF-HFP-LiClO$_4$-LLZTO/Li battery shows outstanding electrochemical performance in the temperature range of 30 to 60 °C. The ionic conductivity of the composite electrolyte membrane at 30 °C and 60 °C is 5.5×10^{-5} S cm^{-1} and 1.0×10^{-5} S cm^{-1}, respectively. At a current density of 0.2 C, the LiFePO$_4$/PVDF-HFP-LiClO$_4$-LLZTO/Li battery shows a high initial specific discharge capacity of 133.3 and 167.2 mAh g^{-1} at 30 °C and 60 °C, respectively. After 50 cycles, the reversible electrochemical capacity of the battery is 121.5 and 154.6 mAh g^{-1} at 30 °C and 60 °C; the corresponding capacity retention rates are 91.2% and 92.5%, respectively. Therefore, this work provides an effective strategy for the design and preparation of solid-state lithium-ion batteries.

Keywords: lithium-ion battery; solid-state electrolyte; composite; electrochemical performance

Citation: Liang, X.; Ning, Y.; Lan, L.; Yang, G.; Li, M.; Tang, S.; Huang, J. Electrochemical Performance of a PVDF-HFP-LiClO$_4$-Li$_{6.4}$La$_{3.0}$Zr$_{1.4}$Ta$_{0.6}$O$_{12}$ Composite Solid Electrolyte at Different Temperatures. *Nanomaterials* 2022, 12, 3390. https://doi.org/10.3390/nano12193390

Academic Editor: Cheol-Min Park

Received: 31 August 2022
Accepted: 26 September 2022
Published: 28 September 2022

Publisher's Note: MDPI stays neutral with regard to jurisdictional claims in published maps and institutional affiliations.

Copyright: © 2022 by the authors. Licensee MDPI, Basel, Switzerland. This article is an open access article distributed under the terms and conditions of the Creative Commons Attribution (CC BY) license (https://creativecommons.org/licenses/by/4.0/).

1. Introduction

Lithium-ion batteries (LIBs) exhibit high energy density, a low self-discharge rate, and a long cycle life, and they have been widely used in various portable electronic devices [1–3]. LIBs continue to prove their irreplaceable value as power suppliers in a variety of applications. However, liquid LIBs include potential explosion hazards due to their flammability; thus, its security issues remain unresolved. Solid-state electrolytes have attracted a wide range of research interests because they are highly flame-retardant and have good flexibility. For example, polymer electrolytes [4–10], ceramic electrolytes [11–17], and ceramic/polymer composite electrolytes [18–23] have been intensively investigated in recent years.

Among these, ceramic/polymer composite electrolytes have attracted wide attention due to their high ionic conductivity, elastic stiffness, high shear modulus, high flexibility, and lower density, which can effectively inhibit the growth of lithium dendrite and make better contact with the electrode interface, thus improving the electrochemical performance of solid-state lithium-ion batteries. Zhang et al. [24] prepared a PVDF/LLZTO composite electrolyte by dispersing LLZTO ceramic powder into a polymer matrix. The ionic conductivity of the composite electrolyte is 5×10^{-4} S cm^{-1} at room temperature, and it exhibits high mechanical strength and good thermal stability. In the LiCoO$_2$/Li battery composed of the composite electrolyte, the initial specific discharge capacity reaches 150 mAh g^{-1}, and the capacity retention rate reaches 98% after 120 cycles at 0.4 C. Gu et al. [25] reported a PVDF-HFP/LLZTO composite solid electrolyte membrane (CSE) with a simple synthesis route and complete raw materials. The mixed electrolyte contains a wide electrochemical stability window (~5 V), excellent mechanical properties (tensile strength over 13 MPa, young's modulus > 50 MPa), good ionic conductivity (3.2×10^{-4} S cm^{-1}) at room temperature, and good electrochemical properties (about 150 mAh g^{-1} at 0.1 C; the coulomb efficiency is 99% after 50 cycles). Although good electrochemical performance was achieved in composite solid electrolytes, novel solid electrolytes with better electrochemical performance need to be developed to match the requirements for a practical battery.

Generally, the electrochemical performance of these solid electrolytes is measured at a certain temperature. As an important performance for practical application, the electrochemical performances of the solid electrolytes at wide temperature ranges were rarely reported. However, the applied temperature has an obvious influence on the electrochemical performance. In this work, the PVDF-HFP-LiClO$_4$-Li$_{6.4}$La$_3$Zr$_{1.4}$Ta$_{0.6}$O$_{12}$ (LLZTO) composite solid electrolyte was prepared using the solution pouring method, and its electrochemical performance was investigated systematically in the temperature range of 30 to 60 °C. It is found that the PVDF-HFP-LiClO$_4$-LLZTO composite solid electrolyte shows excellent electrochemical performance in the temperature range of 30 to 60 °C. By assembling this electrolyte into the battery, the LiFePO$_4$/PVDF-HFP-LiClO$_4$-LLZTO/Li battery shows outstanding electrochemical performance at 30 and 60 °C. It is noted that the electrolyte and the assembled battery show better electrochemical performance at 60 °C than at 30 °C.

2. Experimental Section

2.1. Preparation of PVDF-HFP-LiClO$_4$-LLZTO Composite Solid Electrolyte

CSE was prepared using the solution casting method, and the preparation process is shown in Figure 1. Typically, 4 g of PVDF-HFP (Mn = 600,000, Aladdin, Shanghai, China) and a certain amount of LiClO$_4$ (99%, Aladdin, Shanghai, China) were dissolved/dispersed in N-N dimethylformamide (DMF, 99.8%, Aladdin, Shanghai, China). LLZTO (99%, Kocrystal, Shenzhen, China) was introduced under moderate stirring, and the temperature was set at 50 °C. Then the slurry was transferred into the PTFE mold for static flow casting. PVDF-HFP-LiClO$_4$-LLZTO CSE can be exfoliated after vacuum-drying at 60 °C for 12 h. Finally, the electrolyte membrane was cut into small discs with a diameter of 18 mm and stored in a glove box (<0.01 PPM H$_2$O and O$_2$).

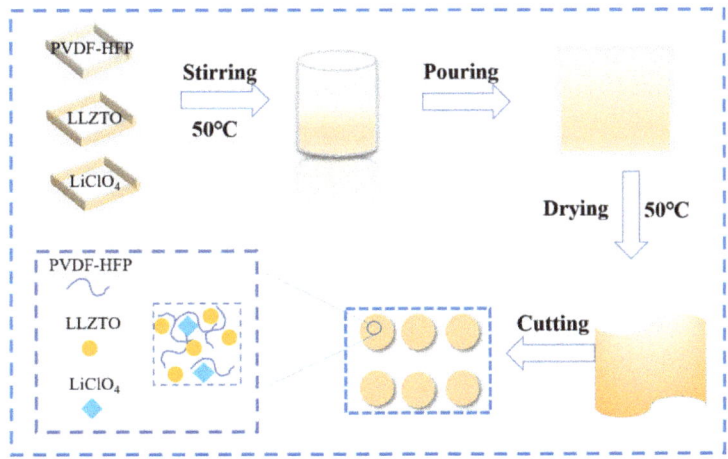

Figure 1. Schematic pattern of the preparation process of CSE.

2.2. Physical Characterization

The phase analysis of the materials was measured on an X-ray diffractometer (XRD, DX-2700, Dandong, China, Cu-Kα, 40 kV × 40 mA). The surface structure of the solid electrolyte films was characterized by scanning electron microscopy (SEM, Phenom Pharos G2, Shanghai, China) equipped with energy dispersive spectrometry (EDS). Raman spectra were collected by a Raman microscope (ATR8000, Fujian, China). Thermogravimetric analysis (TGA) was conducted using a TG analysis system (NetzschF3Tarsus, Bayern, Germany) from 30 to 800 °C with a heating rate of 10 °C min^{-1}.

2.3. Electrochemical Tests

High ionic conductivity at operating temperature is one of the prerequisites for the application of CSE in solid-state batteries. The electrochemical impedance spectroscopy (EIS) of CSE was measured on an electrochemical workstation (DH 7000, Donghua, Jiangsu, China) using button-mounted symmetrical cells assembled from stainless steel (SS)/CSE/SS. The EIS was conducted from 10^6 Hz to 0.01 Hz with an amplitude of 10 mV. The ionic conductivity σ was calculated according to the following formula [24]:

$$\sigma = \frac{L}{RS} \quad (1)$$

where σ is the ionic conductivity (S/cm), L is the thickness of the film (cm), R is the local minimal resistance over all the impedance spectrum (Ω), and S is the contact area between the film and the SS (cm^2).

The electrochemical stability window of CSE was determined by linear sweep voltammetry (LSV) using an SS/CSE/Li semi-symmetric cell with a scanning rate of 5 mV S^{-1} in the voltage range from 2.5 V to 6.0 V. The lithium-ion migration number (T_{Li^+}) of the CSE at 30 °C and 60 °C was measured by the combined measurement of EIS and DC polarization on the same electrochemical station using a symmetric Li/CSE/Li battery. T_{Li^+} is calculated according to the following formula [26,27]:

$$T_{Li^+} = \frac{I_{SS}(\Delta V - I_0 R_0)}{I_0(\Delta V - I_{SS} R_{SS})} \quad (2)$$

where T_{Li^+} represents the number of lithium-ion migrations in the electrolyte, I_0 and I_{SS} are the current values at the beginning and after the DC polarization is stabilized, R_0 and R_{SS} are the impedance values before and after the DC polarization, and ΔV is the voltage applied to both ends of the battery at 50 mV.

Electrochemical measurements of the cells were carried out by a battery test system (Neware, Dongguan, China). Commercial LiFePO$_4$ (LFP, Macklin, Shanghai, China) is used as the cathode active material without further modification, and a lithium plate is used as the anode. The cathode is made of a paste of 80 wt.% LFP, 10 wt.% PVDF, and 10 wt.% conductive carbon black cast on aluminum foil. The active material loading of the prepared cathode is about 1.8 mg cm^{-2}. The assembled LFP/CSE/Li cells were tested in the voltage range of 2.8 to 4.0 V. Liquid electrolyte (1.0 M LiPF$_6$ ethyl carbonate (EC), diethyl carbonate (DEC) and diethyl carbonate (DMC) (V:V:V = 4:3:3)) was added to the surface of the electrolyte membrane to improve the interface contact between the membrane and the electrode.

3. Results and Discussion

The addition of LLZTO inorganic filler can effectively reduce the crystallinity of the polymer, introducing a large number of Li$^+$ migration channels, thus improving the ionic conductivity of the electrolyte membrane and enhancing its mechanical properties [28]. However, excessive LLZTO makes it difficult to completely peel the electrolyte membrane from the mold, and the ionic conductivity of the electrolyte membrane decreases due to the agglomeration of the LLZTO filler. Considering the influence of LLZTO on the ionic conductivity of the electrolyte membrane, the CES were prepared with different amounts of LLZTO (0~20 wt.%), namely 0 wt.% LLZTO, 5 wt.% LLZTO, 10 wt.% LLZTO, 15 wt.% LLZTO, and 20 wt.% LLZTO, respectively. Figure 2a–e shows the EIS of CSE with different LLZTO content at different temperatures; corresponding calculated ionic conductivity of the electrolyte membranes is shown in Figure 2f. It is found that the ionic conductivity of CES increases with the increase in LLZTO content at first, then shows a rapid decline after the LLZTO content is more than 15 wt.%. Therefore, the CSE with 15 wt.% LLZTO was used as the studied object for the following discussion. Obviously, the resistance of the CSE decreases with the increase in temperature, and the calculated ionic conductivity of CSE at 30 °C, 40 °C, 50 °C, 60 °C, and 70 °C are 5.5×10^{-5}, 6.9×10^{-5}, 8.4×10^{-5}, 9.6×10^{-5}, and 1.0×10^{-4} S cm^{-1}, respectively. It should be pointed out that there is little difference between 60 °C and 70 °C, so 60 °C is chosen as the upper-temperature limit for the following discussion.

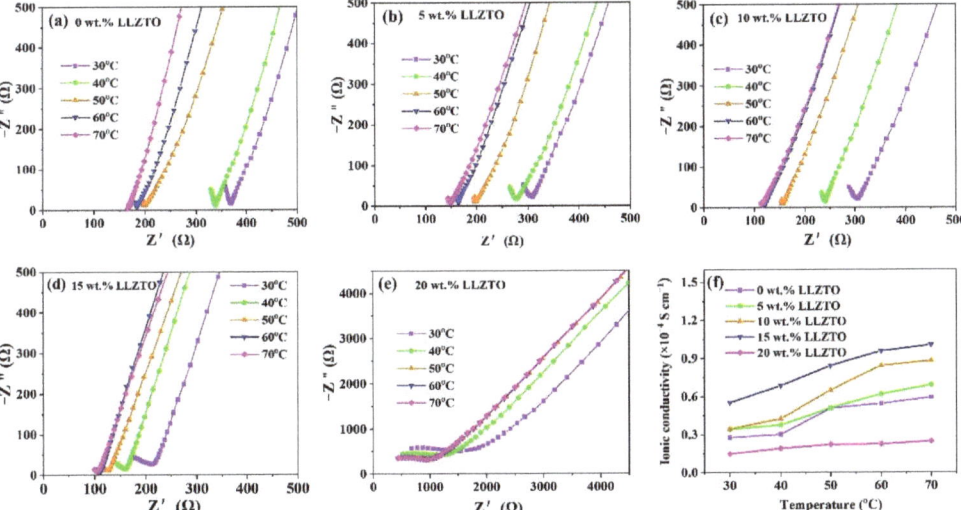

Figure 2. (a–e) EIS and (f) corresponding calculated ionic conductivity of CSE with different LLZTO content at different temperatures.

Figure 3a shows the XRD patterns of the samples. For LLZTO powder, the diffraction peaks are matched with cubic garnet Li$_5$La$_3$Nb$_2$O$_{12}$ (PDF#45-0109), indicating that LLZTO is a typical cubic phase. With respect to CSE, the diffraction peak of LLZTO can also be observed. Figure 3b shows the Raman curves of the materials; it can be seen that the PVDF-HFP peak appeared in the PVDF-HFP/LiClO$_4$ membrane, while there was none in the CSE, indicating that PVDF-HFP reacted with LLZTO. Thermal stability is an important indicator of composite polymer electrolytes. Figure 3c shows the TG analysis of the materials; the test temperature ranges from 30 °C to 800 °C. It can be seen that the weight loss temperature of PVDF-HFP is about 450 °C, PVDF-HFP/LiClO$_4$ starts to decompose at 350 °C, and CSE loses weight at 456 °C. The weight loss of the PVDF-HFP, PVDF-HFP/LiClO$_4$ and CSE at 800 °C are 98.67%, 94.61%, and 71.63%, respectively, indicating that the weight loss of CSE decreases after the addition of LLZTO, which is beneficial to improve the thermal stability of CSE.

The morphology of the CSE was characterized by a scanning electron microscope (SEM). Figure 4a–c shows SEM images of the front, back, and cross-section of the CSE sample, respectively. The positive surface of the electrolyte film has a thickness of about 200 μm, and there are some holes, which may be due to the uneven volatilization of the solvent. Some solid particles and roughness can be observed on the back of the electrolyte film, which may be caused by part of ceramic oxide LLZTO sinking into the bottom of the mold during the drying process of the slurry. The roughness on the back of the electrolyte film is caused by the surface roughness of the PTFE mold. It can be seen from the cross-section that the ceramic oxide LLZTO filler is evenly distributed in the CSE. It is also explained that this heterostructure contributes to the high performance of all-solid-state lithium batteries (ASSLB), as smooth surfaces in contact with the cathode (polymer-rich) will help to reduce interface resistance due to their soft properties, while rough surfaces in contact with the lithium anode (LLZTO-rich) will be beneficial. The EDS mapping of C, Cl, F, La, Ta, and Zr are shown in Figure 4d–i. It can be observed that the LLZTO ceramic particles are uniformly dispersed in the electrolyte membrane. Figure 4j–m shows photos of the electrolyte membranes in different physical states; obviously, the prepared CSE shows excellent mechanical flexibility and can be bent and wound. The excellent mechanical flexibility should be beneficial to inhibit the formation of lithium dendrite.

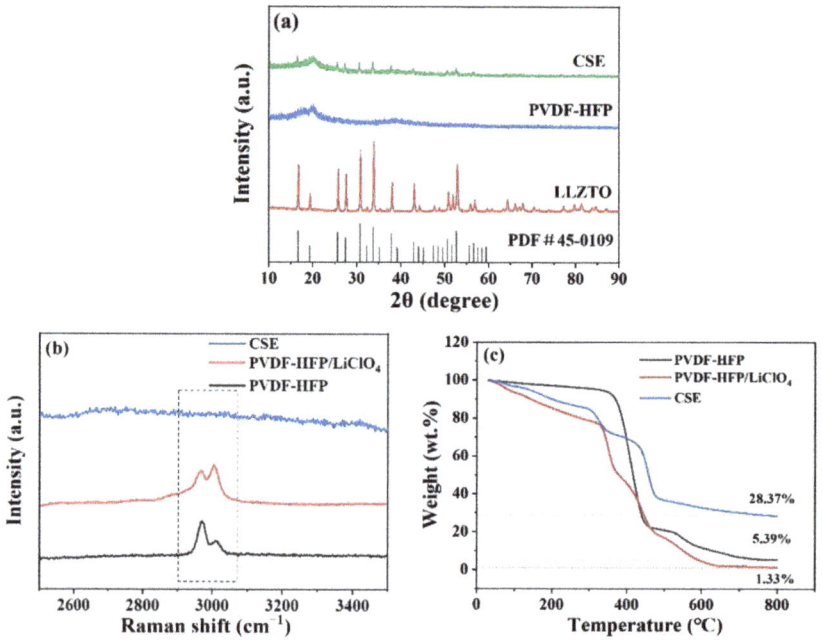

Figure 3. (a) XRD, (b) Raman, and (c) TG of PVDF-HFP, PVDF-HFP-LiClO$_4$ and CSE.

As mentioned above, the ionic conductivity of CSE at 60 °C is 1.0×10^{-4} S cm^{-1}, which is higher than that at 30 °C, being 5.5×10^{-5} S cm^{-1}; the corresponding EIS is shown in Figure 5a,b shows the electrochemical stability windows at the temperatures of 30 °C and 60 °C. The results show that the electrochemical stability window of the CSE at 60 °C is 4.4 V, while the electrochemical stability window at 30 °C is 4.7 V. Li$^+$ migration number is an important parameter to evaluate ion mobility [29]; Figure 5c,d shows the polarization curves and initial/steady impedance curves of the electrolyte membrane at 30 °C and 60 °C. The T_{Li^+} of the composite solid electrolyte at 60 °C is 0.32, higher than that that at 30 °C (0.21). The addition of LLZTO powder as an inorganic filler not only improves the ionic conductivity of the material, but also has a positive impact on the electrochemical window and the migration number of lithium ions [30]. Moreover, the addition of LLZTO ceramic fillers has been reported to slack the polymer chain. Therefore, the interaction between the inorganic filler and the polymer chain promotes the movement of the chain segments and accelerates the dynamic process between them [31].

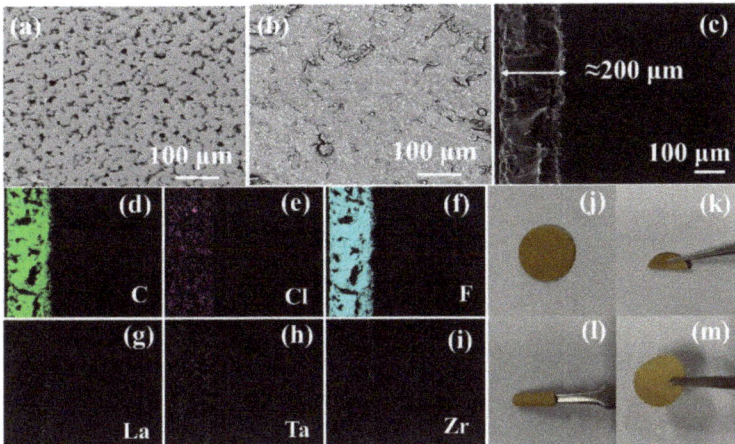

Figure 4. (a,b) SEM images and (c) cross-sectional SEM images of CSE, (d–i) corresponding EDS mapping of CSE and (j–m) macro picture of CSE in different physical states.

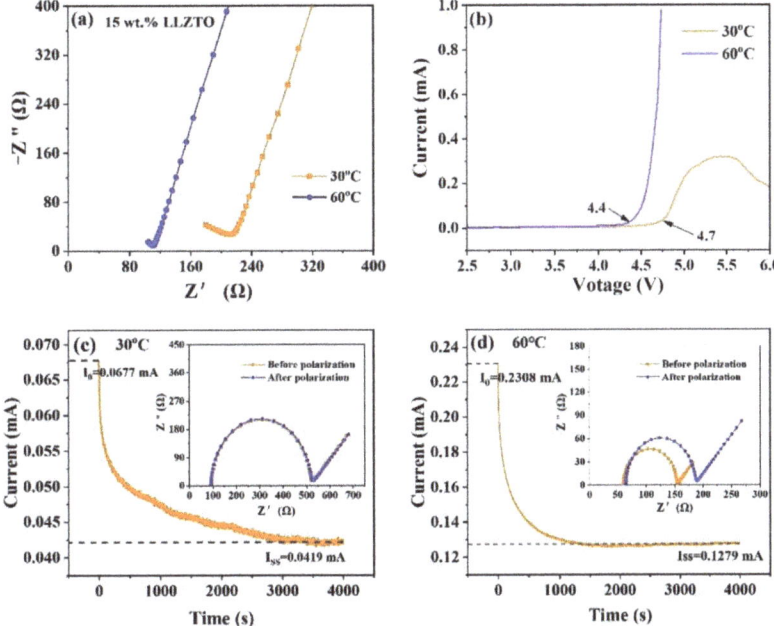

Figure 5. (a) Impedance diagram, (b) LSV diagram, and (c,d) DC polarization and AC impedance curves of CSE at 30 and 60 °C.

In order to explore the effect of CSE on the stability of lithium metal and the transport capacity of lithium ions, a symmetric battery composed of a lithium metal anode and a lithium metal cathode was charged and discharged at 0.5 h in a constant current. Figure 6a,b shows the process of lithium plating/stripping of symmetric cells at different temperatures. After 100 h of cycling at 0.1 mA cm^{-2} and 0.2 mA cm^{-2}, the batteries exhibit stable lithium plating/stripping processes at all given current densities, and the composite solid electrolyte battery exhibits very stable polarization voltage in the 200 h cycle, at either 30 °C or 60 °C. This result shows that the CSE can achieve reversible electroplating and stripping of lithium metal without obvious lithium dendrites.

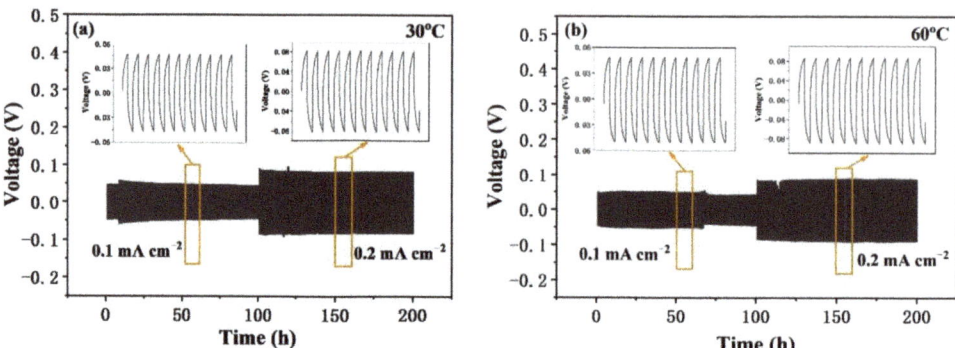

Figure 6. Constant current charge and discharge cycling of Li/CSE/Li symmetric batteries at (**a**) 30 °C and (**b**) 60 °C.

To explore the charge transfer resistance of the battery at different temperatures, EIS measurement of the half-cell assembled by CSE, cathode LFP, and anode Li at different temperatures was conducted, as shown in Figure 7a. Each EIS spectrum consists of a semicircle, caused by charge transfer resistance (Rct), and a diagonal line corresponding to the diffusion of Li$^+$ [32]. It can be seen that the semicircle at 60 °C is smaller than that at 30 °C, indicating that the charge transfer resistance of the battery decreases with the increase in temperature. Figure 7b shows the cyclic voltammetry (CV) of CSE at different temperatures between 2.5 V and 4 V, at a scanning rate of 0.5 mV s^{-1}. In the CV curves shown in Figure 7b, a remarkable reduction peak appears at ~3.0 V for all the cells, and a strong oxidation peak appears at ~3.8 V, corresponding to the lithium ion dilithium and stripping from the LFP. When observed carefully, it can be seen that the CV curve for 60 °C has a larger redox peak area, indicating that it has a larger specific capacity [33].

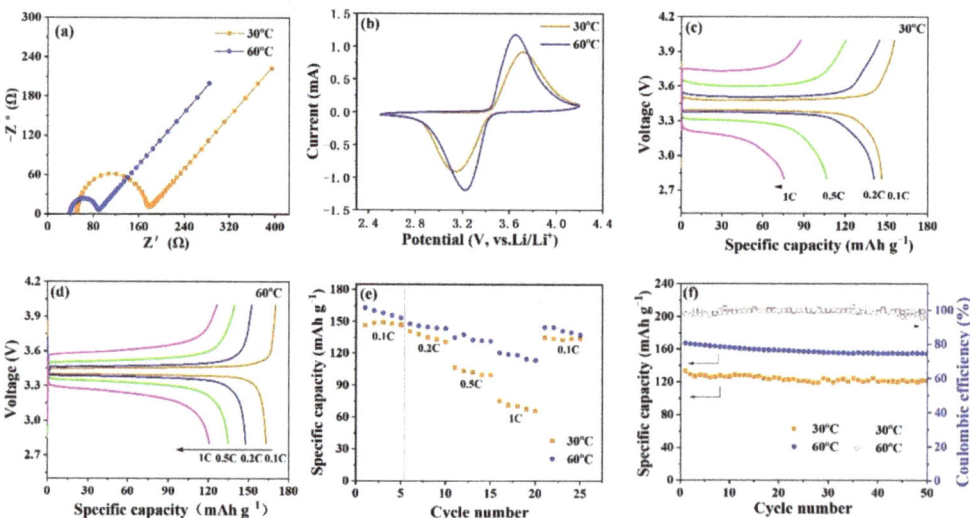

Figure 7. (**a**) Impedance diagram, (**b**) CV curves, (**c**,**d**) initial charge–discharge curves, (**e**) rate capability and (**f**) cycling life of LFP/CSE/Li battery at 30 and 60 °C.

To verify the applicability of the prepared electrolyte membranes at different temperatures, the LFP/CSE/Li battery was assembled, and its electrochemical performances were measured at 30 °C and 60 °C. Figure 7c,d shows the initial charge and discharge (CD) curves of the assembled ASSLB, with different rates at different temperatures, in the voltage range of 2.8~4.0 V. The initial CD curve presents a typically smooth and monotonous voltage plateau due to the extraction/insertion of Li$^+$ from the LFP active substance. The voltage platform gap of the battery at 30 °C is larger than that at 60 °C, indicating that there is a bigger polarization in the charging and discharging process at 30 °C. In addition, for all applied current densities, the battery shows a larger charge-discharge capacity at 60 °C than that at 30 °C, fitting well with the results of the CV.

It is noteworthy that the rate performance of the assembled lithium-ion battery is closely related to the ionic conductivity of the prepared CSE. The CSE has higher ionic conductivity at 60 °C than at 30 °C, so the battery assembled using this membrane shows better rate performance at 60 °C, as shown in Figure 7e. At a current density of 0.1 C, the assembled battery has a specific discharge capacity of 147.9 mAh g^{-1} and 158.2 mAh g^{-1} at 30 °C and 60 °C, respectively. After experiencing the cycle of 0.2, 0.5, and 1 C and recovering to 0.1 C, the specific discharge capacity of the battery can be restored to 133.9 mAh g^{-1} and 141.8 mAh g^{-1}, respectively, indicating that the battery assembled by this composite solid electrolyte has good reversible performance.

Figure 7f shows the cycling life of the LFP/CSE/Li battery with 0.2 C at 30 °C and 60 °C. It can be seen that the initial specific discharge capacities at 30 °C and 60 °C are 133.3 and 167.2 mAh g^{-1}, respectively. After 50 cycles, the specific discharge capacities are 121.5 and 154.6 mAh g^{-1}, respectively; the corresponding retention rates are 91.2 and 92.5%, respectively. The results show that the battery has better cycling life at 60 °C because the high temperature accelerates the relative motion of Li$^+$ and promotes the migration of Li$^+$. In short, the results of electrochemical measurements obtained above indicate that the prepared CSE enables the solid-state battery to show good rate capability, high reversible capacity, and good cycling life at 30 °C and 60 °C.

4. Conclusions

The PVDF-HFP-LiClO$_4$-LLZTO composite solid electrolyte was prepared using the solution pouring method, and its electrochemical performance was investigated systematically in the temperature range of 30 to 60 °C. The ionic conductivity of the composite electrolyte membrane at 30 °C and 60 °C is 5.5×10^{-5} S cm^{-1} and 1.0×10^{-4} S cm^{-1}, respectively, and the electrochemical stability window of the composite electrolyte membrane at 30 °C and 60 °C is 4.7 V and 4.4 V, respectively. By assembling this electrolyte into the battery, the LiFePO$_4$/PVDF-HFP-LiClO$_4$-LLZTO/Li battery shows outstanding electrochemical properties at the temperatures of 30 and 60 °C. At a current density of 0.2 C, the LiFePO$_4$/PVDF-HFP-LiClO$_4$-LLZTO/Li battery shows a high initial specific discharge capacity of 133.3 and 167.2 mAh g^{-1}, respectively, at 30 °C and 60 °C. After 50 cycles, the reversible electrochemical capacity of the battery at 30 °C and 60 °C is 121.5 and 154.6 mAh g^{-1}, respectively, and the corresponding capacity retention rates are 91.2% and 92.5%, respectively. In addition, the battery shows excellent rate capability, especially at 60 °C. We believe this work stimulates further interest in solid composite electrolytes for high-energy density lithium batteries.

Author Contributions: Conceptualization, X.L. and L.L.; methodology, X.L. and Y.N.; software, Y.N.; validation, X.L., Y.N., L.L., G.Y., M.L., S.T. and J.H.; formal analysis, M.L.; investigation, Y.N.; resources, X.L.; data curation, S.T.; writing—original draft preparation, Y.N.; writing—review and editing, G.Y. and J.H.; visualization, X.L.; supervision, X.L.; project administration, X.L.; funding acquisition, X.L. All authors have read and agreed to the published version of the manuscript.

Funding: This research was funded by the Guangxi Natural Science Foundation (No. 2020GXNS-FAA297082), the Guangxi Innovation Driven Development Project (No. AA18242036-2), the National Natural Science Foundation of China (No. 52161033), the Fund Project of the Key Lab of Guangdong for Modern Surface Engineering Technology (No. 2018KFKT01), the Featured Innovation Projects (Natural Science) of General Universities in Guangdong Province (Grant No. 2018GKTSCX082), and the Guangxi University Young and Middle-Aged Teachers' Basic Scientific Research Ability Improvement Project (2021KY1715).

Data Availability Statement: Not applicable.

Conflicts of Interest: The authors declare no conflict of interest.

References

1. Das, A.; Goswami, M.; Illath, K.; Ajithkumar, T.; Arya, A.; Krishnan, M. Synthesis and characterization of LAGP-glass-ceramics-based composite solid polymer electrolyte for solid-state Li-ion battery application. *J. Non-Cryst. Solids* **2021**, *558*, 120654. [CrossRef]
2. Tarascon, J.-M.; Armand, M. Issues and challenges facing rechargeable lithium batteries. *Nature* **2001**, *414*, 359–367. [CrossRef] [PubMed]
3. Lu, Q.; Jie, Y.; Meng, X.; Omar, A.; Mikhailova, D.; Cao, R.; Jiao, S.; Lu, Y.; Xu, Y. Carbon materials for stable Li metal anodes: Challenges, solutions, and outlook. *Carbon Energy* **2021**, *3*, 957–975. [CrossRef]
4. Zhang, M.; Li, M.; Chang, Z.; Wang, Y.; Gao, J.; Zhu, Y.; Wu, Y.; Huang, W. A Sandwich PVDF/HEC/PVDF Gel Polymer Electrolyte for Lithium Ion Battery. *Electrochim. Acta* **2017**, *245*, 752–759. [CrossRef]
5. Wen, Z.Y.; Wu, M.M.; Itoh, T.; Kubo, M.; Lin, Z.X.; Yamamoto, O. Effects of alumina whisker in $(PEO)_8$-$LiClO_4$-based composite polymer electrolytes. *Solid State Ion.* **2002**, *148*, 185–191. [CrossRef]
6. Yang, C.-C.; Lin, S.-J.; Wu, G.-M. Study of ionic transport properties of alkaline poly(vinyl) alcohol-based polymer electrolytes. *Mater. Chem. Phys.* **2005**, *92*, 251–255. [CrossRef]
7. Huang, B.Y.; Wang, Z.X.; Li, G.B.; Huang, H.; Xue, R.J.; Chen, L.Q.; Wang, F.S. Lithium-ion conduction in polymer electrolytes based on PAN. *Solid State Ion.* **1996**, *85*, 79–84. [CrossRef]
8. Kobayashi, K.; Pagot, G.; Vezzù, K.; Bertasi, F.; Di Noto, V.; Tominaga, Y. Effect of plasticizer on the ion-conductive and dielectric behavior of poly(ethylene carbonate)-based Li electrolytes. *Polym. J.* **2020**, *53*, 149–155. [CrossRef]
9. Jacob, M.M.E.; Prabaharan, S.R.S.; Radhakrishna, S. Effect of PEO addition on the electrolytic and thermal properties of PVDF-$LiClO_4$ polymer electrolytes. *Solid State Ion.* **1997**, *104*, 267–276. [CrossRef]
10. Liu, T.; Chang, Z.; Yin, Y.; Chen, K.; Zhang, Y.; Zhang, X. The PVDF-HFP gel polymer electrolyte for Li-O 2 battery. *Solid State Ionics* **2018**, *318*, 88–94. [CrossRef]
11. Huang, X.; Lu, Y.; Song, Z.; Rui, K.; Wang, Q.; Xiu, T.; Badding, M.E.; Wen, Z. Manipulating Li2O atmosphere for sintering dense Li7La3Zr2O12 solid electrolyte. *Energy Storage Mater.* **2019**, *22*, 207–217. [CrossRef]
12. Bi, Z.; Zhao, N.; Ma, L.; Fu, Z.; Xu, F.; Wang, C.; Guo, X. Interface engineering on cathode side for solid garnet batteries. *Chem. Eng. J.* **2020**, *387*, 124089. [CrossRef]
13. MJia, M.; Zhao, N.; Huo, H.; Guo, X. Comprehensive Investigation into Garnet Electrolytes Toward Application-Oriented Solid Lithium Batteries. *Electrochem. Energy Rev.* **2020**, *3*, 656–689.
14. Lu, W.; Xue, M.; Zhang, C. Modified Li7La3Zr2O12 (LLZO) and LLZO-polymer composites for solid-state lithium batteries. *Energy Storage Mater.* **2021**, *39*, 108–129. [CrossRef]
15. Quartarone, E.; Mustarelli, P. Electrolytes for solid-state lithium rechargeable batteries: Recent advances and perspectives. *Chem. Soc. Rev.* **2011**, *40*, 2525–2540. [CrossRef]
16. Xu, R.; Han, F.; Ji, X.; Fan, X.; Tu, J.; Wang, C. Interface engineering of sulfide electrolytes for all-solid-state lithium batteries. *Nano Energy* **2018**, *53*, 958–966. [CrossRef]
17. Yan, B.; Li, X.F.; Xiao, W.; Hu, J.H.; Zhang, L.L.; Yang, X.L. Design, synthesis, and application of metal sulfides for Li-S batteries: Progress and prospects. *J. Mater. Chem. A* **2020**, *8*, 17848–17882. [CrossRef]
18. Dirican, M.; Yan, C.; Zhu, P.; Zhang, X. Composite solid electrolytes for all-solid-state lithium batteries. *Mater. Sci. Eng. R Rep.* **2018**, *136*, 27–46. [CrossRef]
19. Keller, M.; Varzi, A.; Passerini, S. Hybrid electrolytes for lithium metal batteries. *J. Power Sources* **2018**, *392*, 206–225. [CrossRef]
20. Li, S.; Zhang, S.; Shen, L.; Liu, Q.; Ma, J.; Lv, W.; He, Y.; Yang, Q. Progress and Perspective of Ceramic/Polymer Composite Solid Electrolytes for Lithium Batteries. *Adv. Sci.* **2020**, *7*, 1903088. [CrossRef]
21. Li, L.S.; Deng, Y.F.; Duan, H.H.; Qian, Y.X.; Chen, G.H. LiF and $LiNO_3$ as synergistic additives for PEO-PVDF/LLZTO-based composite electrolyte towards high-voltage lithium batteries with dual interfaces stability. *J. Energy Chem.* **2022**, *65*, 319–328. [CrossRef]
22. Li, Y.; Xu, B.; Xu, H.; Duan, H.; Lu, X.; Xin, S.; Zhou, W.; Xue, L.; Fu, G.; Manthiram, A.; et al. Hybrid Polymer/Garnet Electrolyte with a Small Interfacial Resistance for Lithium-Ion Batteries. *Angew. Chem. Int. Ed.* **2016**, *56*, 753–756. [CrossRef] [PubMed]
23. Xu, L.; Li, G.; Guan, J.; Wang, L.; Chen, J.; Zheng, J. Garnet-doped composite polymer electrolyte with high ionic conductivity for dendrite-free lithium batteries. *J. Energy Storage* **2019**, *24*, 100767. [CrossRef]
24. Zhang, X.; Liu, T.; Zhang, S.F.; Huang, X.; Xu, B.Q.; Lin, Y.H.; Xu, B.; Li, L.L.; Nan, C.W.; Shen, Y. Ynergistic coupling between $Li_{6.75}La_3Zr_{1.75}Ta_{0.25}O_{12}$ and poly (vinylidenefluoride) induces high ionic conductivity, mechanical strength, and thermal stability of solid composite electrolytes. *J. Am. Chem. Soc.* **2017**, *139*, 13779–13785. [CrossRef]
25. Gu, Y.; She, S.; Hong, Z.; Huang, Y.; Wu, Y. Enabling lithium metal battery with flexible polymer/garnet type solid oxide composite electrolyte. *Solid State Ion.* **2021**, *368*, 115710. [CrossRef]

26. Lu, J.; Liu, Y.; Yao, P.; Ding, Z.; Tang, Q.; Wu, J.; Ye, Z.; Huang, K.; Liu, X. Hybridizing poly(vinylidene fluoride-co-hexafluoropropylene) with Li6.5La3Zr1.5Ta0.5O12 as a lithium-ion electrolyte for solid state lithium metal batteries. *Chem. Eng. J.* **2019**, *367*, 230–238. [CrossRef]
27. Huggins, R.A. Simple method to determine electronic and ionic components of the conductivity in mixed conductors a review. *Ionics* **2002**, *8*, 300–313. [CrossRef]
28. Tao, X.Y.; Liu, Y.Y.; Liu, W.; Zhou, G.M.; Zhao, J.; Lin, D.C.; Zu, C.X.; Sheng, O.W.; Zhang, W.K.; Lee, H.W.; et al. Solid-State Lithium−Sulfur Batteries Operated at 37 °C with Composites of Nanostructured $Li_7La_3Zr_2O_{12}$/Carbon Foam and Polymer. *Nano Lett.* **2017**, *17*, 2967–2972. [CrossRef]
29. Jiang, G.S.; Qu, C.Z.; Xu, F.; Zhang, E.; Lu, Q.Q.; Cai, X.R.; Hausdorf, S.; Wang, H.Q.; Kaskel, S. Glassy metal-organic-framework-based quasi-solid-state electrolyte for high-performance Lithium-metal batteries. *Adv. Funct. Mater.* **2021**, *31*, 2104300. [CrossRef]
30. Chen, F.; Yang, D.; Zha, W.; Zhu, B.; Zhang, Y.; Li, J.; Gu, Y.; Shen, Q.; Zhang, L.; Sadoway, D.R. Solid polymer electrolytes incorporating cubic Li7La3Zr2O12 for all-solid-state lithium rechargeable batteries. *Electrochim. Acta* **2017**, *258*, 1106–1114. [CrossRef]
31. Liang, Y.F.; Deng, S.J.; Xia, Y.; Wang, X.L. A superior composite gel polymer electrolyte of $Li_7La_3Zr_2O_{12}$-poly (vinylidenefluoride-hexafluoropropylene)(PVDF-HFP) for re-chargeable solid-state lithium ion batteries. *Mater. Res. Bull.* **2018**, *102*, 412–417. [CrossRef]
32. Yan, H.; Huang, L.B.; Huang, Z.Y.; Wang, C.A. Enhanced mechanical strength and ionic conductivity of LLZO solid electrolytes by oscillatory pressure sintering. *Ceram. Int.* **2019**, *45*, 18115–18118.
33. Liu, L.X.; Wang, J.W.; Oswald, S.; Hu, J.P.; Tang, H.M.; Wang, J.H.; Yin, Y.; Lu, Q.Q.; Liu, L.F.; Argibay, E.C.; et al. Decoding of oxygen network distortion in a layered high-rate anode by in situ investigation of a single microelectrode. *ACS Nano* **2020**, *14*, 11753–11764. [CrossRef] [PubMed]

Article

Secondary-Heteroatom-Doping-Derived Synthesis of N, S Co-Doped Graphene Nanoribbons for Enhanced Oxygen Reduction Activity

Bing Li [†], Tingting Xiang [†], Yuqi Shao, Fei Lv, Chao Cheng, Jiali Zhang, Qingchao Zhu, Yifan Zhang and Juan Yang *

School of Materials Science and Engineering, Jiangsu University, Zhenjiang 212013, China
* Correspondence: yangjuan6347@ujs.edu.cn
† These authors contributed equally to this work.

Abstract: The rareness and weak durability of Pt-based electrocatalysts for oxygen reduction reactions (ORRs) have hindered the large-scale application of fuel cells. Here, we developed an efficient metal-free catalyst consisting of N, S co-doped graphene nanoribbons (N, S-GNR-2s) for ORRs. GNRs were firstly synthesized via the chemical unzipping of carbon nanotubes, and then N, S co-doping was conducted using urea as the primary and sulfourea as the secondary heteroatom sources. The successful incorporation of nitrogen and sulfur was confirmed by elemental mapping analysis as well as X-ray photoelectron spectroscopy. Electrochemical testing revealed that N, S-GNR-2s exhibited an E_{onset} of 0.89 V, $E_{1/2}$ of 0.79 V and an average electron transfer number of 3.72, as well as good stability and methanol tolerance. As a result, N, S-GNR-2s displayed better ORR property than either N-GNRs or N, S-GNRs, the control samples prepared with only a primary heteroatom source, strongly clarifying the significance of secondary-heteroatom-doping on enhancing the catalytic activity of carbon-based nanomaterials.

Keywords: secondary-heteroatom-doping; graphene nanoribbons; unzipping; oxygen reduction reaction

1. Introduction

The World Energy Outlook 2021 report, released by the International Energy Agency (IEA), pointed out that the growth speed of new energy power is not fast enough to support the goal of net zero emissions by 2050 [1]. Therefore, new energy conversion and storage devices (such as fuel cells [2–4], metal air batteries [5–7], lithium-ion batteries [8–10], supercapacitors [11–13], etc.) continue to draw researchers' attention. A fuel cell is a device that can convert the chemical energy of a fuel and an oxidizer directly into electric energy with the help of catalysts. It is an environmentally friendly technology with a high energy density and conversion rate [14].

The oxygen reduction reaction (ORR) is a vital electrochemical process which occurs at the cathode side of fuel cells. The sluggish kinetics of ORRs result in serious cathode polarization and energy loss, making them one of the major challenges of the large-scale implementation and commercialization of fuel cells [15,16]. To date, Pt-based nanomaterials still play an important role in catalyzing ORRs as they are regarded as the most efficient catalysts [17]. However, numbers of research works have been conducted on exploring alternative ORR electrocatalysts with a low price and outstanding catalytic performance due to the prohibitive cost, poor stability and methanol tolerance of commercial Pt-based catalysts [18].

Carbon nanomaterials are a hot topic in recent years because of their special structure, physical and chemical properties, and thus their wide range of applications [19,20]. Heteroatoms (e.g., nitrogen, sulfur, boron, phosphorus) doped into carbon nanomaterials could induce electron modulation, rendering the charge distribution and facilitating the adsorption of O_2 to enhance the ORR activity [21]. Nitrogen has an atomic size similar to that of carbon but a different electron configuration, such that nitrogen atoms can change carbon nanomaterials' electronic structures while minimizing the lattice mismatch [22]. N-doped materials, including carbon nanotubes (CNTs) [23–25], graphene [26–28] as well as porous carbons [29–31], have been studied for years in order to regulate nanomaterials' electronic structures and some other properties. In addition, sulfur has also been confirmed to have a beneficial effect on the ORR activity of carbon nanomaterials. The electronegativity of sulfur atoms is similar to that of carbon atoms (C = 2.55, S = 2.58), leading to the change in the spin density of carbon atoms as well as more defect sites [32]. These characteristics make sulfur the second-most efficiently doped element after nitrogen. Given the facts above, the co-doping of nitrogen and sulfur atoms has the potential to further regulate the electronic structure of the adjacent carbon atoms, resulting in a synergistic effect on enhancing ORR properties [33].

Graphene nanoribbons (GNRs) have drawn growing research attention for their unique structure with a large length–width ratio and abundant edge content [34]. They can be regarded as the products of the longitudinal unzipping of CNTs. In addition, GNRs show characteristics such as high chemical stability, low weight, decent specific surface area and low price, making themselves possible to be applied in the field of electrocatalysis [35]. In our previous work, N-doped GNRs (N-GNRs) were successfully prepared and electrochemical tests were conducted [36]. The synergistic effect between N-doping and carbon edges on catalyzing ORRs has been demonstrated, and as a result, as-obtained catalysts exhibited superior ORR property. On the basis of the discussion above, if an additional heteroatom of sulfur is introduced into N-GNRs, the obtained new catalyst can take full advantage of the synergy among the edge structure, nitrogen doping and sulfur doping, and thus the ORR catalytic activity is expected to be further improved [37–39].

Herein, GNRs were firstly prepared via the chemical unzipping of multi-walled CNTs, and N, S co-doped GNRs (N, S-GNR-2s) were obtained by high-temperature annealing with urea as the primary and sulfourea as the secondary heteroatom precursors. Structural characterizations of the sample have verified the successful incorporation of nitrogen and sulfur. N, S-GNR-2s showed better ORR activity than other GNR-based samples doped only with a primary heteroatom source, demonstrating the importance of secondary-heteroatom-doping. As a result, N, S-GNR-2s mainly displayed a 4-electron catalytic pathway towards an ORR, rendering themselves potential ORR electrocatalyst alternatives for the application on fuel cells.

2. Materials and Methods

2.1. Synthesis of N-GNRs, N, S-GNRs and N, S-GNR-2s

GNRs were obtained by the unzipping of multi-walled CNTs (full details are given in the Supplementary Materials) [40]. GNRs and urea (mass ratio of 1:20) were mixed and grinded in an agate grinding bowl [41,42]. The mixture was slowly transferred into a quartz boat and put in a tube furnace, and then underwent thermal annealing in an Ar atmosphere at 900 °C for 2 h to obtain N-GNRs. N-GNRs and sulfourea (mass ratio of 1:20) followed the same steps above but pyrolyzed at different temperatures (800 °C, 900 °C and 1000 °C) for 2 h to obtain N, S-GNR-2s.

To synthesize the control sample of N, S-GNRs, GNRs and sulfourea (1:20) were also mixed and grinded in a crucible and then annealed in Ar at 900 °C for 2 h.

2.2. Structural Characterization

Scanning electron microscopy (SEM, JEOL JSM-7800F; Tokyo, Japan) and transmission electron microscopy (TEM, JEOL JEM-2100; Tokyo, Japan) were applied to observe the morphology of the catalysts. Element mapping analysis was detected to know the distribution of the elements. X-ray photoelectron spectroscopy (XPS, ESCALAB 250Xi; Thermo Fisher, Waltham, MA, USA) was conducted to analyze the bond configuration of the elements.

Raman spectra were collected to monitor the defect level of the samples. All instrument parameters are shown in the Supplementary Materials.

2.3. Electrochemical Characterization

A 2 mg catalyst was uniformly dispersed in a 1 mL mixed solvent of isopropanol and nafion (19.88:0.12) to form a homogeneous ink. Then, a 20 µL catalyst ink was loaded onto a rotating disk electrode (RDE) or rotating ring-disk electrode (RRDE) and dried. A concentration of 0.1 M of KOH was used as the electrolyte with O_2 or N_2 saturation.

All of the electrochemical characterization was performed in a standard three-electrode cell (a Pt wire as the counter electrode, an Ag/AgCl as the reference electrode and an RDE or RRDE as the working electrode). The potentials in this study were converted according to the equation of $E_{vs.\ RHE} = E_{vs.\ Ag/AgCl} + (0.059\ pH + 0.197)$ V [43]. Full details of the electrochemical characterization including cyclic voltammetry (CV), RDE and RRDE measurements are given in the Supplementary Materials.

3. Results and Discussion

Figure 1a illustrates the fabrication process of N, S-GNR-2s. Urea was utilized as the primary heteroatom source to form the N-GNRs via thermal annealing. Then, sulfourea was used as the secondary heteroatom source to incorporate sulfur as well as additional nitrogen to form N, S-GNR-2s. TEM and SEM were applied to observe the morphology of our sample. After chemical oxidation and N, S co-doping, the ribbon structure can be clearly seen (Figure 1b,c), indicating that the CNTs were successfully unzipped and GNRs were formed. However, some side walls of CNTs still existed, implying that the oxidative unzipping was not 100% complete. An SEM image of N, S-GNR-2s is shown in Figure 1d, and the corresponding energy dispersive spectrometer (EDS) elemental mapping proved that nitrogen and sulfur were successfully doped and uniformly distributed on N, S-GNR-2s (Figure 1e).

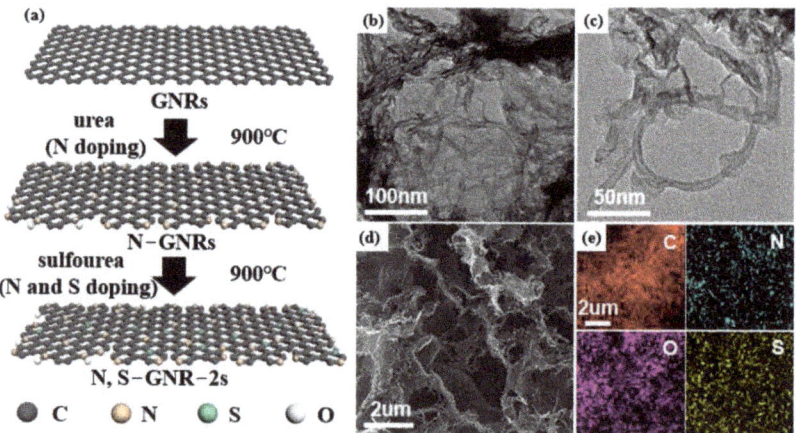

Figure 1. (a) The synthesis process of N, S-GNR-2s. (b,c) TEM images of N, S-GNR-2s. (d) The SEM image and (e) the corresponding EDS element mappings of N, S-GNR-2s showing the distribution of C, N, O and S.

A variety of structural characterizations were used to examine the physical properties of our samples. The existence of C, N, O and S was detected by the XPS analysis of N, S-GNR-2s (Figure 2a), further proving the successful incorporation of nitrogen and sulfur. In the C 1s spectrum, C–C, C–S, C–N, C–O and O = C–O bonds could be directly observed (Figure 2b), demonstrating that parts of N, S-GNR-2s were oxidized with N and S co-doping. The N 1s spectrum can be divided into four peaks: pyridinic N, pyrrolic N, graphitic N and oxidized N (Figure 2c). Since the GNRs possessed abundant edge structures and pyridine N was located at the edges of the carbon materials, the content of pyridine N was high (48.78%) as expected. Figure 2d is the S 2p spectrum of N, S-GNR-2s, in which the fitting peaks of 162.8 eV (S $2p_{3/2}$) and 163.9 eV (S $2p_{1/2}$) are characteristic peaks of thiophene S with different spin orbital coupling. There is also a small peak of S–O at 167.1 eV. Generally, thiophene S is considered as the key coordination type to enhance ORR performance [44,45]. The results confirmed that N and S replaced a small amount of C in the material and acted as dopants, which was beneficial to improving the ORR performance of the catalysts.

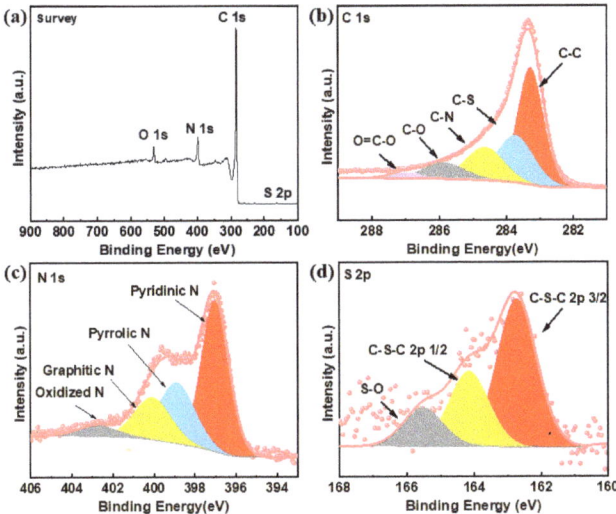

Figure 2. (a) The XPS survey, (b) C 1s, (c) N 1s and (d) S 2p spectra of N, S-GNR-2s.

The D band (located at ~1360 cm^{-1}) and G band (located at ~1580 cm^{-1}) are shown in the Raman spectra of N-GNRs, N, S-GNRs and N, S-GNR-2s (Figure 3). The I_D/I_G ratio of N-GNRs was calculated to be 1.28, owing to N-doping and the formation of the porous structure caused by the release of gas generated during the thermal decomposition of urea. Compared with the N-GNRs, both the N, S-GNRs and N, S-GNR-2s had more defects caused by additional S-doping. As a result, the I_D/I_G ratios of the N, S-GNRs and N, S-GNR-2s were increased to 1.36 and 1.37, respectively [46].

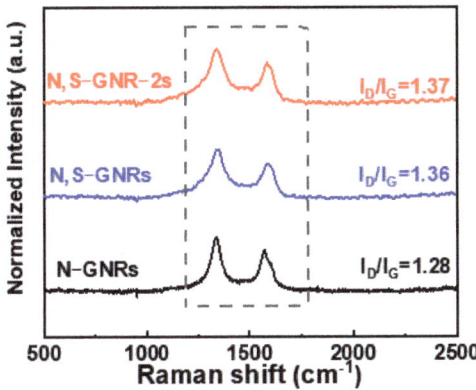

Figure 3. The Raman spectra of N-GNRs, N, S-GNRs and N, S-GNR-2s.

Electrochemical tests of N, S-GNR-2s were carried out in an alkaline medium of 0.1 M KOH. Three samples of N, S-GNR-2s with different thermal annealing temperatures (800 °C, 900 °C and 1000 °C) were examined, among which, N, S-GNR-2s at 900 °C showed the best ORR activity (Figure S1). For this reason, this sample was selected in the following experiments for the performance analysis. Figure 4a displays CV curves of N, S-GNR-2s in O_2 and N_2-saturated solution. No evident peaks appeared when tested in N_2-saturated solution, whereas a distinct ORR peak in O_2-saturated solution was clearly observed at 0.80 V, suggesting good ORR catalytic activity for N, S-GNR-2s. RDE measurement was carried out at different rotating speeds from 625 to 2500 rpm (Figure 4b), and Koutecky–Levich (K–L) plots are correspondingly displayed in Figure 4c. The electron transfer number was extracted to be 3.66~3.67, ranging from 0.3~0.6 V. In addition, RRDE testing was further performed in order to accurately detect the generation of peroxide (HO_2^-) during the reaction (Figure 4d). The average HO_2^- yield was measured and calculated to be about 13.69%, and the average electron transfer number was calculated to be about 3.72 from 0 to 0.8 V (Figure 4e), in correspondence with the result in Figure 4c. The electrocatalytic properties tested by RDE and RRDE distinctly verified that N, S-GNR-2s mainly displayed a 4-electron catalytic pathway towards an ORR. RDE curves at 1600 rpm before and after 2000 CV cycles with a scan rate of 100 mV s^{-1} and potential range of 0.56~0.96 V are shown to evaluate the stability of N, S-GNR-2s (Figure 4f). The diffusion-limited current density after cycling stayed almost unchanged, and the half-wave potential ($E_{1/2}$) slightly shifted from 0.79 V to 0.71 V, demonstrating the decent ORR stability of our catalysts.

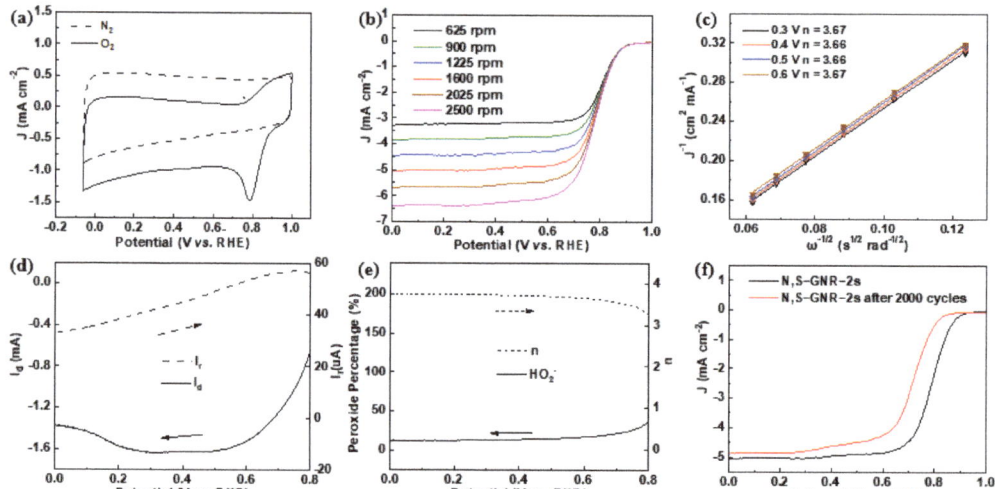

Figure 4. Electrocatalytic property tests of N, S-GNR-2s. (**a**) The CV test in 0.1 M KOH solution. (**b**) The RDE test at rotating speeds from 625 to 2500 rpm. (**c**) K–L plots based on the RDE test. (**d**) RRDE curves at a rotating speed of 1600 rpm. (**e**) The calculated HO_2^- yield and electron transfer number. (**f**) The stability test before and after 2000 cycles.

In order to compare the catalytic performance of different samples (N-GNRs, N, S-GNRs and N, S-GNR-2s) and verify the significance of secondary-heteroatom-doping, various electrochemical analyses were carried out. A CV test was performed, and the curves were recorded and plotted together (Figure 5a). The CV curve of N, S-GNR-2s in O_2 exhibited a more positive oxygen reduction peak than those of the other two samples, revealing the superior catalytic activity of N, S-GNR-2s. Figure 5b shows the RDE curves at 1600 rpm of the three samples, and the corresponding values of onset potential (E_{onset}) and $E_{1/2}$ were plotted in Figure 5c. Among all the samples, N, S-GNR-2s showed the highest E_{onset} of 0.89 V and $E_{1/2}$ of 0.79 V along with a diffusion-limited current density of up to 5.06 mA cm^{-2}, while N-GNRs displayed an E$_{onset}$ of 0.81 V and E$_{1/2}$ of 0.72 V, and N, S-GNRs displayed an E$_{onset}$ of 0.84 V and E$_{1/2}$ of 0.67 V. Tafel plots of the catalysts are displayed in Figure 5d on the basis of the RDE test. N, S-GNR-2s presented a much lower Tafel slope of 74.34 mV dec^{-1} than other catalysts, suggesting the most desirable ORR activity of N, S-GNR-2s. The values of electron transfer number are shown in Figure 5e on the basis of the RDE tests (Figure 4, Figures S2 and S3). N, S-GNR-2s exhibited similar but more steady values than N, S-GNRs, and both of their performances surpassed that of N-GNRs. Chronoamperometric measurement was conducted to test the methanol tolerance of these samples at 1600 rpm at 0.5 V (Figure 5f). When 1 M methanol was added to the test system, they exhibited different levels of performance decay. Interestingly, N, S-GNRs were found to display a high current retention of 95.03% after 500 s, slightly higher than 93.64% for N, S-GNR-2s and 93.16% for N-GNRs, certifying a good methanol tolerance property for all the catalysts. We compared the electrocatalytic property of N, S-GNR-2s with some relevant studies (Table S1), and the results revealed that N, S-GNR-2s showed competitive or even better performance compared to the reported works.

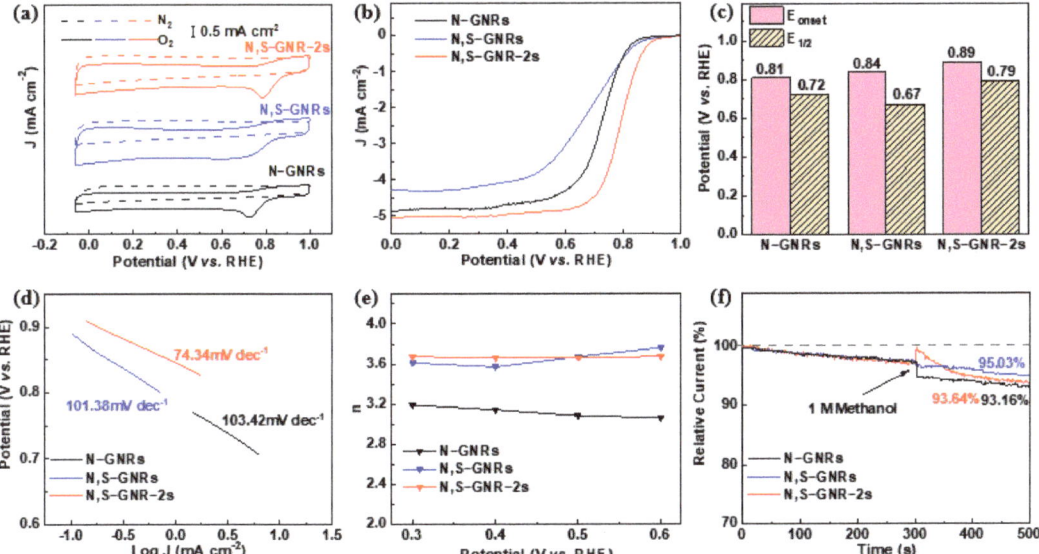

Figure 5. Electrocatalytic property tests of N-GNRs, N, S-GNRs and N, S-GNR-2s. (**a**) The CV test in 0.1 M KOH. (**b**) The RDE test at a rotating speed of 1600 rpm. (**c**) The corresponding values of E$_{onset}$ and E$_{1/2}$. (**d**) The corresponding Tafel plots based on the RDE tests. (**e**) The values of electron transfer number (n) at certain potentials. (**f**) The methanol tolerance test.

To sum up the electrochemical performance above, interesting conclusions can be drawn. Firstly, comparing the performance of N, S-GNR-2s with N-GNRs and N, S-GNRs, the significance of secondary-heteroatom-doping was remarkably certified. Both N-GNRs and N, S-GNRs were synthesized using only a primary heteroatom source, and thus they presented inferior activity to N, S-GNR-2s with secondary-heteroatom-doping. In order to understand the effect of secondary-heteroatom-doping, we further conducted the XPS analysis of N-GNRs and N, S-GNRs. As shown in Figure S4 and Table S2, N, S-GNR-2s displayed a N content of 10.93%, much higher than that of either N-GNRs (5.43%) or N, S-GNRs (5.67%). Thus, it can be concluded that the heteroatom content was greatly increased by secondary-heteroatom-doping, leading to more reaction sites and thus improved ORR property. Secondly, the performance difference between N, S-GNR-2s and N-GNRs also proved that the synergistic effect between N and S induced by N, S co-doping contributed to the enhancement of catalytic activity. Consequently, the highest electrocatalytic performance was obtained for our well-designed catalyst of N, S-GNR-2s.

4. Conclusions

In summary, N, S-GNR-2s were prepared via the N, S co-doping of GNRs using urea and sulfourea as the primary and secondary heteroatom precursors, respectively. The ribbon structure of N, S-GNR-2s was confirmed by TEM observation, suggesting the successful unzipping of CNTs. The introduction of both nitrogen and sulfur was verified by various structural characterizations. Electrochemical testing showed that N, S-GNR-2s exhibited an E_{onset} of 0.89 V, $E_{1/2}$ of 0.79 V and an average electron transfer number of 3.72, as well as good stability and methanol tolerance. As a result, N, S-GNR-2s revealed better ORR property than N-GNRs and N, S-GNRs, demonstrating that secondary-heteroatom-doping contributes greatly to the improvement of electrocatalytic performance for carbon nanomaterials.

Supplementary Materials: The following supporting information can be downloaded at: https://www.mdpi.com/article/10.3390/nano12193306/s1, Figure S1. Electrochemical characterizations of N, S-GNR-2s synthesized at different temperatures in 0.1 M KOH. Figure S2. Electrochemical characterizations of N-GNRs. Figure S3. Electrochemical characterizations of N, S-GNRs. Figure S4. An XPS survey of N-GNRs, N, S-GNRs and N, S-GNR-2s. Table S1. An electrocatalytic property comparison between our work and some relevant studies. Table S2. The elemental contents of N-GNRs, N, S-GNRs and N, S-GNR-2s based on XPS analysis in Figure S4. References [43,47–55] are cited in the supplementary materials.

Author Contributions: Conceptualization, B.L.; data curation, Q.Z. and Y.Z.; formal analysis, Y.Z.; funding acquisition, B.L.; investigation, T.X., F.L., C.C. and J.Z.; methodology, T.X.; supervision, J.Y.; validation, J.Y.; visualization, Y.S.; writing—original draft, T.X.; writing—review and editing, B.L. All authors have read and agreed to the published version of the manuscript.

Funding: This research was funded by the National Natural Science Foundation of China (Grant No. 51972150), the China Postdoctoral Science Foundation (Grant No. 2020M671357) and the Natural Science Foundation of the Jiangsu Province (Grant No. BK20210780).

Data Availability Statement: The data presented in this study are available upon request from the corresponding author.

Conflicts of Interest: The authors declare no conflict of interest.

References

1. Fazendeiro, L.M.; Simoes, S.G. Historical Variation of IEA Energy and CO_2 Emission Projections: Implications for Future Energy Modeling. *Sustainability* **2021**, *13*, 27. [CrossRef]
2. Yuan, X.Z.; Nayoze-Coynel, C.; Shaigan, N.; Fisher, D.; Zhao, N.N.; Zamel, N.; Gazdzicki, P.; Ulsh, M.; Friedrich, K.A.; Girard, F.; et al. A review of functions, attributes, properties and measurements for the quality control of proton exchange membrane fuel cell components. *J. Power Sources* **2021**, *491*, 39. [CrossRef]
3. Fan, J.T.; Chen, M.; Zhao, Z.L.; Zhang, Z.; Ye, S.Y.; Xu, S.Y.; Wang, H.; Li, H. Bridging the gap between highly active oxygen reduction reaction catalysts and effective catalyst layers for proton exchange membrane fuel cells. *Nat. Energy* **2021**, *6*, 475–486. [CrossRef]
4. Tao, Z.W.; Wang, C.Y.; Zhao, X.Y.; Li, J.; Guiver, M.D. Progress in High-Performance Anion Exchange Membranes Based on the Design of Stable Cations for Alkaline Fuel Cells. *Adv. Mater. Technol.* **2021**, *6*, 14. [CrossRef]
5. Borchers, N.; Clark, S.; Horstmann, B.; Jayasayee, K.; Juel, M.; Stevens, P. Innovative zinc-based batteries. *J. Power Sources* **2021**, *484*, 22. [CrossRef]
6. Yang, D.; Chen, D.; Jiang, Y.; Ang, E.H.X.; Feng, Y.Z.; Rui, X.H.; Yu, Y. Carbon-based materials for all-solid-state zinc-air batteries. *Carbon Energy* **2021**, *3*, 50–65. [CrossRef]
7. Kundu, A.; Mallick, S.; Ghora, S.; Raj, C.R. Advanced Oxygen Electrocatalyst for Air-Breathing Electrode in Zn-Air Batteries. *ACS Appl. Mater. Interfaces* **2021**, *13*, 40172–40199. [CrossRef]
8. Galos, J.; Pattarakunnan, K.; Best, A.S.; Kyratzis, I.L.; Wang, C.H.; Mouritz, A.P. Energy Storage Structural Composites with Integrated Lithium-Ion Batteries: A Review. *Adv. Mater. Technol.* **2021**, *6*, 19. [CrossRef]
9. Zhang, X.H.; Li, Z.; Luo, L.A.; Fan, Y.L.; Du, Z.Y. A review on thermal management of lithium-ion batteries for electric vehicles. *Energy* **2022**, *238*, 12. [CrossRef]

10. Murdock, B.E.; Toghill, K.E.; Tapia-Ruiz, N. A perspective on the sustainability of cathode materials used in lithium-ion batteries. *Adv. Energy Mater.* **2021**, *11*, 17. [CrossRef]
11. Zhou, Y.; Qi, H.L.; Yang, J.Y.; Bo, Z.; Huang, F.; Islam, M.S.; Lu, X.; Dai, L.; Amal, R.; Wang, C.H.; et al. Two-birds-one-stone: Multifunctional supercapacitors beyond traditional energy storage. *Energy Environ. Sci.* **2021**, *14*, 1854–1896. [CrossRef]
12. Park, J.; Kim, W. History and Perspectives on Ultrafast Supercapacitors for AC Line Filtering. *Adv. Energy Mater.* **2021**, *11*, 28. [CrossRef]
13. Li, C.; Zhang, X.; Wang, K.; Su, F.Y.; Chen, C.M.; Liu, F.Y.; Wu, Z.-S.; Ma, Y. Recent advances in carbon nanostructures prepared from carbon dioxide for high-performance supercapacitors. *J. Energy Chem.* **2021**, *54*, 352–367. [CrossRef]
14. Neatu, S.; Neatu, F.; Chirica, I.M.; Borbath, I.; Talas, E.; Tompos, A.; Somacescu, S.; Osiceanu, P.; Folgado, M.A.; Chaparro, A.M.; et al. Recent progress in electrocatalysts and electrodes for portable fuel cells. *J. Mater. Chem. A* **2021**, *9*, 17065–17128. [CrossRef]
15. Hu, B.B.; Xia, C.R. Factors influencing the measured surface reaction kinetics parameters. *Asia-Pac. J. Chem. Eng.* **2016**, *11*, 327–337. [CrossRef]
16. Li, M.R.; Zhou, W.; Zhu, Z.H. Recent development on perovskite-type cathode materials based on $SrCoO_3$-delta parent oxide for intermediate-temperature solid oxide fuel cells. *Asia-Pac. J. Chem. Eng.* **2016**, *11*, 370–381. [CrossRef]
17. Li, C.L.; Tan, H.B.; Lin, J.J.; Luo, X.L.; Wang, S.P.; You, J.; Kang, Y.-M.; Bando, Y.; Yamauchi, Y.; Kim, J. Emerging Pt-based electrocatalysts with highly open nanoarchitectures for boosting oxygen reduction reaction. *Nano Today* **2018**, *21*, 91–105. [CrossRef]
18. Ly, A.; Asset, T.; Atanassov, P. Integrating nanostructured Pt-based electrocatalysts in proton exchange membrane fuel cells. *J. Power Sources* **2020**, *478*, 9. [CrossRef]
19. Devi, N.; Sahoo, S.; Kumar, R.; Singh, R.K. A review of the microwave-assisted synthesis of carbon nanomaterials, metal oxides/hydroxides and their composites for energy storage applications. *Nanoscale* **2021**, *13*, 11679–11711. [CrossRef]
20. Chen, Y.F.; Wang, M.Y.; Zhang, J.T.; Tu, J.G.; Ge, J.B.; Jiao, S.Q. Green and sustainable molten salt electrochemistry for the conversion of secondary carbon pollutants to advanced carbon materials. *J. Mater. Chem. A* **2021**, *9*, 14119–14146. [CrossRef]
21. Liu, X.; Dai, L.M. Carbon-based metal-free catalysts. *Nat. Rev. Mater.* **2016**, *1*, 12. [CrossRef]
22. Liu, M.J.; Lee, J.Y.; Yang, T.C.; Zheng, F.Y.; Zhao, J.; Yang, C.M.; Lee, L.Y.S. Synergies of Fe Single Atoms and Clusters on N-Doped Carbon Electrocatalyst for pH-Universal Oxygen Reduction. *Small Methods* **2021**, *5*, 10. [CrossRef]
23. Xiong, W.; Wang, Z.N.; He, S.L.; Hao, F.; Yang, Y.Z.; Lv, Y.; Zhang, W.; Liu, P.; Luo, H. Nitrogen-doped carbon nanotubes as a highly active metal-free catalyst for nitrobenzene hydrogenation. *Appl. Catal. B Environ.* **2020**, *260*, 118105. [CrossRef]
24. Yi, S.J.; Qin, X.P.; Liang, C.H.; Li, J.S.; Rajagopalan, R.; Zhang, Z.J.; Song, J.; Tang, Y.; Cheng, F.; Wang, H.; et al. Insights into $KMnO_4$ etched N-rich carbon nanotubes as advanced electrocatalysts for Zn-air batteries. *Appl. Catal. B Environ.* **2020**, *264*, 118537. [CrossRef]
25. Chen, Z.; Higgins, D.; Tao, H.S.; Hsu, R.S.; Chen, Z.W. Highly Active Nitrogen-Doped Carbon Nanotubes for Oxygen Reduction Reaction in Fuel Cell Applications. *J. Phys. Chem. C* **2009**, *113*, 21008–21013. [CrossRef]
26. Kim, H.W.; Bukas, V.J.; Park, H.; Park, S.; Diederichsen, K.M.; Lim, J.; Cho, Y.H.; Kim, J.; Kim, W.; Han, T.H.; et al. Mechanisms of Two-Electron and Four-Electron Electrochemical Oxygen Reduction Reactions at Nitrogen-Doped Reduced Graphene Oxide. *ACS Catal.* **2019**, *10*, 852. [CrossRef]
27. Gasnier, A.; Luguet, M.; Pereira, A.G.; Troiani, H.; Zampieri, G.; Gennari, F.C. Entanglement of N-doped graphene in resorcinol-formaldehyde: Effect over nanoconfined $LiBH_4$ for hydrogen storage. *Carbon* **2019**, *147*, 284–294. [CrossRef]
28. Wu, X.X.; Chen, K.Q.; Lin, Z.P.; Zhang, Y.M.; Meng, H. Nitrogen doped graphitic carbon from biomass as non noble metal catalyst for oxygen reduction reaction. *Mater. Today Energy* **2019**, *13*, 100–108. [CrossRef]
29. Wang, S.H.; Yan, X.; Wu, K.H.; Chen, X.M.; Feng, J.M.; Lu, P.Y.; Feng, H.; Cheng, H.-M.; Liang, J.; Dou, S.X. A hierarchical porous Fe-N impregnated carbon-graphene hybrid for high-performance oxygen reduction reaction. *Carbon* **2019**, *144*, 798–804. [CrossRef]
30. Liu, Y.M.; Xu, Q.; Fan, X.F.; Quan, X.; Su, Y.; Chen, S.; Yu, H.; Cai, Z. Electrochemical reduction of N-2 to ammonia on Co single atom embedded N-doped porous carbon under ambient conditions. *J. Mater. Chem. A* **2019**, *7*, 26358–26363. [CrossRef]
31. Panomsuwan, G.; Saito, N.; Ishizaki, T. Nitrogen-Doped Carbon Nanoparticle-Carbon Nanofiber Composite as an Efficient Metal-Free Cathode Catalyst for Oxygen Reduction Reaction. *ACS Appl. Mater. Int.* **2016**, *8*, 6962–6971. [CrossRef] [PubMed]
32. Shen, H.J.; Gracia-Espino, E.; Ma, J.Y.; Zang, K.T.; Luo, J.; Wang, L.; Gao, S.; Mamat, X.; Hu, G.; Wagberg, T.; et al. Synergistic Effects between Atomically Dispersed Fe-N-C and C-S-C for the Oxygen Reduction Reaction in Acidic Media. *Angew. Chem. Int. Edit.* **2017**, *56*, 13800–13804. [CrossRef] [PubMed]
33. Lu, G.P.; Shan, H.B.; Lin, Y.M.; Zhang, K.; Zhou, B.J.; Zhong, Q.; Wang, P. A Fe single atom on N,S-doped carbon catalyst for performing N-alkylation of aromatic amines under solvent-free conditions. *J. Mater. Chem. A* **2021**, *9*, 25128–25135. [CrossRef]
34. Chen, Z.P.; Narita, A.; Mullen, K. Graphene Nanoribbons: On-Surface Synthesis and Integration into Electronic Devices. *Adv. Mater.* **2020**, *32*, 26. [CrossRef] [PubMed]
35. Wang, H.M.; Wang, H.S.; Ma, C.X.; Chen, L.X.; Jiang, C.X.; Chen, C.; Xie, X.; Li, A.-P.; Wang, X. Graphene nanoribbons for quantum electronics. *Nat. Rev. Phys.* **2021**, *3*, 791–802. [CrossRef]
36. Xiang, T.; Wu, Z.R.; Sun, Z.T.; Cheng, C.; Wang, W.W.; Liu, Z.Z.; Yang, J.; Li, B. The Synergistic Effect of Carbon Edges and Dopants Towards Efficient Oxygen Reduction Reaction. *J. Colloid Inter. Sci.* **2021**, *610*, 486–494. [CrossRef]

37. Yazdi, A.Z.; Roberts, E.P.L.; Sundararaj, U. Nitrogen/sulfur co-doped helical graphene nanoribbons for efficient oxygen reduction in alkaline and acidic electrolytes. *Carbon* **2016**, *100*, 99–108. [CrossRef]
38. Chen, Y.Y.; Xu, C.X.; Hou, Z.H.; Zhou, M.J.; He, B.H.; Wang, W.; Ren, W.; Liu, Y.; Chen, L.; Xu, W. 3D N, S-co-doped carbon nanotubes/graphene/MnS ternary hybrid derived from Hummers' method for highly efficient oxygen reduction reaction. *Mater. Today Energy* **2020**, *16*, 9100402. [CrossRef]
39. Gong, Y.J.; Fei, H.L.; Zou, X.L.; Zhou, W.; Yang, S.B.; Ye, G.L.; Liu, Z.; Peng, Z.; Lou, J.; Vajtai, R.; et al. Boron- and Nitrogen-Substituted Graphene Nanoribbons as Efficient Catalysts for Oxygen Reduction Reaction. *Chem. Mater.* **2015**, *27*, 1181–1186. [CrossRef]
40. Kosynkin, D.V.; Higginbotham, A.L.; Sinitskii, A.; Lomeda, J.R.; Dimiev, A.; Price, B.K.; Tour, J. Longitudinal unzipping of carbon nanotubes to form graphene nanoribbons. *Nature* **2009**, *458*, 872–876. [CrossRef]
41. Zhang, C.K.; Lin, W.Y.; Zhao, Z.J.; Zhuang, P.P.; Zhan, L.J.; Zhou, Y.H.; Cai, W. CVD synthesis of nitrogen-doped graphene using urea. *Sci. China-Phys. Mech. Astron.* **2015**, *58*, 107801. [CrossRef]
42. Zhang, Z.W.; Jiang, X.M.; Hu, J.H.; Yue, C.J.; Zhang, J.T. Controlled Synthesis of Mesoporous Nitrogen-Doped Carbon Supported Ni-Mo Sulfides for Hydrodesulfurization of Dibenzenethiophene. *Catal. Lett.* **2017**, *147*, 2515–2522. [CrossRef]
43. Li, W.; Min, C.G.; Tan, F.; Li, Z.P.; Zhang, B.S.; Si, R.; Xu, M.; Kiu, W.; Zhou, L.; Yang, X.K. Bottom-Up Construction of Active Sites in a Cu-N_4-C Catalyst for Highly Efficient Oxygen Reduction Reaction. *ACS Nano* **2019**, *13*, 3177–3187. [CrossRef]
44. Abdelkader-Fernandez, V.K.; Domingo-Garcia, M.; Lopez-Garzon, F.J.; Fernandes, D.M.; Freire, C.; de la Torre, M.D.L.; Melguizo, M.; Godino-Salido, M.L.; Pérez-Mendoza, M. Expanding graphene properties by a simple S-doping methodology based on cold CS_2 plasma. *Carbon* **2019**, *144*, 269–279. [CrossRef]
45. Poh, H.L.; Simek, P.; Sofer, Z.; Pumera, M. Sulfur-Doped Graphene via Thermal Exfoliation of Graphite Oxide in H_2S, SO_2, or CS_2 Gas. *ACS Nano* **2013**, *7*, 5262–5272. [CrossRef] [PubMed]
46. Zhao, H.; Weng, C.C.; Ren, J.T.; Ge, L.; Liu, Y.P.; Yuan, Z.Y. Phosphonate-derived nitrogen-doped cobalt phosphate/carbon nanotube hybrids as highly active oxygen reduction reaction electrocatalysts. *Chin. J. Catal.* **2020**, *41*, 259–267. [CrossRef]
47. Li, R.R.; Liu, F.; Zhang, Y.H.; Guo, M.M.; Liu, D. Nitrogen, Sulfur Co-Doped Hierarchically Porous Carbon as a Metal-Free Electrocatalyst for Oxygen Reduction and Carbon Dioxide Reduction Reaction. *ACS Appl. Mater. Inter.* **2020**, *12*, 44578–44587. [CrossRef]
48. Liu, J.T.; Wei, L.L.; Wang, H.Q.; Lan, G.J.; Yang, H.J.; Shen, J.Q. In-situ synthesis of heteroatom co-doped mesoporous dominated carbons as efficient electrocatalysts for oxygen reduction reaction. *Electrochim. Acta* **2020**, *364*, 11137335. [CrossRef]
49. Patil, I.M.; Reddy, V.; Lokanathan, M.; Kakade, B. Nitrogen and Sulphur co-doped multiwalled carbon nanotubes as an efficient electrocatalyst for improved oxygen electroreduction. *Appl. Surf. Sci.* **2018**, *449*, 697–704. [CrossRef]
50. Li, Y.Q.; Xu, H.B.; Huang, H.Y.; Gao, L.G.; Zhao, Y.Y.; Ma, T.L. Facile synthesis of N, S co-doped porous carbons from a dual-ligand metal organic framework for high performance oxygen reduction reaction catalysts. *Electrochim. Acta* **2017**, *254*, 148–154. [CrossRef]
51. Zhang, H.H.; Liu, X.Q.; He, G.L.; Zhang, X.X.; Bao, S.J.; Hu, W.H. Bioinspired synthesis of nitrogen/sulfur co-doped graphene as an efficient electrocatalyst for oxygen reduction reaction. *J. Power Sources* **2015**, *279*, 252–258. [CrossRef]
52. Huang, B.B.; Hu, X.; Liu, Y.C.; Qi, W.; Xie, Z.L. Biomolecule-derived N/S co-doped CNT-graphene hybrids exhibiting excellent electrochemical activities. *J. Power Sources* **2019**, *413*, 408–417. [CrossRef]
53. Wang, S.T.; Liu, Y.; Liu, X.P.; Chen, Y.; Zhao, Y.L.; Gao, S.Y. Fabricating N, S Co-Doped Hierarchical Macro-Meso-Micro Carbon Materials as pH-Universal ORR Electrocatalysts. *ChemistrySelect* **2022**, *7*, e202200044. [CrossRef]
54. Zhang, X.R.; Wang, Y.Q.; Du, Y.H.; Qing, M.; Yu, F.; Tian, Z.Q.; Shen, P.K. Highly active N,S co-doped hierarchical porous carbon nanospheres from green and template-free method for super capacitors and oxygen reduction reaction. *Electrochim. Acta* **2019**, *318*, 272–280. [CrossRef]
55. Nong, J.; Zhu, M.; He, K.; Zhu, A.S.; Xie, P.; Rong, M.Z.; Zhang, M.Q. N/S co-doped 3D carbon framework prepared by a facile morphology-controlled solid-state pyrolysis method for oxygen reduction reaction in both acidic and alkaline media. *J. Energy Chem.* **2019**, *34*, 220–226. [CrossRef]

Article

Preparation and Study of a Simple Three-Matrix Solid Electrolyte Membrane in Air

Xinghua Liang [1,*], Xingtao Jiang [1], Linxiao Lan [1,*], Shuaibo Zeng [2,*], Meihong Huang [3] and Dongxue Huang [1]

1. Guangxi Key Laboratory of Automobile Components and Vehicle Technology, Guangxi University of Science and Technology, Liuzhou 545006, China
2. China School of Automotive and Transportation Engineering, Guangdong Polytechnic Normal University, Guangzhou 510632, China
3. Guangdong Polytechnic of Industry and Commerce, Guangzhou 510550, China
* Correspondence: lxh18589873093@163.com (X.L.); 15506749886@163.com (L.L.); zengshuaibo@gpnu.edu.cn (S.Z.); Tel.: +86-18589873093 (X.L.); +86-18107785376 (L.L.); +86-18819264279 (S.Z.)

Abstract: Solid-state lithium batteries have attracted much attention due to their special properties of high safety and high energy density. Among them, the polymer electrolyte membrane with high ionic conductivity and a wide electrochemical window is a key part to achieve stable cycling of solid-state batteries. However, the low ionic conductivity and the high interfacial resistance limit its practical application. This work deals with the preparation of a composite solid electrolyte with high mechanical flexibility and non-flammability. Firstly, the crystallinity of the polymer is reduced, and the fluidity of Li$^+$ between the polymer segments is improved by tertiary polymer polymerization. Then, lithium salt is added to form a solpolymer solution to provide Li$^+$ and anion and then an inorganic solid electrolyte is added. As a result, the composite solid electrolyte has a Li$^+$ conductivity (3.18 × 10^{-4} mS cm^{-1}). The (LiNi$_{0.5}$Mn$_{1.5}$O$_4$)LNMO/SPLL (PES-PVC-PVDF-LiBF$_4$-LAZTP)/Li battery has a capacity retention rate of 98.4% after 100 cycles, which is much higher than that without inorganic oxides. This research provides an important reference for developing all-solid-state batteries in the greenhouse.

Keywords: composite solid electrolyte; lithium ion conductivity; capacity retention; flame retardancy

Citation: Liang, X.; Jiang, X.; Lan, L.; Zeng, S.; Huang, M.; Huang, D. Preparation and Study of a Simple Three-Matrix Solid Electrolyte Membrane in Air. *Nanomaterials* **2022**, *12*, 3069. https://doi.org/10.3390/nano12173069

Academic Editors: Christian M. Julien and Henrich Frielinghaus

Received: 3 July 2022
Accepted: 25 August 2022
Published: 3 September 2022

Publisher's Note: MDPI stays neutral with regard to jurisdictional claims in published maps and institutional affiliations.

Copyright: © 2022 by the authors. Licensee MDPI, Basel, Switzerland. This article is an open access article distributed under the terms and conditions of the Creative Commons Attribution (CC BY) license (https://creativecommons.org/licenses/by/4.0/).

1. Introduction

All-solid-state batteries have received extensive attention due to the advantages of high safety and high energy density in recent years. The use of electrolyte membranes replaces traditional electrolytes and separators, acting as Li$^+$ transport bridges in batteries [1,2]. The quality of the electrolyte membrane determines the cycle performance of the battery. Solid electrolytes for lithium battery development can generally be divided into two categories: ceramic-based solid-state electrolytes and polymer-based solid-state electrolytes [3,4]. In ceramic solid electrolytes, cation groups and metal cations form a unit cell skeleton, which provides a channel for the transmission of lithium ions. Each unit cell is connected to each other to form a network structure, which enables lithium ions to perform vacancy migration and interstitial migration inside to complete the transport of lithium ions [5]. Inorganic polymer electrolytes have thin films that can be prepared into thinner shapes—which can reduce interfacial impedance with positive and negative electrodes and can have good mechanical ductility—that can adapt to changes in battery conditions during charging and discharging. These materials have been studied for many years, but each has its own advantages and disadvantages [6,7]. The integration of material advantages is an important aspect of innovation. Xiang et al. prepared a Li/LiPON/LiCoO$_2$ all-solid-state battery with a capacity loss of less than 2% after 4000 cycles. Subsequently, organic polymer solid electrolytes, inorganic solid electrolytes, and organic/inorganic composite solid electrolytes appeared in succession [8]. Polyethylene oxide (PEO) based polymer solid electrolyte has attracted much attention [9]. In the PEO-based polymer solid electrolyte, the conductive Li$^+$ in the amorphous region coordinates with the ether oxide (EO) on the PEO segment and realizes Li$^+$ migration through the complexation and decomplexation process of the lithium-oxygen bond (Li-O) movement on the PEO chain segment [10–12]. However, PEO has higher crystallinity and less lithium ion migration at room temperature, resulting in extremely low ion conductivity (only 10^{-8}–10^{-7} S cm^{-1}) [13]. The inorganic solid electrolyte has distinctive merits, including high chemical stability, environmentally friendly, high safety, high conductivity, wide electrochemical window, and good thermal stability. Li et al. [14], prepared a high-performance three-dimensional cross-linked electrolyte based on polyvinylidene fluoride (PVDF) and polyethylene oxide (PEO), which has good performance at room temperature, but the cycle performance needs to be improved. However, shortcomings, such as preparation and storage requirements and flexibility, restrict their development [15–17].

The organic/inorganic composite solid electrolyte was developed to have the conductivity of inorganic solid electrolytes and the flexibility of PEO-based polymer electrolytes, opening up new development prospects for all-solid-state lithium-ion batteries with high safety and a long cycle life [18,19]. However, its electrochemical performance is still worse than traditional liquid batteries [20–22]. Therefore, preparing a greenhouse with high ion channels and solving interface problems has high prospects.

Here, we propose the strategy of combining organic electrolytes and inorganic electrolytes. Firstly, the crystallinity of the polymer is reduced, and the fluidity of Li$^+$ between the polymer segments is improved by tertiary polymer polymerization. Then, lithium salt is added to form a sol polymer solution to provide Li$^+$ and anion and then an inorganic solid electrolyte is added. By adding inorganic fillers, the regular arrangement of polymer segments can be destroyed and made in an amorphous state to increase the amorphous region conducive to lithium ion transport. In addition, inorganic solid electrolytes have high greenhouse conductivity and a wide electrochemical window. Inorganic fillers can also reduce the activation energy of lithium ion migration to form channels conducive to lithium ion migration.

2. Materials and Methods

2.1. Preparation of LATP Powder

The solid phase method uses lithium carbonate (Li$_2$CO$_3$), aluminum oxide (Al$_2$O$_3$), titanium dioxide (TiO$_2$), and ammonium dihydrogen phosphate (NH$_3$H$_2$PO$_4$) zinc oxide (ZnO) as precursors. Each material is weighed according to Li$_{1.3}$Al$_{0.3}$Ti$_{1.7}$(PO$_4$)$_3$(LATP) and Li$_{1.3}$Al$_{0.1}$Zn$_{0.1}$Ti$_{1.8}$(PO$_4$)$_3$(LAZTP). Calculate the molar ratio to prepare 40 g of LATP and LAZTP materials, respectively, for use in subsequent experimental tests. After mixing uniformly, ball mill for 8 h, use ethanol as the dispersant, and the speed of the ball mill is 280 rpm. After ball milling, the powder was dried in a drying oven at 80 °C for 10 h. After the powder was dry, it was processed by grinding. The powder was annealed and calcined in an atmosphere furnace at 950 °C for 4 h to obtain two white LATP and LAZTP precursor powders.

2.2. Preparation of Electrolyte Sheet

After the powder is ground through the grinding table, the two powders are pressed under a pressure of 18 MP for 5 min and pressed into a disc with a diameter of 16 mm and a thickness of 1 mm. The layer powder is placed in an atmosphere furnace, calcined at 900 °C for 4 h at a heating rate of 5 °C/min, and naturally cooled to room temperature to obtain high-temperature sintered LATP and LAZTP solid electrolyte sheets, which are then polished to the required thickness by sandpaper. All LATP and LAZTP electrolyte sheet sintered samples were ultrasonically cleaned in absolute ethanol for 10 min and then dried for 2 h. Conductive silver paste was applied to both sides and steel sheets were pasted and placed in a drying oven for 3 h for testing.

2.3. Preparation of SPLL Composite Electrolytes

Preparation of three-matrix solid electrolyte SPLL (PES-PVC-PVDF-LiBF$_4$-LAZTP). First, add 30 mL of DMF solution in a 100 m beaker, stir it electrically at a temperature of 60 °C, then add 2 g of PES until the solution becomes a clear sol-like solution and then add 1g of PVA. When the mixture is stirred until there are no air bubbles, add 5% PVDF to form a copolymerized base, sonicate for 20 min to mix the three completely, then add LiBF$_4$ (add with PES es: Li=8:1), and stir for 4 h. Then, add 10% LAZTP, ultrasonic for 30 min, stir for 10 h, pour it into a polytetrafluoroethylene container, dry in a blast drying oven at 60 °C for 12 h, then remove and cut it to a size of 16 mm, and put it in a glove box for later use.

2.4. Preparation of Solid-State LFP/SPLL/Li Cell

The CR2032 button cell is assembled in an argon box using the above composite SSE in an argon-filled glove box. The LNMO cathode was prepared by mixing 80 wt.% of commercial LNMO (Shanghai Ales Co., Ltd.), 10 wt.% conductive carbon black, and 10 wt.% PVDF dissolved in NMP solvent. After thorough stirring, the cathode slurry was evenly cast onto the aluminum foil. Subsequently, the aluminum foil was dried under vacuum at 110 °C and then cut into circular electrodes with a diameter of 16 mm. In addition, lithium metal is used as the anode of assembled button batteries. The electrochemical performance of all assembled batteries is measured in the voltage range of 3.5–5 V.

2.5. Physical Characterizations

X-ray diffraction (XRD, D8-Advabce, Bruker, Frankfurt, Germany, in the range of 10–90 °) and Raman spectroscopy were used to analyze the phase structure of the positive electrode material. The surface structure of the material is observed through a field emission scanning electron microscope (SEM, Sigma04-55, ZEISS, HORIBA, Longjumeau, France). The electrochemical window test is carried out by linear sweep voltammetry (LSV) at a scanning rate of 0.1 mV·s^{-1}. A stainless steel sheet (SS) is used as the working electrode, and a lithium sheet is used as the counter electrode. The LAGP glass and GC structural studies were carried out using 13 C Solid State Nuclear Magnetic Resonance (NMR) using BRUKER 700 MHz HD spectrometer on a 2.5 mm Trigamma probe at a spinning frequency of 32 kHz. The SPLL and SPL anode surface was characterized by X-ray photoelectron spectroscopy (XPS, Thermo Escalab 210 system, Dreieich, Germany).

2.6. Electrochemical Measurements

The ionic conductivity of the electrolyte can be obtained by the following equation:

$$\sigma = \frac{L}{(R \times S)}$$

where σ represents the ionic conductivity, L is the thickness of the sample, S represents the contact area between the electrolyte and the test electrode (SS), and R is the resistance measured by impedance spectroscopy.

The lithium ion transference number (t_{Li^+}) is an important parameter for evaluating polymer electrolyte membranes. A higher number shows the lithium ion transfer strength of the membrane, and the transfer is more important during the cycle. Usually, the test is performed by the timing method on a lithium symmetrical battery (such as Li/SPLL/Li) at a voltage of 1 mV for 4000 s, and the value is calculated by the following equation.

$$t_{Li^+} = \frac{I_s(\Delta V - I_0 R_0)}{I_0(\Delta V - I_s R_s)}$$

where I_0 and Is are the current values at the beginning and after the DC polarization is stabilized, and R_0 and R_s are the impedance values before and after DC polarization, respectively, and ΔV is the voltage value acting on both ends of the battery [23].

The electrochemical stability window of the electrolyte membrane is obtained by linear scanning voltammetry. The lithium sheet is used as the counter electrode and the reference electrode, the stainless steel sheet is used as the working electrode (SS), and the electrolyte membrane is in the middle (Li/SPLL/SS). The test range is 0–6 V, and the sweep speed is 0.5 mV s^{-1}.

The cycle rate test is a standard for evaluating the quality of a battery. The assembled battery is tested for the cycle and rate performance between the 3.5–5 V electrochemical window, using the Xinwei tester and the DH7000 (Shenzhen, China) tester for impedance and CV testing.

3. Results

It can be seen from Figure 1b that the diffraction peaks of the two samples are consistent with the standard card (35-0754) of the NASICON (Na$^+$ superionic conductor) structure (R-3c). The three prominent diffraction peaks are sharp, indicating that the synthesized samples are of high crystallinity. Negligible spurious peaks indicate some purity of the LATP and LAZTP samples. The intensity of the diffraction peak at 1 is higher than at 2 in Figure 1b, meaning the successful incorporation of Zn element into the lattice system, and it causes vacancies or defects in the lattice in the LAZTP sample. The SEM-EDS images in Figure S1 further confirmed that the Zn element is successfully doped with a small amount in the LAZTP sample.

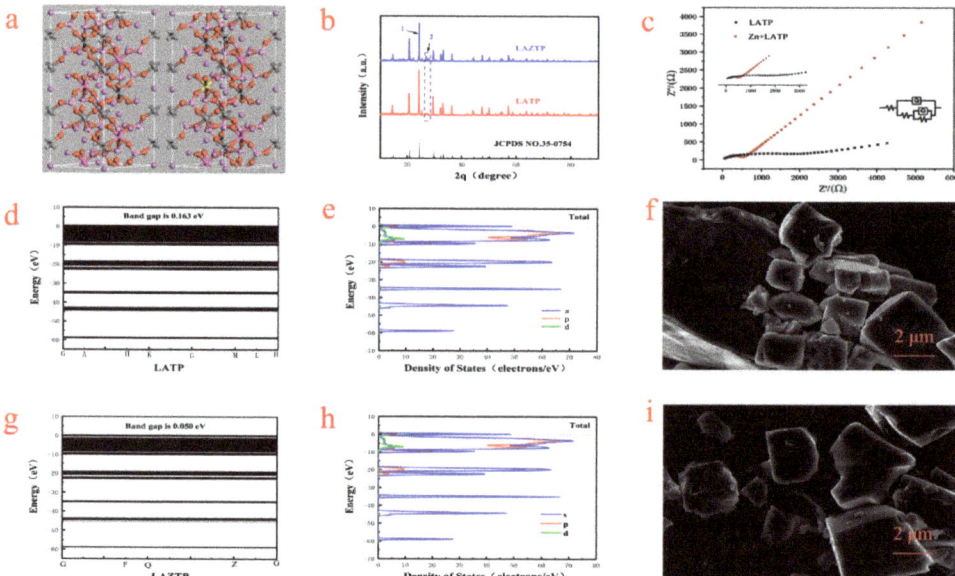

Figure 1. (**a**) LATP and LAZTP unit cell model. (**b**) XRD patterns of LATP and LAZTP. (**c**) Nyquist patterns of LATP and LAZTP. (**d**,**e**) Band structure and total density of states diagrams of LATP. (**g**,**h**) Band structure and total density and states diagrams of LAZTP. (**f**) SEM image of LAZTP. (**i**) SEM image of LATP.

As shown in Figure 1c, the conductivity of LAZTP is much higher than LATP (1.69 × 10^{-3} S cm^{-1} vs. 4.02 × 10^{-4} S cm^{-1}). Theoretical calculations are used to confirm the further conductivity of LAZTP. The calculation results show that the energy bandgap of LATP is 0.165 eV (Figure 1d), while the energy bandgap of LAZTP (Figure 1g) is only 0.05 eV. Such a small bandgap is beneficial for electrons to transition from the valence band into the conduction band and improve the electronic conductivity of LAZTP.

The density of states of the two samples is shown in Figure 1e,h. LAZTP is metalized after doping. The energy transition decreases and the conductivity increases, which is consistent with the experimental results, indicating that the doping of Zn element greatly improves the conductivity of the material.

As shown in Figure 1f,i, the particle distribution of the samples is relatively uniform with low agglomeration and clear boundary. Compared with LATP, the LAZTP has a smaller grain size and a smoother surface, which is conducive to the migration of Li$^+$.

The interaction mechanism between lithium ions and electrolyte skeleton was further revealed by XRD and Raman spectroscopy. As shown in Figure 2a, the XRD patterns of (Polyethersulfone)PES, (Polyvinyl chloride)PVC, and PVDF are different, and their characteristic peaks are displayed in the SP (PVDF-PES-PVC) polymer. The diffraction peak at ≈20° is broad, indicating that amorphous regions are formed in the SP polymer, conducive to the transmission of Li$^+$. No impurity peaks can be observed in the XRD patterns, suggesting the high purity of the composite electrolyte. Figure 2b shows the Raman spectra of the polymers in the range of 2000–3000 cm^{-1}. The characteristic peaks of the SP polymer are only the superposition of the peaks of the three polymers, indicating that the SP polymer is only copolymerized of the three polymers, which is in good agreement with XRD results. Figure 2c shows the XRD patterns of SP, SPL, and SPLL. Compared with SP, the peaks at 2θ ≈ 20° and ≈33° of SPLL are significantly weakened, indicating that with the addition of LiBF$_4$, the heterogeneous salt doping process changes the locally ordered polymer-ion assembly, filling the defective LiBF$_4$ nanocrystalline grains into the intercrystalline network, changing the lattice structure, and improving the conductivity of the copolymer substrate. The peak intensities of SPLL are different from SPL, which is due to the addition of LAZTP in SPLL. Figure 2d shows the Raman spectra of SP, SPL(PVDF-PES-PVC-LiBF$_4$), and SPLL, which reflects that the structure of SP changes with the addition of LiBF$_4$, and the structure of SPL changes with the addition of LAZTP.

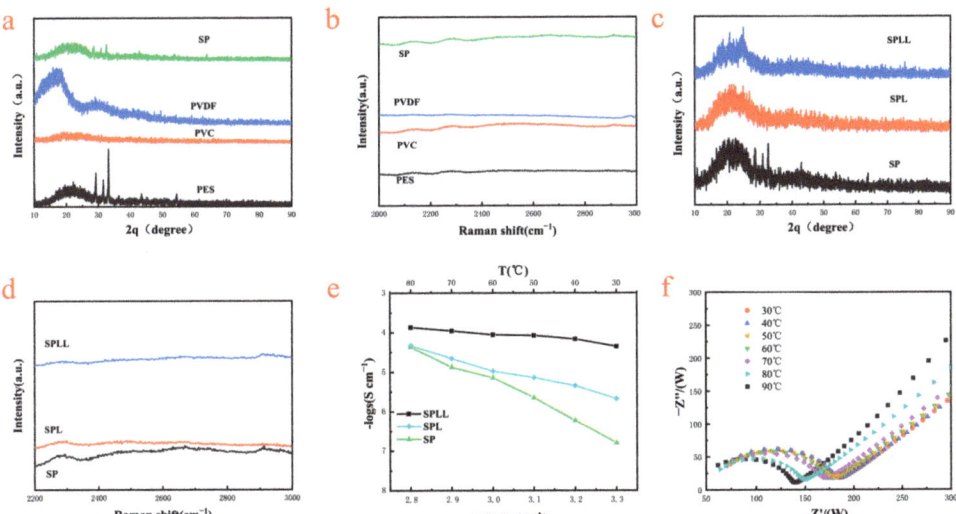

Figure 2. (**a,c**) XRD patterns of different polymer. (**b,d**) Raman patterns of different polymer. (**e**) Conductivity of polymer electrolyte at different temperatures. (**f**) SPLL impedance diagram at different temperatures.

Electrochemical impedance spectroscopy (EIS) was utilized to explore the conductivity of the composite electrolytes [24]. As shown in Figure 2e, the conductivity of SPL polymer electrolytes increases as the temperature rises. However, for SPLL polymer electrolyte, the conductivity did not dramatically change, indicating that the electrolyte membrane is adapted to a broad temperature environment and is more stable than SPL. Figure 2f shows that the interface impedance decreases as the temperature changes for SPLL. Still, the phase does not change, indicating that the prepared electrolyte membrane can maintain stable cycling at different temperatures.

Figure 3 shows the preparation flow chart of the polymer electrolyte. PES, PVC, and PVDF polymers constitute a polyelectrolyte and then lithium salt LiBF$_4$ is added to form a stable polymer. LAZTP is used as an additive to improve lithium ion transport channels, and inorganic metal oxides are added to increase the amorphous area of the polymer. After mixing uniformly, the composite is poured on a polytetrafluoroethylene template. Subsequently, the composite is cut to the size of a steel sheet after drying. Afterward, the product is assembled into a quasi-solid battery. Then, electrolyte was added to the battery to improve the humidity of the composite interface. The battery was taken out from the glove box and put in a dry box to stabilize polymerization. Finally, the battery was tested under air conditions.

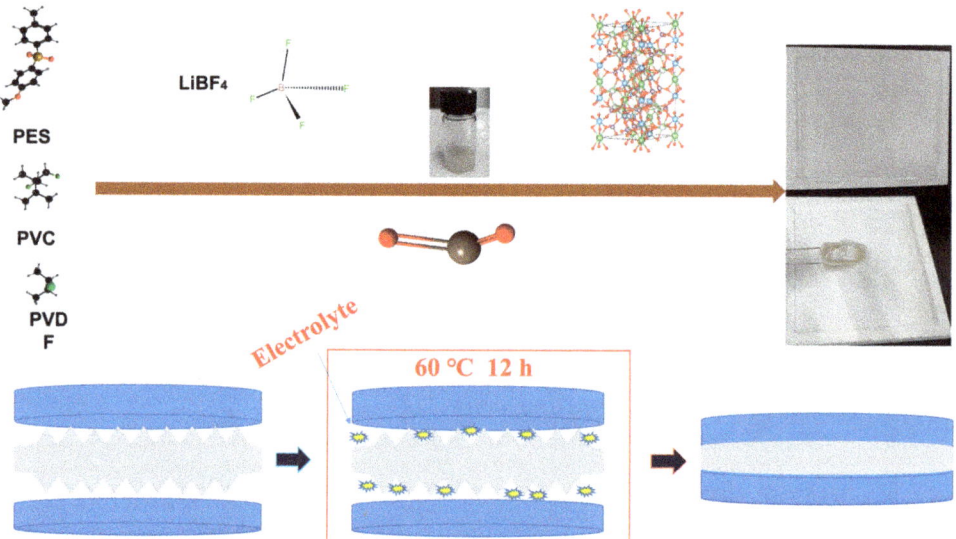

Figure 3. Schematic diagram of preparation of PES-PVC-PVD-LiBF$_4$-LATP polymer electrolyte and assembly of solid-state LiNi$_{0.5}$Mn$_{1.5}$O$_4$/SPLL/Li lithium battery.

Figure 4a shows the assembly sequence of the prepared electrolytes in the solid-state battery. The lithium ion migration number of SPL is calculated by chronoamperometry combined with AC impedance spectroscopy. At the initial stage of polarization, both Li$^+$ and LiBF$_4$ provide current. At the end of polarization, only Li$^+$ are transferred from one lithium electrode to another, and the current reaches a constant value. In SPL polymer electrolytes, the mobility of Li$^+$ is lower than that of the corresponding anions due to the high complexation of Li$^+$ with the three polymer substrates. As shown in Figure 4b,e, the Li+ transference number of SPL is 0.32, while the lithium ion migration number of SPLL is 0.5, which is significantly higher than that of SPL. It was believed that the high lithium ions mobility of SPLL is due to the binding effect of LAZTP on anions.

The electrochemical stability window of the composite electrolytes was carried out by linear sweep voltammetry. As shown in Figure 4d, when the voltage rises to 5.1 V, the SPL is oxidized and decomposed, while the electrochemical stability window of SPLL can increase to more than 5.2 V, indicating that adding LAZTP to the polymer can increase the electrochemical stability window of polymer electrolyte. The wide electrochemical stability window of the composite electrolyte system is due to the strong interaction between small molecules and the trapping effect of a large number of micropores [25]. Therefore, SPLL can match the high voltage cathode of high-energy-density lithium-ion batteries.

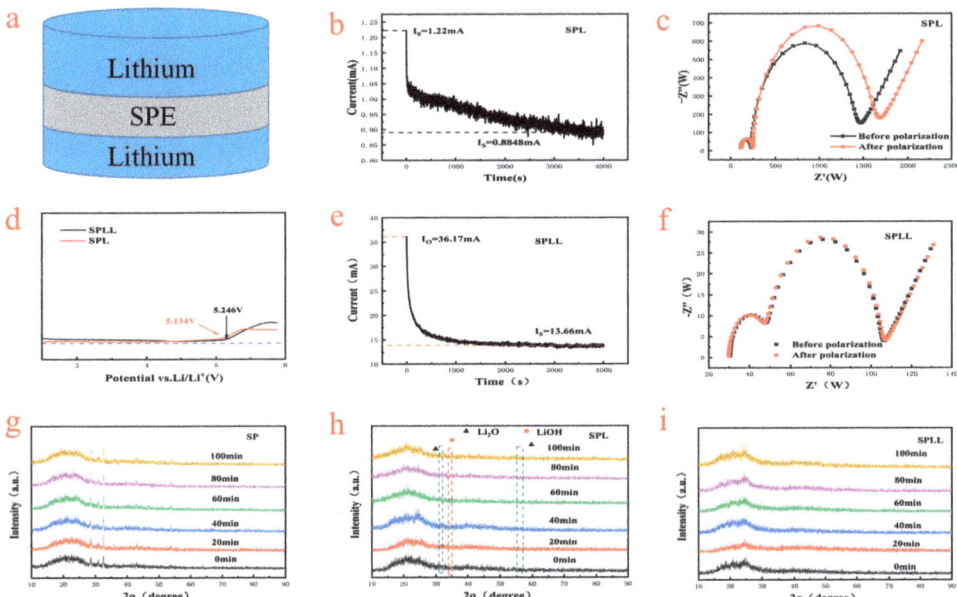

Figure 4. (**a**) Polymer electrolyte membrane assembly test. (**b**,**c**) SPL timing current and impedance before and after timing. (**d**) SPL and SPLL linear voltammetry test. (**e**,**f**) SPLL timing current and impedance before and after timing. (**g–i**) XRD patterns after exposing to air (40% relative humidity).

The structural stability of the composite electrolytes was evaluated in the air. The indoor relative humidity is 40%. XRD was used to analyze the change of crystal structure. As shown in Figure 4g, new peaks at 2θ ≈ 36° and ≈55° were produced after 40 min, indicating that slight water decomposition occurred in the SP polymer. As shown in Figure 4h, new peaks appeared after 40 min, corresponding to Li_2O and LiOH, indicating that the crystal structure of SPL polymer electrolyte changed. As shown in Figure 4i, the peak strength of SPLL polymer does not change significantly, indicating that the addition of LAZTP improves the stability of the composite electrolyte.

Figure 5 presents the chemical composition of the polymer electrolyte by X-ray photoelectron spectroscopy (XPS). Figure 5a represents the C element of SPL and SPLL. The peaks of SPLL at the low value are higher than SPL, indicating that the internal C structure of the polymer has changed after adding LAZTP and SiO_2, resulting in highly stable under air circulation. As shown in Figure 5b, for SPLL, two peaks at 402.3 eV and 399.8 eV in the pristine N1s spectrum correspond to the N in $[bmim]^+$ cation (Ncation) and $[Tf_2N]^-$ anion (Nanion). Compared with SPL, the N element peaks of SPLL increased significantly, indicating that the number of anion groups increased, which accelerated the transport of lithium ions. As shown in Figure 5c, the intensity of the Li^+ peak is weak, suggesting an increase of the amorphous region area in the polymer, thereby increasing the Li^+ mobility of the polymer.

To prove that the addition of LAZTP and SiO_2 reduces the crystallinity of polymer electrolyte and provides a lithium ion migration channel, we analyzed the local chemical environment of lithium ion by ^{13}C solid-state nuclear magnetic resonance (NMR). As shown in Figure 5d, the peaks of SPL are scattered, and there are many impurity peaks. In contrast, for SPLL, the peaks near 50 ppm and 140 ppm gradually become sharp, indicating that the addition of LAZTP and SiO_2 reduces the crystallinity of SPLL polymer and enhances the conductivity of the material, which is consistent with the EIS results [26].

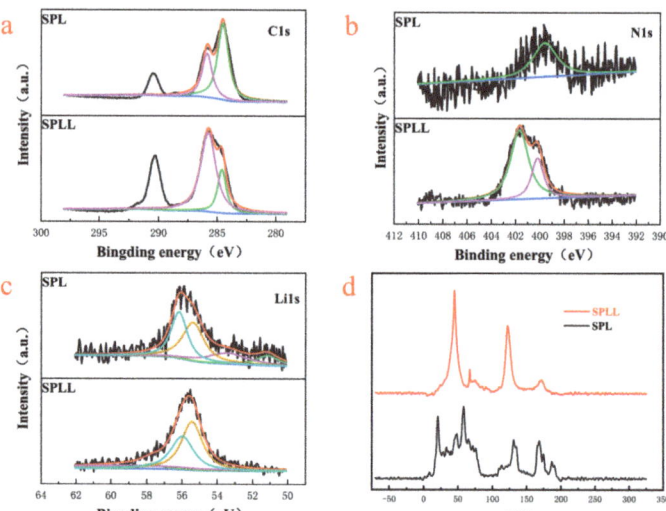

Figure 5. (**a**) Li1s, (**b**) N1s, and (**c**) Li1s XPS spectra of the SPL and SPLL. (**d**) Solid state NMR of the SPL and SPLL.

Constant current charge and discharge (GCD) were used to evaluate the influence of the addition of LAZTP and SiO$_2$ on the electrochemical stability of the polymer electrolytes under the current density of 0.05, 0.1 and 0.2 mA cm^{-2}. Figure 6c shows that the Li/SPLL/Li battery exhibits a stable Li electroplating/stripping under the corresponding current density. Moreover, it displays excellent stability even after 1000 cycles of long-term plating/stripping, demonstrating a good rate performance. On the contrary, the SPL polymer appears flocculent at 0.2 mA cm^{-2} during the plating/stripping process (Figure S2) [27]. Figure 6b shows that the interface impedance increases after the cycling test, indicating that a stable SEI film is formed at the interface, reducing the loss at the interface and improving the Li$^+$ transmission rate. According to literature reports, adding inorganic metal oxide LAZTP [28] and non-metal oxide SiO$_2$ [29] can increase the conductivity and amorphous region of the polymer electrolyte and improve the lithium ion migration.

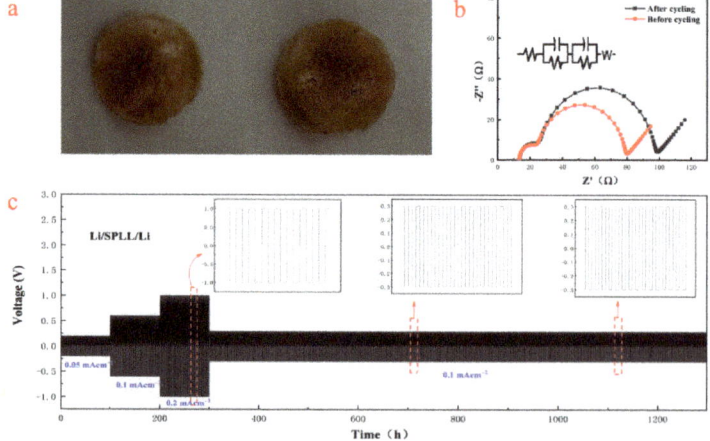

Figure 6. (**a**) Two-sided photo after GCD cycled 1000 times. (**b**) impedance spectra before and after GCD cycle. (**c**) GCD cycling of the Li/SPLL/Li cells at 0.05, 0.1 and 0.2 mA cm^{-2}.

To evaluate the electrochemical stability of the electrolytes, the LNMO/SPLL/Li and LNMO/SPL/Li batteries were assembled with SPLL and SPL electrolyte membranes, respectively [30]. Figure 7a,d shows the rate performance of the two batteries at 0.1 C, 0.2 C, 0.3 C, 0.5 C, 1.0 C, and 2.0 C. Compared with the SPL polymer battery, the SPLL polymer electrolyte battery exhibits a better rate performance. Figures S3 and 7b are the cycle performance and CV curves of the LNMO/SPLL/Li battery. The CV curves change slightly after five cycles, indicating that the battery has a stable cycle performance [31]. Figure 7c displays the discharge/charge curves of the first 100 cycles at a rate of 0.1 C at room temperature. It has a high voltage discharge platform of 4.7 V and a high cycle stability. The charge and discharge specific capacity of the first cycle are 140 mAh/g and 122 mAh/g, respectively, which are close to the theoretical specific capacity of LNMO, suggesting an excellent cycle stability of the SPLL polymer [32]. Figure 7e shows the Nyquist plots of the LNMO/SPLL/Li battery after various charging/discharging cycles at the fully charged state. The charge transfer impedance decreases and finally stabilizes with the increase of the cycle number, indicating that the SPLL polymer has a good cycle performance. As shown in Figure 7f, the capacity retention rate of the LNMO/SPLL/Li battery is still 98.4% after 100 cycles. On the contrary, the LNMO/SPL/Li battery displays a poor cycle performance (Figure S4), indicating that adding inorganic metal oxide LAZTP and non-metallic oxide SiO_2 could improve the stability of the polymer electrolyte.

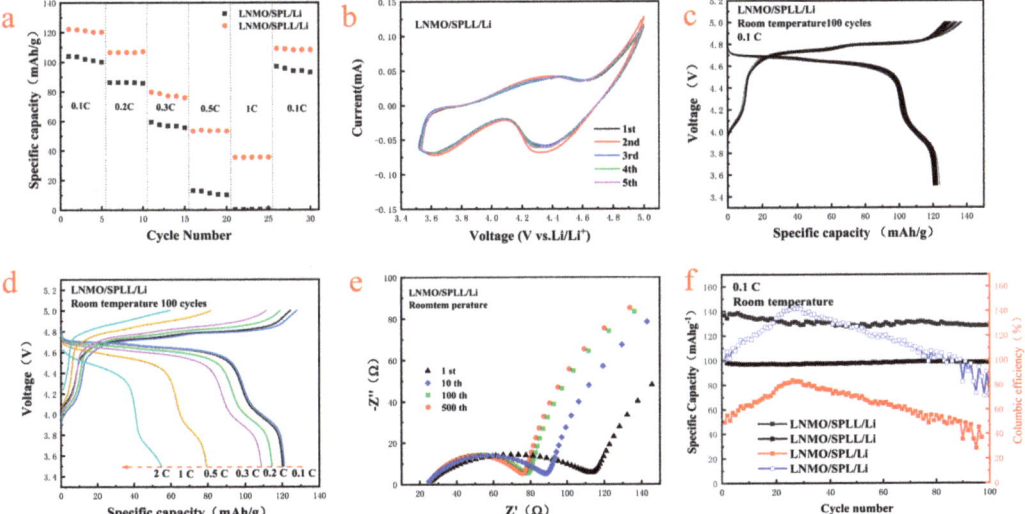

Figure 7. Electrochemical performance of the all-solid-state cell with LNMO cathode and SPLL: (**a**) rate capability; (**b**) CV curves of LNMO/SPLL/Li; (**c**) typical discharge–charge profiles at 0.1 C rate; (**d**) charge/discharge curves under various current densities ranging from 0.1 to 2.0 C; (**e**) Nyquist plots; and (**f**) the cycling performance between 3.5 and 5 V at 0.1 C rate.

The morphology of the SPLL polymer electrolyte was characterized by SEM before and after cycling. The electrolyte exhibits a smooth surface morphology before cycling (Figure S5a,b) and shows a cross-networking morphology after 100 cycles, indicating that the polymer electrolyte membrane reacts to form an ion conduction channel for lithium ion transmission. It can also be seen from the cross-section that the uncirculated electrolyte cross-section is neatly arranged and dense. When electrochemical charging and discharging are carried out, a network structure appears inside the electrolyte, which indicates that a chemical reaction occurs internally during the charging and discharging process, forming a new type of electrolyte. The conductive network provides a powerful channel for lithium ion transmission.

Figure S6 shows the morphology of the SPLL polymer electrolyte membrane before and after cycling. The electrolyte displays a white morphology before cycling. However, the surface of the electrolyte turns black after 100 cycles, which is the residue of the LNMO cathode. At the polymerization temperature (60 °C), SPLL is in close contact with LNMO. The good interface contact ability is conducive to the shuttle of Li+ in the interface. Figure S7 shows the XRD patterns of the polymer electrolyte membrane before and after cycling. The characteristic peak is very sharp before cycling, and no by-products were generated after 20 cycles, indicating that no side reactions occurred during cycling.

The cycling stability of the LNMO/SPLL/Li batteries is evaluated at a current density of 0.25 C. As shown in Figure 8, the battery can reach up to 117.5 mAhg^{-1} in the first cycle. After 500 cycles, the capacity can still reach 85.6 mAhg^{-1} with a capacity retention rate of 82.5% [33]. For such a good performance, we attribute it to several aspects. First, the LAZTP can provide lithium ion transport channels to improve the ionic conductivity of the polymer. On the other hand, the non-metal oxide SiO_2 can increase the amorphous area of the polymer and expand the lithium ion transport channel. Finally, we add a small amount of electrolyte on both sides of the polymer SPLL to reduce interface impedance.

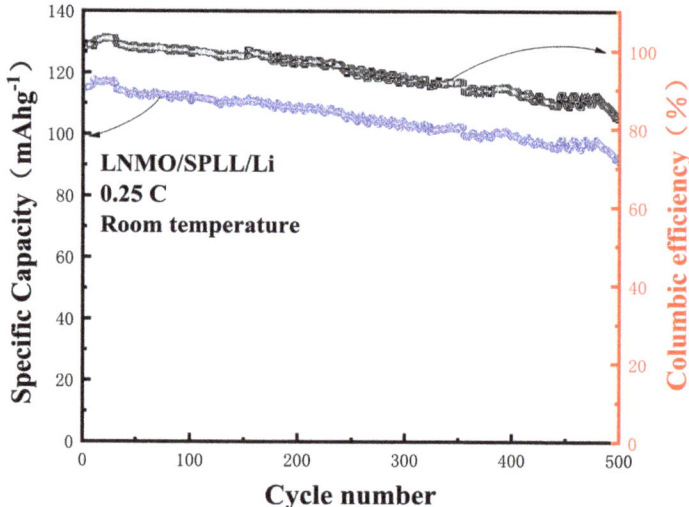

Figure 8. Cycling performance of the assembled all-solid-state LNMO/SPLL/Li battery at 0.25 C.

The polymer electrolyte is ignited (Figure S8) to test its stability in the air. The electrolyte membrane burns when the fire source touches the electrolyte for 2 s. Subsequently, the electrolyte membrane goes out when being removed from the fire source. When we again put it on the polymer electrolyte for 4 s, the electrolyte membrane burned up, and the electrolyte membrane extinguished immediately as the fire source was removed. These results show that the electrolyte has a good air stability.

4. Conclusions

This work prepared a composite solid electrolyte with high mechanical flexibility and non-flammability. Firstly, the crystallinity of the polymer is reduced, and the fluidity of Li+ between the polymer segments is improved by tertiary polymer polymerization. The composite solid electrolyte has an excellent Li+ conductivity (3.18×10^{-4} mS cm^{-1}). The LNMO/SPLL/Li battery has a capacity retention rate of 98.4% after 100 cycles, which is much higher than that without inorganic oxides. This research provides an important reference for developing all-solid-state batteries in a greenhouse.

Supplementary Materials: The following supporting information can be downloaded at: https://www.mdpi.com/article/10.3390/nano12173069/s1, Figure S1: SEM-EDS of the LAZTP. (a) 10 μm size LAZTP. (b) Zn element content chart; Figure S2: SPL's GCD at three different current densities; Figure S3: CV test of SPLL under different cycle rates; Figure S4: (a) 50 cycles of LNMO/SPL/Li at 0.1c rate in the greenhouse. (b) The first charge and discharge of LNMO/SPL/Li at different rates; Figure S5: (a,b) Surface map of the SPLL before and after cycle. The SEM cross-sectional view of (c) and (d); Figure S6: Physical image of the surface before and after the SPLL cycle; Figure S7: XRD before and after SPLL cycle; Figure S8: Physical image of the surface before and after the SPLL cycle. SP is an electrolyte membrane formed by stirring and drying three polymers in DMF (among them, DMF 30 mL, PVA 1 g, PES 2 g, PVDF 5% copolymerized base); SPL is added on the basis of SP LiBF4 (PES es:Li = 8:1); SPLL is based on SPL with 10%LAZTP added.

Author Contributions: Conceptualization, X.L.; methodology, L.L.; writing—original draft preparation, X.J.; writing—review and editing, S.Z.; supervision, M.H.; formal analysis, D.H. All authors have read and agreed to the published version of the manuscript.

Funding: This research was supported by the: Guangxi Natural Science Foundation (No. 2020GXNS-FAA297082), Guangxi Innovation Driven Development Project (No. AA18242036-2), National Natural Science Foundation of China (No. 52161033), Fund Project of the GDAS Special Project of Science and Technology Development, Guangdong Academy of Sciences Program (No. 2020GDASYL-20200104030), and Fund Project of the Key Lab of Guangdong for Modern Surface Engineering Technology (No. 2018KFKT01).

Data Availability Statement: Not applicable.

Conflicts of Interest: The authors declare no conflict of interest. The funders had no role in the design of the study; in the collection, analyses, or interpretation of data; in the writing of the manuscript, or in the decision to publish the results.

References

1. Cheng, X.B.; Zhao, C.Z.; Yao, Y.X.; Liu, H.; Zhang, Q. Recent Advances in Energy Chemistry between Solid-State Electrolyte and Safe Lithium-Metal Anodes. *Chem* **2019**, *5*, 74–96. [CrossRef]
2. Ghazi, Z.A.; Sun, Z.H.; Sun, C.G.; Qi, F.L.; An, B.G.; Li, F.; Cheng, H.M. Key Aspects of Lithium Metal Anodes for Lithium Metal Batteries. *Small* **2019**, *15*, 1900687–1900714. [CrossRef] [PubMed]
3. Forsyth, M.; Porcarelli, L.; Wang, X.E.; Goujon, N.; Mecerreyes, D. Innovative electrolytes based on ionic liquids and polymers for next-generation solid-state batteries. *Acc. Chem. Res.* **2019**, *52*, 686–694. [CrossRef] [PubMed]
4. Choudhury, S.; Stalin, S.; Vu, D.; Warren, A.; Deng, Y.; Biswal, P.; Archer, L.A. Solid-state polymer electrolytes for high-performance lithium metal batteries. *Nat. Commun.* **2019**, *410*, 398. [CrossRef]
5. Feng, J.K.; Yan, B.G.; Liu, J.C.; Lai, M.O.; Li, L. All solid state lithium ion rechargeable batteries using NASICON structured electrolyte. *Mater. Technol.* **2013**, *228*, 76–279. [CrossRef]
6. Chen, R.; Qu, W.; Guo, X.; Li, L.; Wu, F. The pursuit of solid-state electrolytes for lithium batteries: From comprehensive insight to emerging horizons. *Mater. Horiz.* **2016**, *3*, 487–516. [CrossRef]
7. Chen, R.; Li, Q.; Yu, X.; Chen, L.; Li, H. Approaching practically accessible solid-state batteries: Stability issues related to solid electrolytes and interfaces. *Chem. Rev.* **2019**, *120*, 6820–6877. [CrossRef]
8. Xiang, J.W.; Yang, L.Y.; Yuan, L.X.; Zhang, Y.; Huang, Y.Y.; Lin, J.; Feng, P.; Huang, Y.H. Alkali-Metal Anodes: From Lab to Market-ScienceDirect. *Joule* **2019**, *3*, 2334–2363. [CrossRef]
9. Zhang, H.T.; Wu, H.; Wang, L.; Xu, H.; He, X.M. Benzophenone as indicator detecting lithium metal inside solid state electrolyte. *J. Power Sources* **2021**, *492*, 229661–229666. [CrossRef]
10. Das, A.; Goswami, M.; Illath, K.; Ajithkumar, T.G.; Arya, A.; Krishnan, M. Synthesis and characterization of LAGP-glass-ceramics-based composite solid polymer electrolyte for solid-state Li-ion battery application. *J. Non-Cryst Solids* **2021**, *558*, 120654–120664. [CrossRef]
11. Mücke, R.; Finsterbusch, M.; Kaghazchi, P.; Fattakhova-Rohlfing, D.; Guillon, O. Modelling electro-chemical induced stresses in all-solid-state batteries: Anisotropy effects in cathodes and cell design optimisation. *J. Power Sources* **2021**, *489*, 229430–229439. [CrossRef]
12. Liang, Y.; Liu, Y.; Chen, D.; Dong, L.; Guang, Z.; Liu, J.; Yuan, B.; Yang, M.; Dong, Y.; Li, Q.; et al. Hydroxyapatite functionalization of solid polymer electrolytes for high-conductivity solid-state lithium-ion batteries. *Mater. Today Energy* **2021**, *20*, 100694–100703. [CrossRef]
13. Li, J.H.; Wang, R.G. Recent advances in the interfacial stability, design and in situ characterization of garnet-type $Li_7La_3Zr_2O_{12}$ solid-state electrolytes based lithium metal batteries. *Ceram. Int.* **2021**, *47*, 13280–13290. [CrossRef]

14. Yi, S.H.; Xu, T.H.; Li, L.; Gao, M.M.; Du, K.; Zhao, H.L.; Bai, Y. Fast ion conductor modified double-polymer (PVDF and PEO) matrix electrolyte for solid lithium-ion batteries. *Solid State Ion.* **2020**, *355*, 115419–115429. [CrossRef]
15. Yu, H.; Han, J.S.; Hwang, G.C.; Cho, J.S.; Kang, D.W.; Kim, J.K. Optimization of high potential cathode materials and lithium conducting hybrid solid electrolyte for high-voltage all-solid-state batteries. *Electrochim. Acta* **2021**, *365*, 137349–137357. [CrossRef]
16. Qu, W.J.; Yan, M.X.; Luo, R.; Qian, J.; Wen, Z.Y.; Chen, N.; Li, L.; Wu, F.; Chen, R.J. A novel nanocomposite electrolyte with ultrastable interface boosts long life solid-state lithium metal batteries. *J. Power Sources* **2021**, *484*, 229195–229203. [CrossRef]
17. Shahi, M.; Hekmat, F.; Shahrokhian, S. Hybrid supercapacitors constructed from double-shelled cobalt-zinc sulfide/copper oxide nanoarrays and ferrous sulfide/graphene oxide nanostructures. *J. Colloid. Interface Sci.* **2021**, *585*, 750–763. [CrossRef]
18. Zhang, J.; Su, Y.; Zhang, Y. Recent advances in research on anodes for safe and efficient lithium–metal batteries. *Nanoscale* **2020**, *12*. [CrossRef]
19. Zhang, Z.; Huang, Y.; Gao, H.; Li, C.; Hang, J.X.; Liu, P.B. MOF-derived multifunctional filler reinforced polymer electrolyte for solid-state lithium batteries. *J. Energy Chem.* **2021**, *60*, 259–271. [CrossRef]
20. Duluard, S.; Paillassa, A.; Lenormand, P.; Taberna, P.L.; Simon, P.; Rozier, P.; Ansart, F. Dense on Porous Solid LATP Electrolyte System: Preparation and Conductivity Measurement. *J. Am. Ceram. Soc.* **2017**, *100*, 1–7. [CrossRef]
21. Liang, J.N.; Hwang, S.; Li, S.; Luo, J.; Sun, Y.P.; Zhao, Y.; Sun, Q.; Li, W.H.; Li, M.S.; Li, R.Y.; et al. Stabilizing and understanding the interface between nickel-rich cathode and PEO-based electrolyte by lithium niobium oxide coating for high-performance all-solid-state batteries. *Nano Energy* **2020**, *78*, 105107–105118. [CrossRef]
22. Jung, W.D.; Jeon, M.; Shin, S.S.; Kim, J.S.; Jung, H.G.; Kim, B.K.; Lee, J.H.; Chung, Y.C.; Kim, H. Functionalized Sulfide Solid Electrolyte with Air-Stable and Chemical-Resistant Oxysulfide Nanolayer for All-Solid-State Batteries. *ACS Omega* **2020**, *5*, 26015–26022. [PubMed]
23. Li, D.; Chen, L.; Wang, T.S.; Fan, L.Z. 3D Fiber-Network-Reinforced Bicontinuous Composite Solid Electrolyte for Dendrite-free Lithium Metal Batteries. *ACS Appl. Mater. Inter.* **2018**, *10*, 7069–7078. [CrossRef]
24. Zhang, Q.Q.; Liu, K.; Ding, F.; Liu, X.J. Recent advances in solid polymer electrolytes for lithium batteries. *Nano Res.* **2017**, *10*, 4139–4174. [CrossRef]
25. Yue, L.P.; Ma, J.; Zhang, J.J.; Zhao, J.J.W.; Dong, S.M.; Liu, Z.H.; Cui, G.L.; Chen, L.Q. All solid-state polymer electrolytes for high-performance lithium-ion batteries. *Energy Storage Mater.* **2016**, *5*, 139–164. [CrossRef]
26. Zhang, Y.; Li, L.Y.; Ding, Z.Y.; Chen, Y.F.; Yuan, Q.H.; Sun, R.T.; Li, K.K.; Liu, C.; Wu, J.W. SnO_2 nanoparticles embedded in 3D hierarchical honeycomb-like carbonaceous network for high-performance lithium ion battery. *J. Alloys Compd.* **2021**, *858*, 157716–157724.
27. Wang, H.; Hu, P.; Liu, X.T.; Shen, Y.; Li, X.Y.; Li, Z.; Huang, Y.H. Sowing Silver Seeds within Patterned Ditches for Dendrite-Free Lithium Metal Batteries. *Adv. Sci.* **2021**, *20*, 2100684–2100692. [CrossRef]
28. Liang, X.H.; Han, D.; Wang, Y.T.; Lan, L.X.; Mao, J. Preparation and performance study of a PVDF–LATP ceramic composite polymer electrolyte membrane for solid-state batteries. *RSC Adv.* **2018**, *8*, 40498–40504.
29. Li, C.; Huang, Y.; Feng, X.; Zhang, Z.; Gao, H.; Huang, J. Silica-assisted cross-linked polymer electrolyte membrane with high electrochemical stability for lithium-ion batteries. *J. Colloid Interface Sci.* **2021**, *594*, 1–8. [CrossRef]
30. Miao, X.Y.; Qin, X.; Huang, S.Y.; Wei, T.Y.; Lei, C.R. Hollow spherical $LiNi_{0.5}Mn_{1.5}O_4$ synthesized by a glucose-assisted hydrothermal method. *Mater. Lett.* **2021**, *289*, 129417. [CrossRef]
31. Wei, Q.L.; DeBlock, R.H.; Butts, D.M.; Choi, C.; Dunn, B. Pseudocapacitive Vanadium-based Materials toward High-Rate Sodium-Ion Storage. *Energy Environ. Mater.* **2020**, *3*, 221–234. [CrossRef]
32. Mitsuharu, T.; Riki, K.; Koji, Y. High-capacity Li-excess lithium nickel manganese oxide as a Co-free positive electrode material. *Mater. Res. Bull.* **2021**, *137*, 111178–111188.
33. Chen, L.J.; Song, K.M.; Shi, J.; Zhang, J.Y.; Mi, L.W.; Chen, W.H.; Liu, C.T.; Shen, C.Y. PAANa-induced ductile SEI of bare micro-sized FeS enables high sodium-ion storage performance. *Sci. China Mater.* **2020**, *64*, 105–114. [CrossRef]

Article

Quasi-Solid-State Lithium-Sulfur Batteries Assembled by Composite Polymer Electrolyte and Nitrogen Doped Porous Carbon Fiber Composite Cathode

Xinghua Liang, Yu Zhang, Yujuan Ning, Dongxue Huang, Linxiao Lan * and Siying Li *

Guangxi Key Laboratory of Automobile Components and Vehicle Technology, Guangxi University of Science and Technology, Liuzhou 545006, China; lxh18589873093@163.com (X.L.); yuzhang5332@163.com (Y.Z.); yujuan996@163.com (Y.N.); hdx877348318@163.com (D.H.)
* Correspondence: l15506749886@163.com (L.L.); lisiying@gxust.edu.cn (S.L.)

Abstract: Solid-state lithium sulfur batteries are becoming a breakthrough technology for energy storage systems due to their low cost of sulfur, high energy density and high level of safety. However, its commercial application has been limited by the poor ionic conductivity and sulfur shuttle effect. In this paper, a nitrogen-doped porous carbon fiber (NPCNF) active material was prepared by template method as a sulfur-host of the positive sulfur electrode. The morphology was nano fiber-like and enabled high sulfur content (62.9 wt%). A solid electrolyte membrane (PVDF/LiClO$_4$/LATP) containing polyvinylidene fluoride (PVDF) and lithium aluminum titanium phosphate (Li$_{1.3}$Al$_{0.3}$Ti$_{1.7}$(PO$_4$)$_3$) was prepared by pouring and the thermosetting method. The ionic conductivity of PVDF/LiClO4/LATP was 8.07 × 10^{-5} S cm^{-1} at 25 °C. The assembled battery showed good electrochemical performance. At 25 °C and 0.5 C, the first discharge specific capacity was 620.52 mAh g^{-1}. After 500 cycles, the capacity decay rate of each cycle was only 0.139%. The synergistic effect between the composite solid electrolyte and the nitrogen-doped porous carbon fiber composite sulfur anode studied in this paper may reveal new approaches for improving the cycling performance of a solid-state lithium-sulfur battery.

Keywords: solid-state lithium-sulfur battery; composite polymer electrolytes; porous carbon

Citation: Liang, X.; Zhang, Y.; Ning, Y.; Huang, D.; Lan, L.; Li, S. Quasi-Solid-State Lithium-Sulfur Batteries Assembled by Composite Polymer Electrolyte and Nitrogen Doped Porous Carbon Fiber Composite Cathode. *Nanomaterials* **2022**, *12*, 2614. https://doi.org/10.3390/nano12152614

Academic Editor: Sergio Brutti

Received: 30 June 2022
Accepted: 25 July 2022
Published: 29 July 2022

Publisher's Note: MDPI stays neutral with regard to jurisdictional claims in published maps and institutional affiliations.

Copyright: © 2022 by the authors. Licensee MDPI, Basel, Switzerland. This article is an open access article distributed under the terms and conditions of the Creative Commons Attribution (CC BY) license (https:// creativecommons.org/licenses/by/ 4.0/).

1. Introduction

With the increasing energy demand for energy storage equipment in the current market, traditional lithium-ion batteries cannot meet the requirements due to low energy density, poor safety and high cost [1,2]. Therefore, it is imperative to develop and research a new battery system with high specific energy and high safety. The theoretical specific capacity and theoretical specific energy of lithium-sulfur battery can reach 1675 mAh g^{-1}, 2600 wh kg^{-1}. What's more, sulfur has obvious advantages in environmental protection, acquisition cost, and so on. It is considered to be the most promising next-generation new energy storage system [3,4]. However, the commercialization of lithium-sulfur batteries still faces some problems [5,6]. Firstly, liquid electrolytes have the safety problem of inflammability and the possibility of explosion, and the lithium dendrite grown from the negative electrode pierces the diaphragm, leading to short circuit [7,8]. Secondly, sulfur has poor electrical conductivity and low utilization rate of active materials. The "shuttle effect" caused by polysulfide dissolution leads to low capacity and coulomb efficiency [1,9].

In order to solve the above problems, more and more attention has been paid to solid-state lithium-sulfur batteries [10,11]. On the one hand, the cathode side requires a high conductivity material to improve contact with low conductivity S. Carbon materials have high electrical conductivity, high specific surface area and excellent mechanical properties, which can provide a conductive network for sulfur and discharge products (Li$_2$S) and improve the electrochemical performance of sulfur cathodes [12]. In addition, the introduction of nitrogen doping cannot only significantly improve the electrical conductivity of carbon materials, but also introduce active sites on the surface of carbon materials [13]. Therefore, nitrogen-doped carbon materials can be used in lithium-sulfur batteries as active materials with high electronic conductivity and strong physical and chemical adsorption. On the other hand, there is no solid electrolyte that can meet all the requirements, such as high ionic conductivity at room temperature, a wide electrochemical stability window, good mechanical properties, etc. [14–17]. The advantages and disadvantages of different solid-state electrolytes are integrated by using composite electrolytes, which provides a new idea for further study of solid-state lithium-sulfur batteries. Polyoxyethylene (PEO) and polyvinylidene fluoride (PVDF) are common polymer electrolyte substrates [18–20]. Currently, polyoxyethylene (PEO) has been widely studied in lithium-sulfur batteries, but its ionic conductivity is low at room temperature, and it can only show good ionic conductivity in the amorphous state of 60~90 °C [21]. Compared with PEO, the PVDF electrolyte has better mechanical strength and a higher melting point. Adding inorganic filler to a polymer electrolyte to form CPEs can effectively improve ionic conductivity and lithium-ion transference number. Common inorganic electrolytes include NASICON type, $Li_{10}GeP_2S_{12}$(LGPS) type, Li_xPON type, Li_2S-P_2S_5 type and $Li_7La_3Zr_2O_{12}$ (LLZO) type [22–25]. Lithium titanium aluminum phosphate (LATP) is a glass ceramic material with NASICON type three-dimensional network structure, which has the advantages of high mechanical strength, high ionic conductivity, high temperature stability and stability to air and water [26,27]. However, the application of LATP as an electrolyte in batteries is limited by its large interfacial impedance and side effects. By adding a certain amount of nano-scale ceramic materials into the polymer electrolyte, the composite polymer-ceramic electrolyte (CPEs) formed has lower interfacial resistance and higher ionic conductivity [28,29]. It can inhibit the formation of lithium dendrite and the shuttle effect of polysulfide and can be effectively applied to solid-state lithium-sulfur batteries.

In this paper, a nitrogen-doped porous carbon fiber active material (NPCNF) with a microporous structure and nanofiber shape was prepared via template method. The NPCNF/S electrode exhibits excellent performance due to the better electrical conductivity and strong physical and chemical adsorption of carbon and nitrogen doped materials. A PVDF/$LiClO_4$/LATP composite solid electrolyte (CPEs) was prepared, which combined the advantages of inorganic electrolytes and polymer electrolytes. It has the characteristics of a wide electrochemical window, high ionic conductivity and stable mechanical properties at room temperature. The assembled quasi-solid lithium sulfur battery was tested at 25 °C and had excellent performance. This study proves that the long cycle performance of a solid-state lithium-sulfur battery is improved at a large magnification rate, which provides ideas for subsequent research.

2. Materials and Methods

2.1. Materials

The raw materials included PEO-PPO-PEO (P123) (99%, Aladdin, Shanghai, China), $C_8H_{20}O_4Si$ (99%, Aladdin, Shanghai, China), HCl (98%, Aladdin, Shanghai, China), $C_2H_4N_4$ (99%, Aladdin, Shanghai, China), HF (40%, Aladdin, Shanghai, China), S (99%, Aladdin, Shanghai, China), polyvinylidene fluoride (PVDF) (Mw = 600,000, Macklin, Shanghai, China), lithium bisimide ($LiClO_4$) (99.99% purity, Aladdin, Shanghai, China), $Li_{1.3}Al_{0.3}Ti_{1.7}(PO_4)_3$ (99.99% purity, Macklin, Shanghai, China).

2.2. Preparation of the NPCNF/S Composite

We dissolved 1 g PEO-PPO-PEO (P123) in 6 mL $C_8H_{20}O_4Si$ under magnetic stirring. Then, we added 32 mL ethanol and 0.583 mL concentrated hydrochloric acid (HCl, 37%) to the solution. After adding 4 mL of deionized water, wefully stirred the hydrolysis for 2 h. Adding 2.8 g dicyandiamide (DCDA) as carbon source and nitrogen source, the semi-solid colloid was obtained by stirring and drying at 80 °C. The powder was dried overnight at

80 °C to obtain a white powder; then, we calcined N_2 in a tubular furnace at 1000 °C for 60 min at a heating rate of 3 °C min^{-1}. After cooling to room temperature, the sintered powder was poured into 5%HF solution to clean the template. Fully cleaned samples were dried at 60 °C for 12 h to obtain the final product NPCNF. The NPCNF was mixed with elemental S at a mass ratio of 1:2 and calcined at 155 °C for 12 h in a tube furnace under a nitrogen atmosphere. Cooling to room temperature to obtain NPCNF/S.

2.3. Preparation of the Composite Solid Electrolyte Membrane (CPEs)

PVDF, LATP and LiClO$_4$ powders were vacuum-dried at 60 °C for 24 h. PVDF, LATP, LiClO$_4$ and DMF were weighed at the mass ratio of 10:1:0.124:80. PVDF was dissolved in 40 mL DMF and stirred at 55 °C for 1 h to form a transparent viscous solution. LATP and LiClO$_4$ were added and stirred for 5~6 h. Finally, the mixed solution was cast into a polytetrafluoroethylene mold and vacuum dried at 60 °C for 24~72 h to obtain flexible electrolyte films with ceramic/polymer composites

2.4. Battery Assembly

NPCNF/S, conducting carbon and polyvinylidene fluoride (PVDF), were dissolved in N-methylpyrrolidone (NMP) at a mass ratio of 7:2:1 and stirred to a obtain uniform slurry. The slurry was coated on aluminum foil and dried in a vacuum drying oven for 12 h. The composite electrolyte was cut into discs with a diameter of 18 mm. The 2025-coin cells were assembled and tested. We then added two drops of electrolyte. The electrolyte was 1.0 mol LiTFSI in DOL:DME = 1:1 vol% with 1.0 wt% LiNO$_3$.

2.5. Characterization

Via X-ray diffraction (XRD, D8-Advance, Bruker, Germany), the material phase was analyzed by measuring the diffraction data in the range of 10~90°. Via thermogravimetric analysis (TGA) measurement in air atmosphere temperature under the condition of 10 °Cmin^{-1}, we performed an analysis of material quality, along with the change of temperature. The cathode material was tested via Raman spectroscopy (XploRA PLUS, HORIBA, France) under a 523 nm Raman microscope. The microscopic morphology of the sample was characterized via scanning electron microscope (SEM, Sigma04-55, ZEISS, Germany). The composition and valence of solid electrolyte elements were determined by X-ray photoelectron spectroscopy (XPS, K-alpha, Thermo, America) at 5 kV.

2.6. Electrochemical Measurements

The timing current of lithium symmetric battery was tested at a voltage of 0.5 mV, lasting 4000 s, and the formula was as follows:

$$t_{Li+} = Is(\Delta V - I_0 R_0)/I_0(\Delta V - I s R s)$$

The lithium-ion transfer rate (t_{Li+}) can be obtained. I_0 and Is are current values after DC polarization starts and stabilizes, R_0 and Rs are the impedance values before and after the DC polarization, and ΔV is the value of the voltage applied to both ends of the battery.

For the ionic conductivity test, battery assembly used SS as a symmetrical battery and electrochemical impedance test together to calculate the ionic conductivity. The frequency range of impedance test is 0.1~106 Hz.

$$\sigma = L/S \times R$$

where σ represents the ionic conductivity, L is the thickness of electrolyte, S represents the contact area between electrolyte and test electrode (SS) and R is the impedance value of battery electrolyte measured by EIS. The battery test system (CT-400, Neware, Hong Kong, China) performed constant current charge–discharge cycle tests between 1.5 and 3 V. At 25 °C, the electrochemical workstation (DH-7000, Donghua, Shanghai, China) was used for cyclic voltammetry (CV) test at 1.5~3 V and 0.2 mV s^{-1}. Linear sweep voltammetry (LSV) was used to perform electrochemical window tests at 2~6 V at a scanning rate of 0.1 mV s^{-1}.

3. Results

Figure 1 shows the manufacturing process of CPEs and NPCNF/S positive poles. The NPCNF material was prepared via the etching template method. Its unique hole structure increased the specific surface area of the material, and it could load more elemental sulfur. After mixing with S, the positive electrode sheet was obtained after the slurry coating. The polymer, lithium salt and inorganic electrolyte were fully dissolved in the mixed solution, and the nanoscale LATP was uniformly combined with PVDF to obtain CPEs.

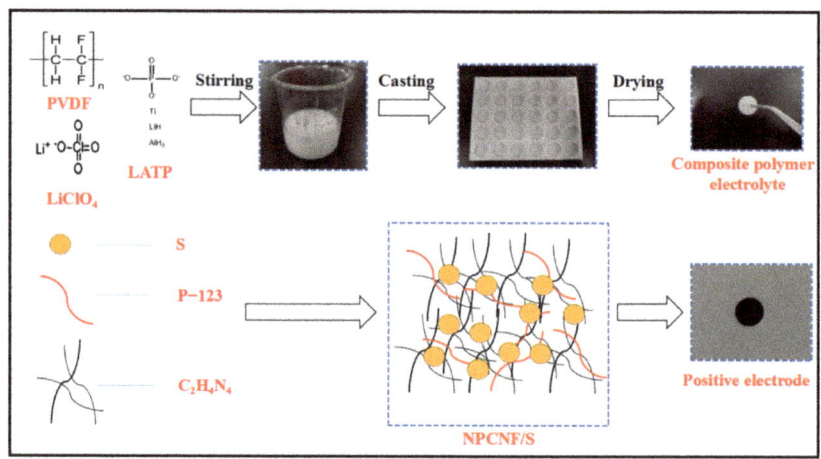

Figure 1. A schematic illustration of the fabrication of the positive electrolyte and CPEs.

XRD patterns of NPCNF and NPCNF/S are shown in Figure 2a. NPCNF/S has an obvious diffraction peak corresponding to elemental S at 23.04°. There is a diffraction peak at 24.8° of NPCNF corresponding to the (002) plane of graphite carbon, which proves that a certain amount of graphite amorphous carbon is formed in the material. The diffraction peak at two places indicates that the characteristics of elemental S and NPCNF are retained in NPCNF/S. NPCNF/S, compared with the diffraction peak of sulfur, was reduced greatly, and this is due to the large amounts of S fully penetrated into the microporous structure of the carbon fiber material [30]. The corresponding morphology can be observed in the SEM figure (Figure 3c,d). Raman spectroscopy was used to test NPCNF and NPCNF/S, as shown in Figure 2b. The D band and G band intensity ratios of NPCNF and NPCNF/S are 1.04 and 1.03, respectively. The differences were small, indicating that the introduction of sulfur particles did not change the graphitization degree of NPCNF. The $I_{D/G}$ values are all greater than 1, indicating that the active material has a high degree of graphite carbonization and good conductivity [31]. In order to determine the sulfur content of the NPCNF/S sample, TGA measurement was carried out, as shown in Figure 2c. The mass change of the sample was measured when the temperature was raised to 800 °C at a heating rate of 10 °C min^{-1} in a nitrogen flow. It can be seen that there was about a 17 wt% amount of weight loss when NPCNF rose to 800 °C, and the elemental sulfur rapidly sublimated to complete disappearance at 250~350 °C. The sulfur content of the NPCNF/S sample is about 62.9 wt%. The sulfur loading and content of the cathode is 0.38 mg cm^{-2}. An experiment on adsorption of polysulfide lithium was carried out using NPCNF, as shown in Figure ??d. Firstly, we added Li$_2$S$_4$ion (Bottle No. 1); 10 mg of NPCNF was added as the No. 2 solution. After standing for 30 min, a clear and transparent liquid was formed in bottle No.3. The pore structure of the NPCNF material had an obvious adsorption and anchoring effect on Li$_2$S$_4$, which results in an inhibiting "shuttle effect" of lithium-sulfur batteries.

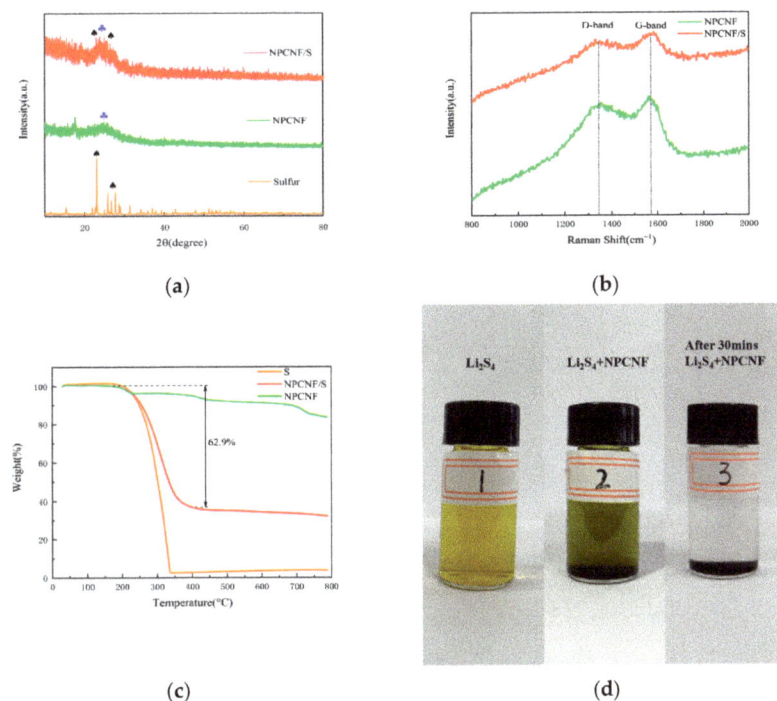

Figure 2. (**a**) XRD tests of the NPCNF and NPCNF/S (**b**) Raman spectra of NPCNF and NPCNF/S (**c**) TGA curves of S, NPCNF and NPCNF/S (**d**) The photograph of the static adsorption test.

SEM characterization tests were conducted for NPCNF and NPCNF/S, as shown in Figure 3. Figure 3a,b shows that the surface of NPCNF presents an irregular reticular structure resembling nanofiber. After being fully etched by the HF solution, an EDS test analysis of NPCNF shows that no Si element was found in the material, and nano-SiO_2 particles generated by tetraethyl orthosilicate hydrolysis were cleaned and removed. The holes leftover increase the specific surface area of the material, which is conducive to the load of S and sulfide in the positive electrode. At the same time, polysulfide can be adsorbed through physical action to provide channels for ion transfer in the battery. After loading S, the sample changes from a nanofiber to porous mesoporous structure, but the original carbon fiber conductive network still remains, as shown in Figure 3c,d. This unique porous mesoporous structure anchors polysulfide, which inhibits the "shuttle effect" and improves the cycling performance of the battery [32]. Figure 3e–h shows the EDS test analysis element map in the specified region of NPCNF/S, where C,N,S elements are evenly distributed, proving the uniformity of material doping.

The XRD results of CPEs can be seen in Figure 4a, 24.2° and 20°, respectively, correspond to characteristic peaks of LATP and PVDF, and another wide peak appears at 38.9°, indicating that PVDF is dominated by γ phase [33]. The characteristic peaks of PVDF and LATP were retained in the samples, indicating that the PVDF and LATP did not combine with each other, but kept their respective characteristics together. Figure 4b shows that the absorption peak of PVDF/$LiClO_4$ complex at 785,910,1131,1438 and 1590 cm^{-1} did not shift with the addition of LATP. The change of peak value at 910 cm^{-1} corresponds to the out-of-plane bending of C-H bond, and the change of peak value at 1590 cm^{-1} is the stretching vibration of C-C bond and C=O, indicating that the addition of LATP is conducive to lithium-ion migration [34]. The TGA tests were conducted for PVDF/$LiClO_4$ and PVDF/$LiClO_4$/LATP. As shown in Figure 4c, rapid weight loss occurred at 400~500 °C. It can be seen from Figure 4d that the temperature of rapid sublimation loss of PVDF/LiClO$_4$ and PVDF/LiClO$_4$/LATP were 471 °C and 433 °C. The weight loss rate of PVDF/LiClO$_4$/LATP was lower than that of PVDF/LiClO$_4$, indicating that the addition of LATP improved the thermal stability of CPEs.

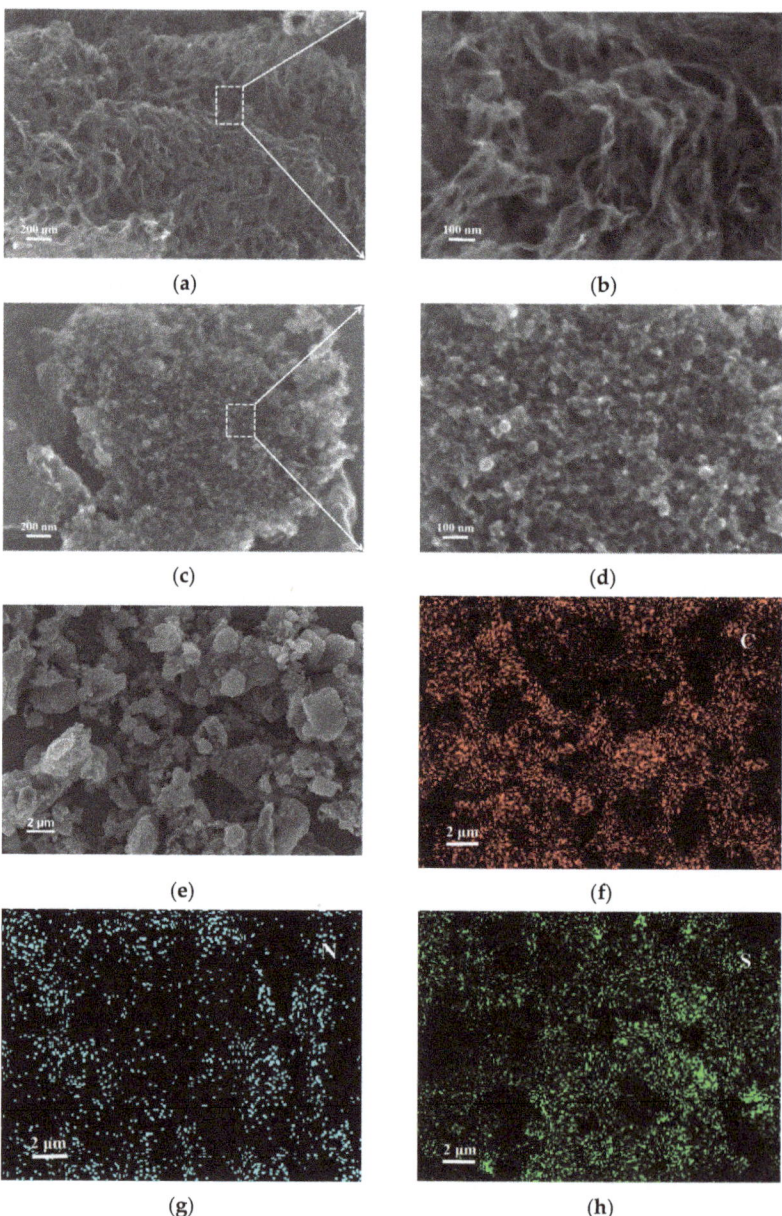

Figure 3. (**a**,**b**) SEM images of NPCNF (**c**,**d**) SEM images of NPCNF/S (**e**) SEM image of NPCNF/S and corresponding elemental mapping images (**f**) C element (**g**) N element (**h**) S element.

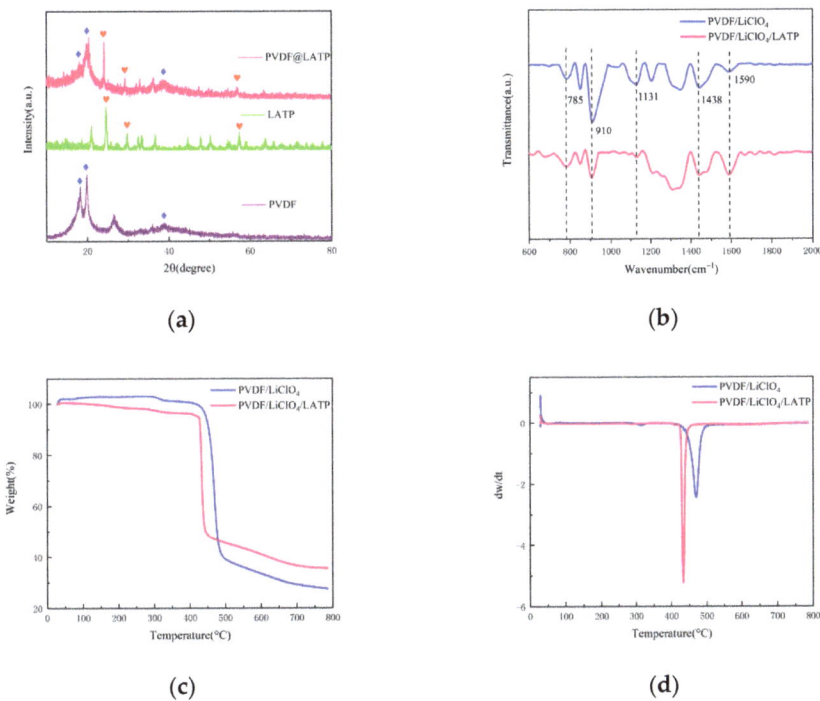

Figure 4. (**a**) XRD tests of the CPEs (**b**) ATR-FTIR (**c**) TGA curves of CPEs (**d**) DTG curves of CPEs.

See Figure 5a for measuring the EIS of CPEs and LATP under 25 °C. It is obvious that the impedance of electrolyte was greatly reduced, which was due to the better flexibility of the membrane made by the combination of PVDF and LATP, as shown in Figure 5e, greatly reducing the interface impedance. The ionic conductivity was at 25 °C is 8.07×10^{-5} S cm^{-1}. Figure 5b shows the Arrhenius diagram of CPEs. With the increase of temperature, the ionic conductivity also increased correspondingly. The increase of temperature promoted the expansion of the polymer and generated free volume in the polymer, which enhanced the segment movement of the polymer and increased the ionic conductivity [35]. The electrochemical window is also an indicator to evaluate the performance of CPEs. Therefore, linear sweep voltammetry (LSV) was used to characterize the electrochemical window. As shown in Figure 5c, the composite solid electrolyte membrane could withstand a voltage of 4.56 V, which is more than sufficient for Li-S batteries. Figure 5d shows the initial impedance spectrum and the impedance spectrum and timing current curve after polarization. The lithium-ion transfer rate of CPEs was calculated to be 0.77. Compared with a traditional liquid electrolyte (t_{Li+} < 0.5) [36], the addition of LATP improved the lithium-ion transfer rate and made the CPEs have better performance. A high lithium-ion transfer rate can generally reduce the concentration of movable anions in CPEs, thus reducing electrode polarization and the accompanying side reactions [37].

Figure 6 shows the CPEs interface and surface SEM characterization tests. Figure 6a,b shows the porous structure of CPEs, which is consistent with the SEM image of the surface in Figure 6c,d. The thickness of CPEs was 183.1 μm, and the porous structure formed by PVDF fiber winding nano-LATP particles was conducive to the transport of lithium-ions [38]. The EDS spectrum in Figure 6e–h shows the existence of element P, proving that LATP was uniformly distributed in CPEs, which itself was conducive to the formation of lithium-ion migration channels [39].

In order to further study the performance of CPEs, XPS tests were carried out on C, F, O and S elements in CPEs after 500 battery cycles. As shown in Figure 7a, the peak at

284.9 eV is the C-C bond peak of organic carbon, and the peak at 286.2, 288.3 and 290.5 eV are carbon–oxygen bonding peaks, indicating the existence of Li_2CO_3 in SEI film [40]. The characteristic peak of -CF_3 appeared at 292.9 eV, corresponding to 684.9 eV in Figure 7b and indicating that -CF_2 in PVDF underwent dehydrogenation to generate LiF [41], which exactly corresponded to the LiF peak at 687.9 eV in Figure 7b. The LiF can inhibit the growth of lithium dendrites and increase the diffusion rate of lithium-ions [42,43]. Figure 7c shows the O1s orbital graph. The characteristic peak of -ClO_4 at 533 eV is the free -ClO_4 in $LiClO_4$. At 532.3 eV, -SO_4 shows the positive S reaction to generate the sulfate salt, which is consistent with the -SO_4 at 169.2 and 170.4 eV in Figure 7d. In Figure 7d, 164.4 and 165.6 eV are natural sulfur [44,45], and no peak bond of polysulfide is found, indicating that the PVDF/$LiClO_4$/LATP electrolyte has a certain inhibitory effect on shuttle effect.

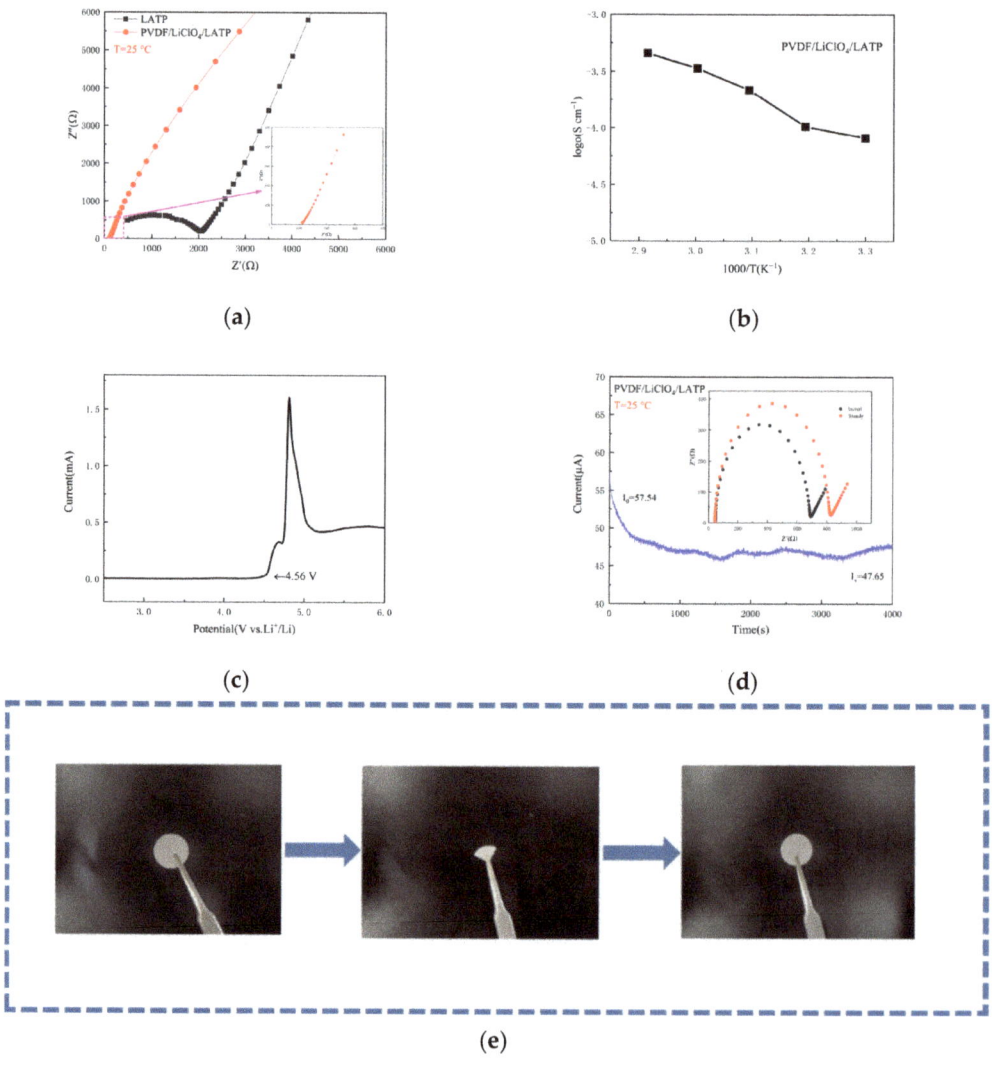

Figure 5. (**a**) EIS of CEPs; (**b**) Arrhenius plots of CEPs; (**c**) electrochemical window test; (**d**) time-ampere measurement symmetrical Li/CEPs/Li battery at 0.5 mV polarization and Nyquist plot initial/steady state internal illustration; (**e**) a photograph showing the flexibility of CEPs.

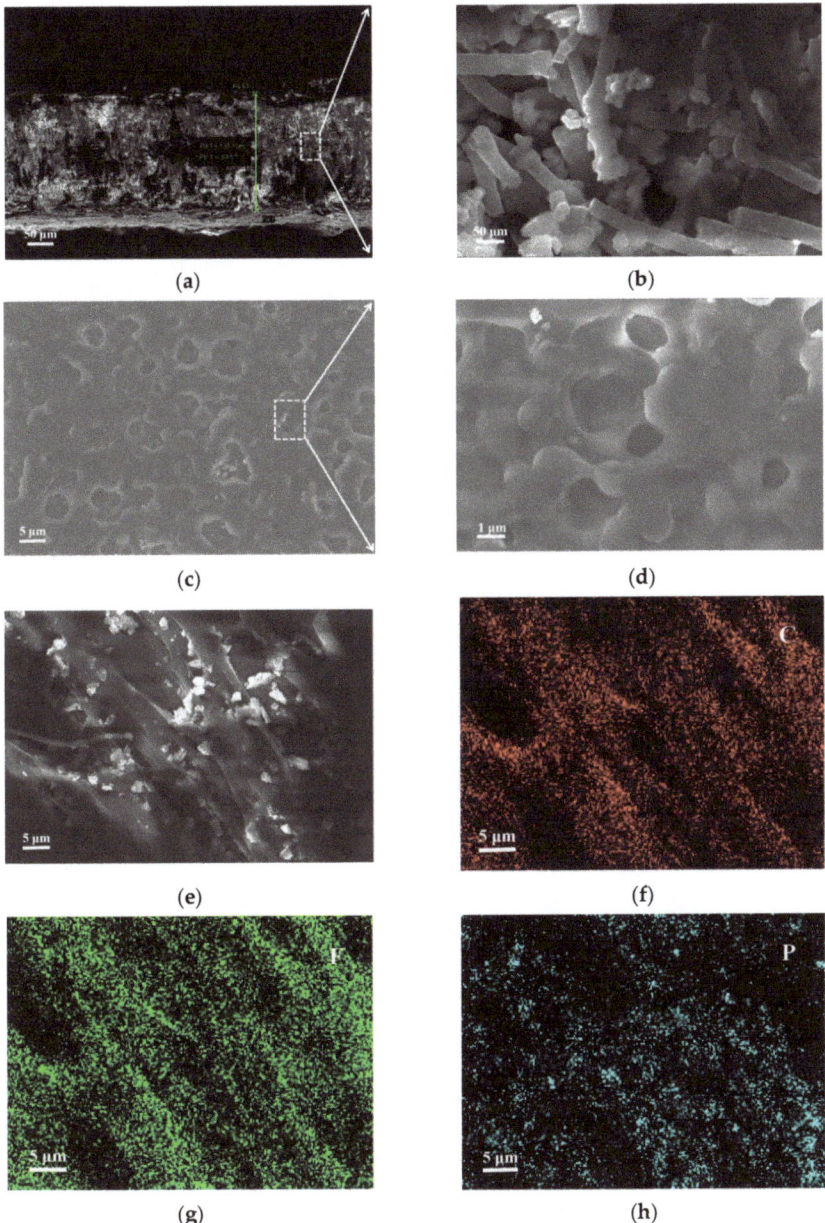

Figure 6. (**a**,**b**) A cross-sectional SEM image of CEPs; (**c**,**d**) SEM images of CEPs; (**e**) an SEM image of CEPs and corresponding elemental mapping images; (**f**) C element; (**g**) F element; (**h**) P element.

Figure 8a presents the CV curves of the Li | PVDF/LiClO$_4$/LATP | NPCNF/S batteries in the voltage range of 1.5~3 V at a scanning rate of 0.2 mV S^{-1} at 25 °C. In the first scan, two reduction peaks appear at 2.3 V and 1.98 V, indicating that S8 is reduced to Li$_2$S$_n$ ($4 \leq n \leq 8$) and Li$_2$S$_2$/Li$_2$S [46] during the discharge process. An oxidation peak that appeared at 2.5 V suggests Li$_2$S$_2$/Li$_2$S oxidized in the process of charging. This is consistent with the phenomenon in the charge–discharge curve of Figure 8c. As the cycle continues, the SEI film tends to be stable and the test curves coincide well, which proves that the polymer electrolyte has good reversible properties. In order to further understand the electrochemical properties of solid-state lithium-sulfur batteries, the AC impedance after different charge–discharge cycles was measured, as shown in Figure 8b. All impedance spectra exhibit at least one semicircular with a Warburg component for the diffusion of lithium-ions through the electrode. After the first charge–discharge cycle, the RCT value was 193.5 Ω, as the number of cycles increased, Rct decreased and finally stabilized after the 10th cycle. It indicates that a stable SEI film is formed in the battery. Research regarding the rate performance test between 0.1 and 1 C is shown in Figure 8d. The discharge specific capacities of the Li | PVDF/LiClO$_4$ LATP | NPCNF/S battery were 595.5 mAh g^{-1} (0.1 C, 1st), 292.3 mAh g^{-1} (0.2 C, 10st), 200.9 mAh g^{-1} (0.5 C, 15st), 141.5 mAh g^{-1} (1 C, 20st). When the rate was restored to 0.1 C, the specific capacity was 390.5 mAh g^{-1}. This shows that the reversible specific capacity of the battery can be maintained after the charging and discharging cycle with a high rate, which proves that the battery has good rate performance. Moreover, the lithium ions diffusion coefficient can be obtained by a series of processing in terms of CV curves at different scan rates, shown in Figure 8e,f. The anodic and cathodic Li$^+$ diffusion rate of $D_{(ALi+)} = 1.06 \times 10^{-8}$, $D_{(BLi+)} = 2.06 \times 10^{-9}$ and $D_{(CLi+)} = 4.36 \times 10^{-9}$ cm^2 s^{-1}. Figure 8g shows the test of 500 long cycles of the battery at 25 °C and 0.5 C. The capacity decay of only 0.139% per cycle, and the coulomb efficiency of the whole cycle is close to 100%, indicating that the battery has good cycle performance. The conductive framework in NPCNF/S and composite electrolytes (CPEs) can provide embedded channels for polysulfide formed in the charge–discharge cycle and inhibit the deposition of polysulfide at the Li-anode interface, thus reducing the influence of "shuttle effect" and improving the performance of batteries.

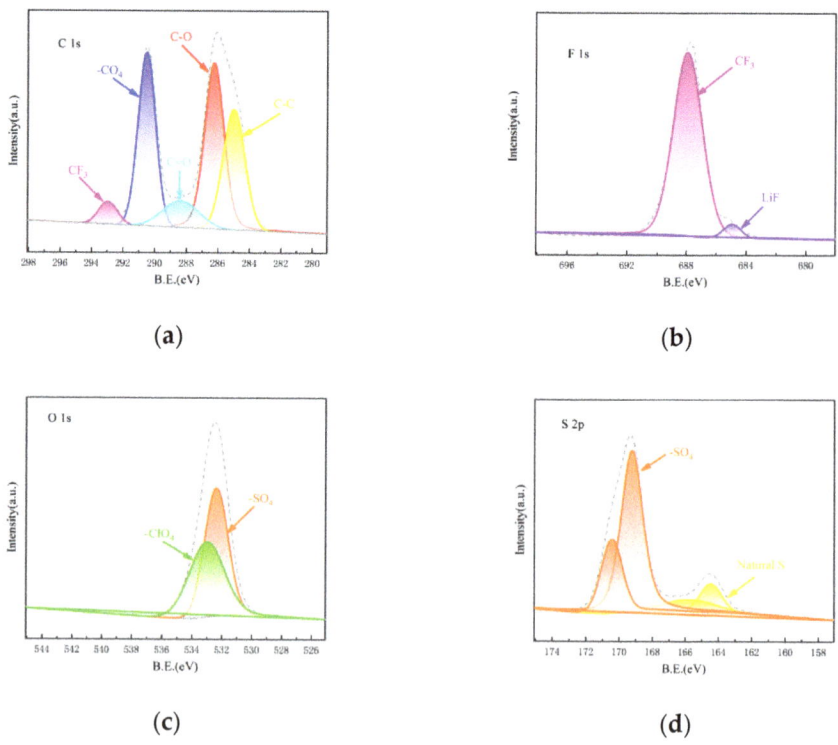

Figure 7. XPS of the CEPs after 500 cycles (**a**) C 1s (**b**) F 1s (**c**) O 1s (**d**) S 2p.

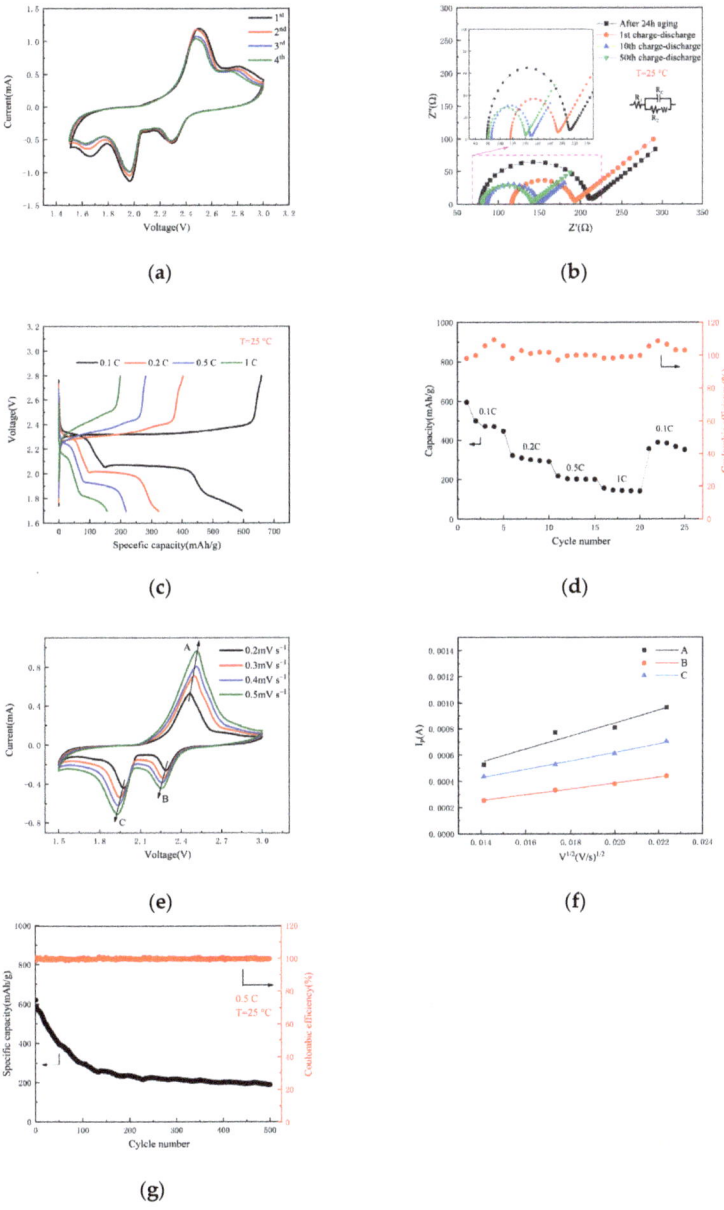

Figure 8. (**a**) CV curves of the Li | PVDF/LiClO$_4$/LATP | NPCNF/S cell at a scan rate of 0.2 mV s^{-1}; (**b**) EIS of the Li | PVDF/LiClO$_4$/LATP | NPCNF/S cell at different cycle numbers; (**c**) voltage-specific capacity curve of the Li | PVDF/LiClO$_4$/LATP | NPCNF/S cell; (**d**) rate performance; (**e**) CV curves of the cell in the range of 0.2–0.5 mV s^{-1}; (**f**) corresponding linear fits of the peak currents of the cell; (**g**) cyclic performance of the Li | PVDF/LiClO$_4$/LATP | NPCNF/S cell at 0.5 C.

4. Conclusions

In this paper, NPCNF as a high-efficiency conductive skeleton of sulfur electrode active material was prepared by template method. The PVDF/LiClO$_4$/LATP electrolyte with good performance was prepared via mixed solution casting method. The NPCNF has a good morphology, such as with nano fiber, having an obvious adsorption effect on polysulfide, and the sulfur content can reach 62.9 wt%. The active material has a high carbonation degree of graphite and good electrical conductivity. At 25 °C, the ionic conductivity of PVDF/LiClO$_4$/LATP electrolyte is 8.07×10^{-5} S cm^{-1}, and the lithium-ion transfer rate can reach 0.77. With the addition of nanoscale LATP, the overall performance of CPEs is better than that of a garnet type and PVDF-based solid electrolytes. The assembled cell has a low impedance, and the RCT value of the first ring is 193.5 Ω. The battery has a good rate performance and can work at 1 C, maintaing a certain specific capacity. At 25 °C and 0.5 C, the specific discharge capacity of 500 cycles is 620.52 mAh g^{-1}, and the capacity decay rate of each cycle is only 0.139%. This method for preparing the excellent sulfur positive electrode, combined with the composite electrolyte membrane, provides a new idea for improving the long cycle performance of solid-state lithium-sulfur batteries at room temperature.

Author Contributions: X.L.: Supervision, resources, conceptualization, writing—review and editing, project administration, and funding acquisition; Y.Z.: Investigation, methodology, writing—original draft, formal analysis, and data curation; Y.N.: data curation; D.H.: investigation; L.L.: funding acquisition; S.L.: writing—review and editing. All authors have read and agreed to the published version of the manuscript.

Funding: This research was supported by Guangxi Natural Science Foundation (No. 2020GXNS-FAA297082), Guangxi Innovation Driven Development Project (No. AA18242036-2); the Fund Project of the Key Lab of Guangdong for ModernSurface Engineering Technology (No. 2018KFKT01); the National Natural Science Foundation of China (No. 52161033).

Data Availability Statement: The data that support the findings of this study are available upon reasonable request.

Acknowledgments: We are thankful to the following authors for their contributions, as well as for the purchase of materials and conducting the relative experiments: Yu Zhang, Xinghua Liang, Yujuan Ning, Dongxue Huang, Linxiao Lan and Siying Li.

Conflicts of Interest: The authors declare no conflict of interest.

References

1. Pang, Q.; Liang, X.; Kwok, C.Y.; Nazar, L.F. Advances in lithium-sulfur batteries based on multifunctional cathodes and electrolytes. *Nat. Energy* **2016**, *1*, 16132. [CrossRef]
2. Fu, A.; Wang, C.; Pei, F.; Cui, J.; Fang, X.; Zheng, N. Recent advances in hollow porous carbon materials for lithium-sulfur batteries. *Small* **2019**, *15*, 1804786. [CrossRef] [PubMed]
3. Fan, L.L.; Deng, N.P.; Yan, J.; Li, Z.H.; Kang, W.M.; Cheng, B.W. The recent research status quo and the prospect of electrolytes for lithium sulfur batteries. *Chem. Eng. J.* **2019**, *369*, 874–897. [CrossRef]
4. Liu, Y.J.; Zhou, P.; Hao, S. Rechargeable solid-state Li-Air and Li-S batteries: Materials, construction, and challenges. *Adv. Energy Mater.* **2018**, *8*, 1701602. [CrossRef]
5. Urbonaite, S.; Poux, T.; Petr, N. Progress towards commercially viable Li-S battery cells. *Adv. Energy Mater.* **2015**, *5*, 1500118. [CrossRef]
6. Skilton, R.A.; Bourne, R.A.; Amara, Z.; Horvath, R.; Jin, J.; Scully, M.J.; Streng, E.; Tang, S.L.Y.; Summers, P.A.; Wang, J. Remote-controlled experiments with cloud chemistry. *Nat. Chem.* **2014**, *7*, 1–5. [CrossRef]
7. Tao, X.; Chen, X.; Xia, Y.; Huang, H.; Gan, Y.; Wu, R.; Chen, F.; Zhang, W. Highly mesoporous carbon foams synthesized by a facile, cost-effective and template-free Pechini method for advanced lithium-sulfur batteries. *J. Mater. Chem. A* **2013**, *1*, 3295–3301. [CrossRef]
8. Zheng, G.; Yang, Y.; Cha, J.J.; Hong, S.S.; Cui, Y. Hollow carbon nanofiber-encapsulated sulfur cathodes for high specific capacity rechargeable lithium batteries. *Nano Lett.* **2011**, *11*, 4462–4467. [CrossRef]
9. Manthiram, A.; Fu, Y.; Chung, S.H.; Zu, C.; Su, Y.S. Rechargeable lithium-sulfur batteries. *Chem. Rev.* **2014**, *114*, 11751–11787. [CrossRef]
10. Mangani, L.R.; Villevieille, C. Mechanical vs. chemical stability of sulphide-based solid-state batteries. Which one is the biggest challenge to tackle? Overview of solid-state batteries and hybrid solid state batteries. *J. Mater. Chem. A* **2020**, *8*, 10150–10167. [CrossRef]
11. Lei, D.; Shi, K.; Ye, H.; Wan, Z.; Wang, Y.; Shen, L.; Li, B.; Yang, Q.H.; Kang, F.; He, Y.B. Solid-state electrolytes: Progress and perspective of solid-state lithium-sulfur batteries. *Adv. Funct. Mater.* **2018**, *28*, 1870272. [CrossRef]

12. Jeong, B.O.; Kwon, S.W.; Kim, T.J.; Lee, E.H.; Jeong, S.H.; Jung, Y. Effect of carbon black materials on the electrochemical properties of sulfur-based composite cathode for lithium-sulfur cells. *J. Nanosci. Nanotechnol.* **2013**, *13*, 7870–7874. [CrossRef]
13. Li, X.; Sun, X.L. Nitrogen-doped carbons in Li-S batteries: Materials design and electrochemical mechanism. *Front. Energy Res.* **2014**, *2*, 49. [CrossRef]
14. Yu, J.; Lee, M.; Kim, Y.; Lim, H.K.; Chae, J.; Hwang, G.S.; Lee, S. Agent molecule modulated low-temperature activation of solid-state lithium-ion transport for polymer electrolytes. *J. Power Sources* **2021**, *505*, 229917–229924. [CrossRef]
15. Chen, X.; He, W.; Ding, L.X.; Wang, S.; Wang, H. Enhancing interfacial contact in all solid state batteries with a cathode-supported solid electrolyte membrane framework. *Energy Environ. Sci.* **2019**, *12*, 938–944. [CrossRef]
16. Zheng, Y.; Yao, Y.; Ou, J.; Li, M.; Luo, D.; Dou, H.; Li, Z.; Amine, K.; Yu, A.; Chen, Z. A review of composite solid-state electrolytes for lithium batteries: Fundamentals, key materials and advanced structures. *Chem. Soc. Rev.* **2020**, *49*, 8790–8839. [CrossRef]
17. Zuo, C.; Yang, M.; Wang, Z.; Jiang, K.; Li, S.; Luo, W.; He, D.; Liu, C.; Xie, X.; Xue, Z. Cyclophosphazene-based hybrid polymer electrolyte via epoxyamine reaction for high-performance all-solid-state lithium-ion batteries. *J. Mater. Chem. A* **2019**, *7*, 18871–18879. [CrossRef]
18. Chen, Y.; Shi, Y.; Liang, Y.; Dong, H.; Hao, F.; Wang, A.; Zhu, Y.; Cui, X.; Yao, Y. Hyperbranched PEO-based hyperstar solid polymer electrolytes with simultaneous improvement of ion transport and mechanical strength. *ACS Appl. Energy Mater.* **2019**, *2*, 1608–1615. [CrossRef]
19. Zhai, P.; Peng, N.; Sun, Z.; Wu, W.; Kou, W.; Cui, G.; Zhao, K.; Wang, J. Thin laminar composite solid electrolyte with high ionic conductivity and mechanical strength towards advanced all-solid-state lithium-sulfur battery. *J. Mater. Chem. A* **2020**, *8*, 23344–23353. [CrossRef]
20. Barbosa, J.C.; Correia, D.M.; Gonçalves, R.; de Zea Bermudez, V.; Silva, M.M.; Lanceros-Mendez, S.; Costa, C.M. Enhanced ionic conductivity in poly(vinylidene fluoride) electrospun separator membranes blended with different ionic liquids for lithium ion batteries. *J. Colloid Interface Sci.* **2021**, *582*, 376–386. [CrossRef]
21. Zhang, Y.; Zhao, Y.; Gosselink, D.; Chen, P. Synthesis of poly(ethylene-oxide)/nanoclay solid polymer electrolyte for all solid-state lithium/sulfur battery. *Ionics* **2015**, *21*, 381–385. [CrossRef]
22. McGrogan, F.P.; Swamy, T.; Bishop, S.R.; Eggleton, E.; Porz, L.; Chen, X.; Chiang, Y.M.; Van Vliet, K.J. Compliant yet brittle mechanical behavior of Li_2S-P_2S_5 lithium-ion-conducting solid electrolyte. *Adv. Energy Mater.* **2017**, *7*, 1602011. [CrossRef]
23. Li, M.; Bai, Z.; Li, Y.; Ma, L.; Dai, A.; Wang, X.; Luo, D.; Wu, T.; Liu, P.; Yang, L.; et al. Electrochemically primed functional redox mediator generator from the decomposition of solid state electrolyte. *Nat. Commun.* **2019**, *10*, 1890–1898. [CrossRef]
24. Li, M.; Frerichs, J.E.; Kolek, M.; Sun, W.; Zhou, D.; Huang, C.J.; Hwang, B.J.; Hansen, M.R.; Winter, M.; Bieker, P. Solid-state lithium-sulfur battery enabled by Thio-LiSICON/Polymer composite electrolyte and sulfurized polyacrylonitrile cathode. *Adv. Funct. Mater.* **2020**, *30*, 1910123. [CrossRef]
25. Deng, S.; Sun, Q.; Li, M.; Adair, K.; Yu, C.; Li, J.; Li, W.; Fu, J.; Li, X.; Li, R.; et al. Insight into cathode surface to boost the performance of solid-state batteries. *Energy Storage Mater.* **2021**, *35*, 661–668. [CrossRef]
26. Schell, K.G.; Bucharsky, E.C.; Lemke, F.; Hoffmann, M.J. Effect of calcination conditions on lithium conductivity in $Li_{1.3}Ti_{1.7}Al_{0.3}(PO_4)_3$ prepared by sol-gel route. *Ionics* **2016**, *23*, 821–827. [CrossRef]
27. Han, J.P.; Zhang, B.; Wang, L.Y.; Zhu, H.L.; Qi, Y.X.; Yin, L.; Li, N.; Lun, N.; Bai, Y.J. $Li_{1.3}Al_{0.3}Ti_{1.7}(PO_4)_3$ behaving as a fast ionic conductor and bridge to boost the electrochemical performance of $Li_4Ti_5O_{12}$. *ACS Sustain. Chem. Eng.* **2018**, *6*, 7273–7282. [CrossRef]
28. Zheng, F.; Kotobuki, M.; Song, S.; Lai, M.O.; Lu, L. Review on solid electrolytes for all-solid-state lithium-ion batteries. *J. Power Sources* **2018**, *389*, 198–213. [CrossRef]
29. Gao, H.; Xue, L.; Xin, S.; Goodenough, J.B. A high-energy-density potassium battery with a polymer-gel electrolyte and a polyaniline cathode. *Angew. Chem. Int. Ed. Engl.* **2018**, *130*, 5449–5453. [CrossRef]
30. Fu, C.; Ma, Y.; Zuo, P.; Zhao, W.; Gao, Y.; Tang, Y.; Yin, G.; Wang, J. In-situ thermal polymerization boosts succinonitrile-based composite solid-state electrolyte for high performance Li-metal battery. *J. Power Sources* **2021**, *496*, 229861. [CrossRef]
31. Cheng, D.; Zhao, Y.; An, T.; Wang, H.; Zhou, H.; Fan, T. 3d interconnected crumpled porous carbon sheets modified with high-level nitrogen doping for high performance lithium sulfur battery. *Carbon* **2019**, *154*, 58–66. [CrossRef]
32. Xu, Z.L.; Kim, J.K.; Kang, K. Carbon nanomaterials for advanced lithium sulfur batteries. *Nano Today* **2018**, *19*, 84–107. [CrossRef]
33. Cai, X.; Lei, T.; Sun, D.; Lin, L. A critical analysis of the α, β and γ phases in poly(vinylidene fluoride) using ftir. *Rsc Adv.* **2017**, *7*, 15382–15389. [CrossRef]
34. Shan, Y.; Li, L.; Yang, X. Solid-state polymer electrolyte solves the transfer of lithium ions between the solid–solid interface of the electrode and the electrolyte in lithium–sulfur and lithium-ion batteries. *ACS Appl. Energy Mater.* **2021**, *4*, 5101–5112. [CrossRef]
35. Baskaran, R.; Selvasekarapandian, S.; Kuwata, N.; Kawamura, J.; Hattori, T. Conductivity and thermal studies of blend polymer electrolytes based on PVAc–PMMA. *Solid State Ion.* **2006**, *177*, 2679–2682. [CrossRef]
36. Cho, Y.G.; Hwang, C.; Cheong, D.S.; Kim, Y.S.; Song, H.K. Gel/solid polymer electrolytes characterized by in situ gelation or polymerization for electrochemical energy systems. *Adv. Mater.* **2019**, *31*, 1804909. [CrossRef]
37. Kim, D.H.; Hwang, S.; Cho, J.J.; Yu, S.; Kim, S.; Jeon, J.; Ahn, K.H.; Lee, C.; Song, H.K.; Lee, H. Toward fast operation of lithium batteries: Ion activity as the factor to determine the concentration polarization. *ACS Energy Lett.* **2019**, *4*, 1265–1270. [CrossRef]
38. Bag, S.; Zhou, C.; Kim, P.J.; Pol, V.G.; Thangadurai, V. Lif modified stable flexible PVDF-garnet hybrid electrolyte for high performance all-solid-state li-s batteries. *Energy Storage Mater.* **2019**, *24*, S2405–S8297. [CrossRef]

39. Zhang, Q.; Wang, Q.; Huang, S.; Jiang, Y.; Chen, Z. Preparation and electrochemical study of PVDF-HFP/LATP/g-C$_3$N$_4$ composite polymer electrolyte membrane. *Inorg. Chem. Commun.* **2021**, *131*, 108793–108799. [CrossRef]
40. Yue, Q.; Shi, S.; Hong, L.; Jr, L. Defect thermodynamics and diffusion mechanisms in Li2CO3 and implications for the solid electrolyte interphase in Li-ion batteries. *J. Phys. Chem. B* **2013**, *117*, 8579–8593.
41. Bai, Y.; Li, L.; Li, Y.; Chen, G.; Zhao, H.; Wang, Z.; Li, H.; Wu, F.; Wu, C. Reversible and irreversible heat generation of nca/si–c pouch cell during electrochemical energy-storage process. *Energy Storage Mater.* **2019**, *16*, 411–418. [CrossRef]
42. Yuan, Y.; Feng, W.; Ying, B.; Li, Y.; Chen, G.; Wang, Z.; Wu, C. Regulating li deposition by constructing lif-rich host for dendrite-free lithium metal anode. *Energy Storage Mater.* **2018**, *16*, 411–418. [CrossRef]
43. Chao, X.; Bing, S.; Gustafsson, T.; Edstrm, K.; Hahlin, M. Interface layer formation in solid polymer electrolyte lithium batteries: An xps study. *J. Mater. Chem. A* **2014**, *2*, 7256–7264.
44. Fang, R.; Xu, H.; Xu, B.; Li, X.; Li, Y.; Goodenough, J.B. Reaction mechanism optimization of solid-state Li-S batteries with a PEO-based electrolyte. *Adv. Funct. Mater.* **2020**, *31*, 2001812. [CrossRef]
45. Liang, X.; Hart, C.; Pang, Q.; Garsuch, A.; Weiss, T.; Nazar, L.F. A highly efficient polysulfide mediator for lithium−sulfur batteries. *Nat. Commun.* **2015**, *6*, 5682–5689. [CrossRef]
46. Zhang, X.; Chen, K.; Sun, Z.; Hu, G.; Xiao, R.; Cheng, H.M.; Li, F. Structure-related electrochemical performance of organosulfur compounds for lithium-sulfur batteries. *Energy Environ. Sci.* **2020**, *13*, 1076–1095. [CrossRef]

Article

Aramid Fibers Modulated Polyethylene Separator as Efficient Polysulfide Barrier for High-Performance Lithium-Sulfur Batteries

Jifeng Gu [1], Jiaping Zhang [1], Yun Su [1] and Xu Yu [2,*]

[1] College of Physics and Electronic Engineering, Xinxiang University, Xinxiang 453003, China; zk_gujifeng@163.com (J.G.); wdsysx@163.com (J.Z.); bit-1@163.com (Y.S.)
[2] School of Chemistry and Chemical Engineering, Yangzhou University, Yangzhou 225002, China
* Correspondence: yxypz15@yzu.edu.cn

Abstract: The separators with high absorbability of polysulfides are essential for improving the electrochemical performance of lithium–sulfur (Li–S) batteries. Herein, the aramid fibers coated polyethylene (AF-PE) films are designed by roller coating, the high polarity of AFs can strongly increase the binding force at AF/PE interfaces to guarantee the good stability of the hybrid film. As confirmed by the microscopic analysis, the AF-PE-6 film with the nanoporous structure exhibits the highest air permeability by the optimal coating content of AFs. The high absorbability of polysulfides for AF-PE-6 film can effectively hinder the migration of polysulfides and alleviate the shuttle effect of the Li–S battery. AF-PE-6 cell shows the specific capacity of 661 mAh g^{-1} at 0.1 C. After 200 charge/discharge cycles, the reversible specific capacity is 542 mAh g^{-1} with the capacitance retention of 82%, implying the excellent stability of AF-PE-6. The enhanced cell performance is attributed to the porous architecture of the aramid layer for trapping the dissolved sulfur-containing species and facilitating the charge transfer, as confirmed by SEM and EDS after 200 cycles. This work provides a facile way to construct the aramid fiber-coated separator for the inhibition of polysulfides in the Li–S battery.

Keywords: nanoporous; aramid fibers; polysulfides; Li–S battery

1. Introduction

Lithium–sulfur (Li–S) batteries with a high energy density (2600 Wh kg^{-1}) and high theoretical specific capacity (1675 mAh g^{-1}) have been considered the promising energy storage system in practical applications [1,2]. However, the poor electrical conductivity, large volume expansion, and the shuttle mechanism of polysulfides are still important factors to affect the performance of Li–S batteries, such as the fast degradation of specific capacity [2–7]. The shutting mechanism is induced by the free migration of polysulfides anions between cathode and anode during the charge/discharge process, which not only reduces the utilization of active materials but also results in the low columbic efficiency of Li–S batteries [8–10].

Recently, the efforts in the modification of host materials have been reviewed to solve the solubility and diffusion of polysulfides [11–13]. The separator is an important component to guarantee the safety of Li–S batteries by separating the cathode and anode without direct contact. The modification of separators is an effective strategy to solve the polysulfide diffusion in an organic electrolyte, such as coating some porous materials on the separator surface, which can act as a physical barrier to block the polysulfide diffusion during cycling. The coating of porous carbon or carbon nanotubes on commercial polypropylene (PP) separators has been reported to show the improved electrochemical performance for Li–S batteries by effectively trapping the polysulfides [14,15]. Meanwhile, many reported lectures demonstrate that the high porosity of metal oxides or sulfides (MnO$_2$, TiO$_2$, Al$_2$O$_3$, and MoS_2) as a coating layer can also enhance the capacitive behavior of Li–S batteries due to their polarity, hydrophilic property, and high absorbability by interacting with polysulfides [15–20]. However, these porous materials are almost nonpolar or weakly polar conductive materials, which only own a single physical barrier to the dissolution of lithium polysulfides. Considering the high polarity surface of the separator, the coating layer may be peeled off from the separator after long-term charge/discharge cycles because of the weak binding force between inorganic coating materials and the separator.

The exploration of high polarity coating materials is necessary to be studied. Aramid fibers (AFs) serving as new building blocks have attracted attention owing to the low cost, high strength, high-temperature resistance, and excellent dimensional stability [21–23], which is favorable to improve the mechanical property of the separator and the ionic conductivity of batteries [22]. Yang and co-workers report an aramid nanofiber/bacterial cellulose (ANFs/BC) composite, which exhibits increased ionic conductivity and interfacial compatibility. The electrochemical performance is significantly improved as the optimal ANFs/BC as the separator in the battery cell [22]. However, the aramid fibers coating on polyethylene (PE) membrane with the nanoporous structures as the separator for Li–S batteries has yet to be explored.

Herein, the aramid fibers coating on PE membrane (AF-PE) was prepared, and the effect of coating content on the porosity of AF-PE films is discussed. AF-PE-6 with the optimal coating content shows better air permeability than other AF-PE films. As the AF-PE-6 film applies as the separator in a lithium–sulfur battery, the electrochemical performance is dramatically enhanced in contrast with pristine PE films, such as the high specific capacity of 661 mAh g^{-1} at 0.1 C, and 38.9% of capacity retention from 0.1 to 1 C. Especially, the specific capacity remained the value of 542 mAh g^{-1} after 200 charge/discharge cycles. This work provides a facile synthetic route to prepare the high polarity separators to further enhance the capacitive performance of the Li–S batter.

2. Experimental Section

2.1. Materials

The PE microporous separators (GRE-20, Green Inc, Xinxiang, China, 20 μm) were used as base membranes. The organic electrolyte was made by dissolving 1M bis-(trifluoromethane)sulfonamide lithium (LiTFSI) into a 10 mL mixed solvent of dimethoxyethane (DME) and 1,3-dioxolane (DOL) in 2:1 volume ratio to test the electrochemical performances. Aramid fibers were purchased from Dongbang special fiber Co., Ltd. (Zhangjiagang, China).

2.2. Preparation of AF-PE Separators

Aramid fibers have a limited solubility in NMP (or DMF, DMAc); however, fast dissolution could be obtained after adding a certain amount of salt (LiCl or $CaCl_2$). Hence 3 g of aramid fibers and 0.3 g of LiCl (Aladdin Industrial Co., Shanghai, China) were mixed with the mass ratio of 10:1 as the co-solvent and subsequently immersed in 50 g of N-Methy pyrrolidone (NMP, Aladdin Industrial Co., Shanghai, China) under magnetic stirring at 65 °C for 8 h. The obtained transparent solution was coated on polyethylene (PE) separators by a simple roller coating technology, and then the separator was immersed in deionized water for about 5 min. Finally, the separator was dried at 55 °C for 12 h under vacuum conditions and named AF-PE-6. As for comparisons, the different mass ratios of aramid fibers and co-solvent (5:1 and 15:1) were treated at the same synthetic process and noted as AF-PE-3 and AF-PE-9.

2.3. Characterization

Surface morphology of AF-PE and PE films with different strains were observed using Zeiss Scanning electron microscopy (SEM, SIGMA, ZEISS, Oberkochen, Germany) and energy dispersive X-ray analysis (EDX) at an accelerating voltage of 5 kV. The average pore size, pore size distribution and porosity were evaluated using a through pore size analyzer instrument (porosimeter 3G, Quantachrome Instruments, Boynton, FL, USA). Permeability test was evaluated using Gurley test instruments 4410N.

2.4. Electrochemical Measurements

The electrochemical performance of AF-PE separators was measured by assembling CR2032 type coin cells in the glove box (Mbraun, M. Braun Inertgas-Systeme GmbH, Garching, Germany) at Argon atmosphere. The sulfur cathodes were prepared by a conventional slurry coating method with a doctor blade, and the detailed process was listed as follows: 80 wt% of pure sulfur (Sigma-Aldrich, St. Louis, MO, USA), 10 wt% of carbon black (Super P), and 10 wt% of PVDF were placed in an agate mortar and ground with adding few drops NMP as the solvent for 40 min. The obtained slurry was pasted onto the aluminum foil and dried in a vacuum oven at 60 °C overnight. The mass loading of the sulfur cathodes was about 2.5 mg cm^{-2}. The sulfur cathode acted as the working electrode, pristine PE and AF-PE-6 films as the separator and lithium metals (Sigma-Aldrich) as the anode electrode, the assembled batteries were named pristine PE cell and AF-PE-6 cell. Galvanostatic charge/discharge (GCD) measurement was performed at different current densities in the voltage range of 1.8–2.8 V with program-controlled battery test equipment (LAND CT2001A, Wuhan LAND Electronic Co.Ltd., Wuhan, China). Ionic conductivities of the membrane with electrolyte were measured by sandwiching it between two stainless steel electrodes, and the ionic conductivity was calculated using formula: $\sigma = d/RA$, where d was the thickness, A was the separator effective area of a membrane and R was the bulk resistance. Ionic conductivities and the electrochemical impedance spectroscopy (EIS) were measured by an electrochemical workstation (CHI660E, Shanghai, China) over a frequency range of 1 Hz–100 kHz with an AC voltage amplitude of 5 mV. For comparison, the assembled CR2032 coin cell with commercial PE film as a separator was measured under the same condition.

3. Results and Discussion

The aramid fibers coating on polyethylene (PE) membrane (AF-PE) were prepared by combining a simple bar coating process and low-temperature vacuum drying methods in Figure 1a. The existence of a co-solvent of aramid/LiCl is favorable to obtain the homogeneous solution, and the treatment in DI water aims to cure the film and remove the residues or co-solvent. Scanning electron microscopy (SEM) was characterized to reveal the morphological structure of AF-PE and pristine PE films. It can be found that the pristine PE film exhibits cross-linked internetworks and porous structures in Figure 1b. The morphological structure of AF-PE films is affected by the coating content of aramid fibers on the PE surface by filling with the porous structure (Figure 1c–e). AF-PE-6 with optimal coating content of AF shows the increased porosity including mesopores and micro-pores in Figure 1d, which is better than AF-PE-3 with the insufficient content of AFs and AF-PE-9 with the over-coating of AFs. The average pore size of all samples is shown in Figure 1f, and the AF-PE-6 owns the value of 98.9 nm, which is larger than that of AF-PE-3 (81.1 nm), AF-PE-9 (59.9 nm), and is smaller than pristine PE film (104.7 nm), respectively. The porous structure is favorable for the fast electrolyte diffusion, while the tortuous pores of AF-PE-6 can localize the polysulfide species diffusing from the cathode to the anode sites. The porous structure of the aramid-coated separator can also be quantitatively characterized by measuring the Gurley value and porosity. The air permeability of AF-PE is affected by the coating content of AF. In comparison to pristine PE film (278 s), the Gurley value of AF-PE-3 increased resulting from the AF coating. However, the Gurley value of AF-PE-6 (440 s) is smaller than these of AF-PE-3 (571 s) and AF-PE-9 (760 s), implying a better air permeability of AF-PE-6. The porosity of AF-PE-6 (49.5 + 1.5%) is higher than pristine PE (37 + 0.5%), AF-PE-3 (43.2 + 0.7%) and AF-PE-9 (41.7 + 1.4%), which is attributing to the optimal coating content of AFs in Figure 1g. This result is attributed to the optimal coating content of AF because the insufficient content of AF blocks the original pores of pristine PE film and the overloading of AFs results in the increased densification of AF-PE-9 film.

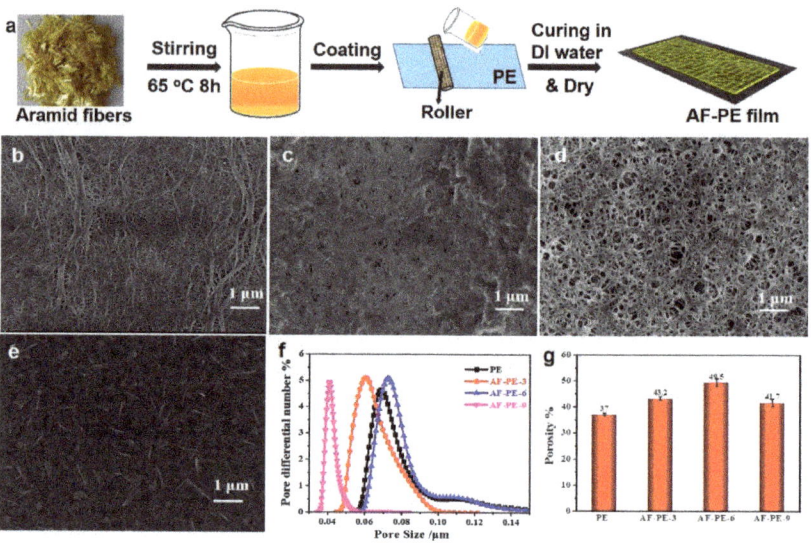

Figure 1. (**a**) Schematic illustration of the preparation of AF-PE films. SEM images of: (**b**) pristine PE film; (**c**) AF-PE-3; (**d**) AF-PE-6; and (**e**) AF-PE-9; (**f**) The pore size distribution of pristine PE, AF-PE-3, AF-PE-6 and AF-PE-9 films; (**g**) Porosity of PE and aramid-coated separator.

To initially evaluate the quality of AF-PE and pristine PE film, the ionic conductivity of separators and the electrochemical performance of the constructed Li–S cells are compared with the AF-PE films and commercial sulfur as the separators and cathode material. As shown in Figure 2a, the ion conductivity value of the AF-PE-6 film can reach up to about 0.57 mS cm^{-1}, which is almost 2.5, 1.6 and 1.7 times larger than that of PE (0.23 mS cm^{-1}), AF-PE-3 (0.36 mS cm^{-1}) and PF-PE-9 (0.33 mS cm^{-1}). The large ion conductivity for AF-PE-6 is assigned to the increased porosity. The resistance of the fresh Li–S cells is confirmed by electrochemical impedance spectroscopy (EIS) in Figure 2b. The semicircle at the high-to-medium frequency and an inclined line at low frequency correspond to the charge transfer resistance (R_{ct}) and mass transfer process, respectively. AF-PE-6 cell shows a smaller diameter of the semicircle and steeper slope line than that of pristine PE cell, implying a faster charge transfer kinetics. The R_{ct} value for AF-PE-6 cell (21.56 Ω) is smaller than that of PE cell (260.1 Ω), attributing to the enhanced affinity and wettability for the accumulation of polar liquid electrolytes by coating optimal content aramid fibers [24,25]. Especially, AF-PE-9 cell shows the highest R_{ct} value, and the overloading of aramid fiber can increase the density of film and decrease the average pore size, which is not favorable for the electrolyte ion passing through the film.

The electrochemical performance of pristine PE cell and AF-PE-6 cell is further confirmed by galvanostatic charge/discharge (GCD) at 0.1 C with the applied potential range from 1.8 to 2.8 V. Figure 3b show the GCD curves of pristine PE cell and AF-PE-6 cell, and both cells display two voltage plateaus arising from the two steps redox reaction of elemental sulfur with metallic lithium during the discharge process. Interestingly, AF-PE-6 cell owns a lower charge plateau potential and higher discharge plateau potential than this pristine PE cell, and a smaller potential separation between charge and discharge plateau indicates a better kinetic behavior for AF-PE-6 cell. For the first cycle, the discharge capacity of AF-PE-6 cell is 731 mAh g^{-1} with the charge capacity of 753 mAh g^{-1} at 0.1 C, which decreases to 687 mAh g^{-1} for the second cycle and 665 mAh g^{-1} for the fifth cycle, respectively. The degradation of the discharge capacity can be attributed to the formation of the SEI layer. It can be found that the discharge capacity is almost stable after 5 charge/discharge cycles. The GCD curves of pristine PE cell and AF-PE-6 cell at

0.1 C are shown in Figure 3b, and the discharge capacity of AF-PE-6 cell is 687 mAh g^{-1} higher than that of pristine PE cell (638 mAh g^{-1}). As the C rate increased by a factor of 10 (Figure 3c), pristine PE cell owns the discharge capacity of 72 mAh g^{-1} with the charge capacity of 74 mAh g^{-1}, and the calculated capacity retention is only 11.2%, respectively. The discharge capacity is 267 mAh g^{-1} for AF-PE-6 cell with a capacity retention of 38.9%, which is higher than that of pristine PE cell (Figure 3d). The high discharge capacity and good capacitance retention for AF-PE-6 cell reflect the enhanced sulfur utilization to tolerate the high charge currents.

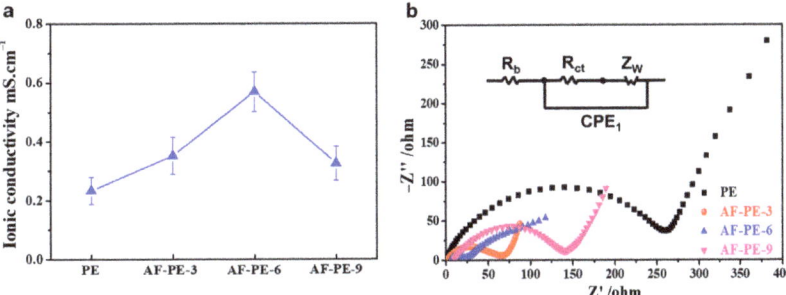

Figure 2. (**a**) The ionic conductivity of PE and AP-PE films, (**b**) Nyquist plots of pristine PE, AF-PE-3, AF-PE-6 and AF-PE-9.

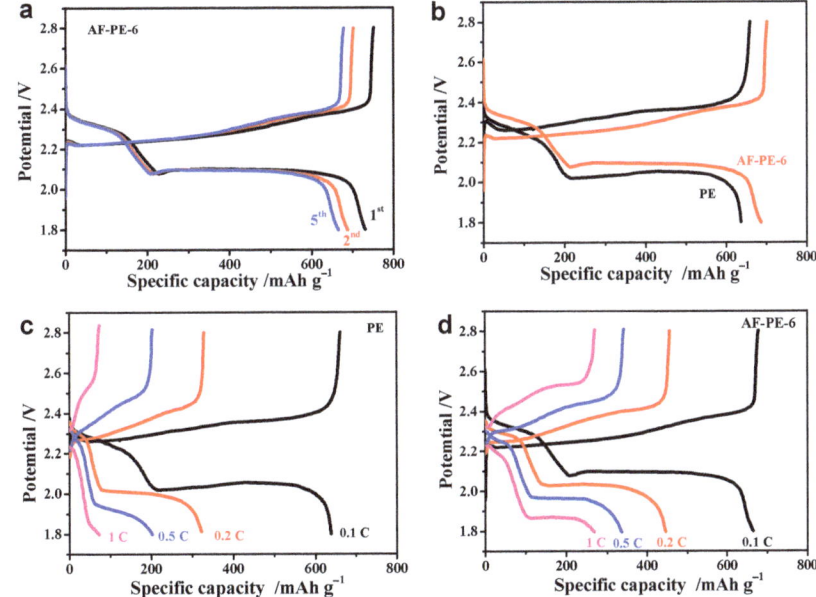

Figure 3. (**a**) The 1st, 2nd and 5th cycle of GCD curves of AF-PE-6 cell; (**b**) The GCD curves of pristine PE and AF-PE-6 cells. The GCD curves at different C rates of (**c**) pristine PE; and (**d**) AF-PE-6 cells.

Furthermore, the stability of pristine PE and AF-PE-6 cells were evaluated at different C rates with 10 cycles for each in Figure 4a. As the C rates increased from 0.1 C to 1 C, the specific discharge capacity is decreased from 661 to 247 mAh g^{-1} for AF-PE-6 cell, and then the discharge capacity keeps at the value of 618 mAh g^{-1} as the C rate returns to 0.1 C. The loss of discharge capacity is only 43 mAh g^{-1}. In comparison, the discharge capacity of pristine PE cell dramatically decreases from 656 mAh g^{-1} at 0.1 C to 61 mAh g^{-1}

at 1 C. The improved electrochemical performance of AF-PE-6 cell can be attributed to the limitation of polysulfides at the sulfur cathode/aramid-coated separator interface by physical absorption and electrochemical deposition. The cycling test of pristine PE and AF-PE-6 cells is carried out by GCD at 0.1 C for 200 cycles. Figure 4b,c shows the GCD curves of pristine PE and AF-PE-6 cells at different cycles. It can be found that the specific discharge capacity of AF-PE-6 cell is only 622, 603 and 542 mAh g^{-1} loss for the 50th cycle, 100th and 200th cycles. After 200 charge/discharge cycles, the capacity retention of AF-PE-6 cell (81.9%) is higher than that of the pristine PE cell (64.8%) in Figure 4d. Meanwhile, the related coulombic efficiency of AF-PE-6 cell is higher than that of the PE cell. The excellent rate capability and cyclic stability for AF-PE-6 cell can be attributed to the blocking effect of aramid coating on the polysulfides.

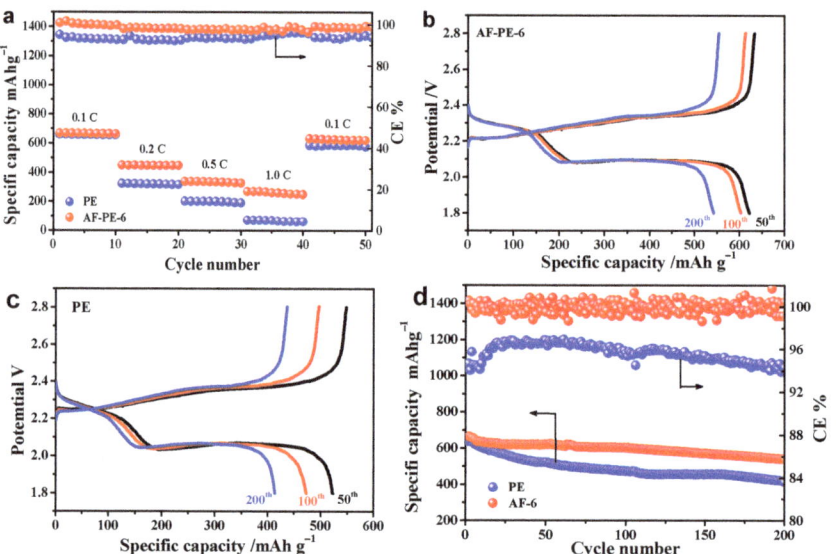

Figure 4. (a) The specific capacity and the related coulombic efficiency of PE and AF-PE-6 cells at various C rates with 10 cycles for each. The GCD curves of: (b) AF-PE-6 and (c) PE cells at the 50th, 100th and 200th cycles; (d) The cyclic performance of PE and AF-PE-6 cells and the related coulombic efficiency.

Figure 5a shows the EIS result of pristine PE and AF-PE-6 cells after 200 GCD cycles. AF-PE-6 cell exhibits two depressed semicircles at high and middle frequency, and an inclined line at low frequency. The R_s value of AF-PE-6 cell is smaller than that of pristine PE cell, implying the efficient inhibition of polysulfides by coating aramid fibers. In comparison to pristine PE cell, the R_{ct} values of AF-PE-6 cell is significantly decreased after 200 cycles, and a significant decrease in charge transfer resistance is attributed to the dissolution and redistribution of the active materials during the chemical activation process [26].

The change in surface morphology of pristine PE and AF-PE-6 films after 200 cycles is probed by SEM. In comparison to pristine PE film, the color of AF-PE-6 film becomes yellow (Figure 5b), arising from the interception and adsorption of soluble polysulfides by the optimal coating content of aramid fibers. Figure 5c,d are the SEM images of pristine PE and AF-PE-6 films after the cycling test. In comparison to pristine PE film, AF-PE-6 film with a small pore size can not only block the polysulfides, but also as a barrier for trapping the polysulfides. The elemental mapping of S for pristine PE and AF-PE-6 films is disclosed in Figure 5e,f. The existence and uniform distribution of S can be found on the surface of films. However, the content of elemental S on AF-PE-6 film is more than that of PE film, indicating that the coating of aramid fiber is favorable for the absorption of polysulfides. Therefore, this work provides a promising strategy to construct the separators to efficiently suppress the shuttling mechanism of polysulfides for Li–S batteries.

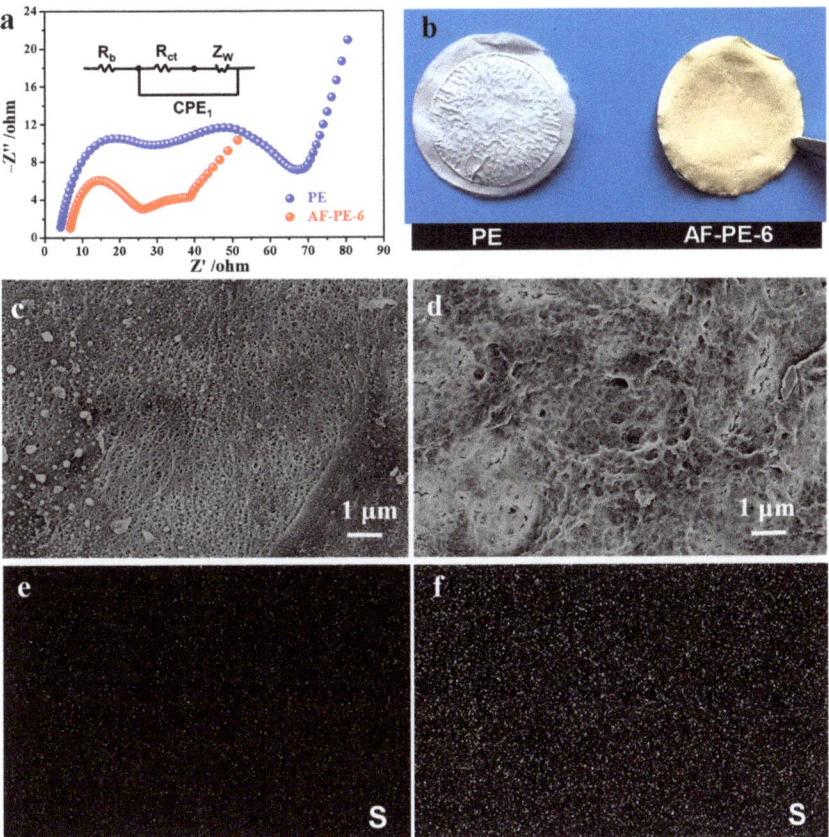

Figure 5. (**a**) Nyquist plots and (**b**) the optical images of pristine PE film and AF-PE-6 film after 200 cycles. SEM images of (**c**) pristine PE and (**d**) AF-PE-6 films after 200 cycles. Elemental mapping of S of (**e**) pristine PE and (**f**) AF-PE-6 films after 200 cycles.

4. Conclusions

The AF-PE-6 film with high porosity has been prepared and acts as the separator for the Li–S battery. The morphological structure of AF-PE-6 is characterized by SEM, and the effect of coating content of AFs on the porosity of hybrid films is discussed. The air permeability of AF-PE-6 is superior to other control samples determined by the optimal coating content of AFs, which showed the enhanced electrochemical performance of the Li–S battery. The specific discharge capacity is 661 mAh g^{-1} at 0.1 C, and 247 mAh g^{-1} of specific capacity is maintained at the C rate increased by a factor of 10, which is better than pristine PE cell. The high specific capacity and good rate capability of AF-PE-6 cell are attributed to the high porosity of the separator and the increased absorbability of polysulfides by coating AFs.

Author Contributions: Conceptualization, X.Y.; formal analysis, J.G., J.Z. and Y.S.; funding acquisition, J.G.; investigation, J.Z. and Y.S.; writing—original draft, J.G.; writing—review and editing, X.Y. All authors have read and agreed to the published version of the manuscript.

Funding: This research was funded by the Key Project of Natural Science of the Education Department of Henan Province (Grant No. 21B430014).

Institutional Review Board Statement: Not applicable.

Informed Consent Statement: Not applicable.

Data Availability Statement: The data that support the findings of this study are available from the corresponding authors upon reasonable request.

Acknowledgments: The authors would like to thank the anonymous reviewers who helped to significantly improve the quality of the research article.

Conflicts of Interest: The authors declare no conflict of interest.

References

1. Balach, J.; Linnemann, J.; Jaumann, T.; Giebeler, L. Metal-based nanostructured materials for advanced lithium-sulfur batteries. *J. Mater. Chem. A* **2018**, *6*, 23127–23168. [CrossRef]
2. Ponraj, R.; Kannan, A.G.; Ahn, J.H.; Lee, J.H.; Kang, J.; Han, B.; Kim, D.W. Effective trapping of lithium polysulfides using a functionalized carbon nanotube-coated separator for lithium-sulfur cells with enhanced cycling stability. *ACS Appl. Mater. Inter.* **2017**, *9*, 38445–38454. [CrossRef] [PubMed]
3. Kong, L.; Li, B.-Q.; Peng, H.-J.; Zhang, R.; Xie, J.; Huang, J.-Q.; Zhang, Q. Porphyrin-derived graphene-based nanosheets enabling strong polysulfide chemisorption and rapid kinetics in lithium-sulfur batteries. *Adv. Energy Mater.* **2018**, *8*, 1800849. [CrossRef]
4. Huang, S.; Lim, Y.V.; Zhang, X.; Wang, Y.; Zheng, Y.; Kong, D.; Ding, M.; Yang, S.A.; Yang, H.Y. Regulating the polysulfide redox conversion by iron phosphide nanocrystals for high-rate and ultrastable lithium-sulfur battery. *Nano Energy* **2018**, *51*, 340–348. [CrossRef]
5. He, Y.; Chang, Z.; Wu, S.; Qiao, Y.; Bai, S.; Jiang, K.; He, P.; Zhou, H. Simultaneously inhibiting lithium dendrites growth and polysulfides shuttle by a flexible MOF-based membrane in Li-S batteries. *Adv. Energy Mater.* **2018**, *8*, 1802130. [CrossRef]
6. Chung, S.-H.; Manthiram, A. Rational design of statically and dynamically stable lithium-sulfur batteries with high sulfur loading and low electrolyte/sulfur ratio. *Adv. Mater.* **2018**, *30*, 1705951. [CrossRef] [PubMed]
7. Cheng, Z.; Pan, H.; Zhong, H.; Xiao, Z.; Li, X.; Wang, R. Porous organic polymers for polysulfide trapping in lithium-sulfur batteries. *Adv. Funct. Mater.* **2018**, *28*, 1707597. [CrossRef]
8. Ji, Y.; Liu, X.; Xiu, Y.; Indris, S.; Njel, C.; Maibach, J.; Ehrenberg, H.; Fichtner, M.; Zhao-Karger, Z. Magnesium-sulfur batteries: Polyoxometalate modified separator for performance enhancement of magnesium-sulfur batteries. *Adv. Funct. Mater.* **2021**, *31*, 2100868. [CrossRef]
9. Zhou, W.; Zhao, D.; Wu, Q.; Fan, B.; Dan, J.; Han, A.; Ma, L.; Zhang, X.; Li, L. Amorphous CoP nanoparticle composites with nitrogen-doped hollow carbon nanospheres for synergetic anchoring and catalytic conversion of polysulfides in Li-S batteries. *J. Colloid Interf. Sci.* **2021**, *603*, 1–10. [CrossRef]
10. Yu, H.; Zeng, P.; Liu, H.; Zhou, X.; Guo, C.; Li, Y.; Liu, S.; Chen, M.; Guo, X.; Chang, B.; et al. Li$_2$S in situ grown on three-dimensional porous carbon architecture with electron/ion channels and dual active sites as cathodes of Li-S batteries. *ACS Appl. Mater. Inter.* **2021**, *13*, 32968–32977. [CrossRef]
11. Wu, Z.; Wang, W.; Wang, Y.; Chen, C.; Li, K.; Zhao, G.; Sun, C.; Chen, W.; Ni, L.; Diao, G. Three-dimensional graphene hollow spheres with high sulfur loading for high-performance lithium-sulfur batteries. *Electrochim. Acta* **2017**, *224*, 527–533. [CrossRef]
12. Deng, C.; Wang, Z.; Wang, S.; Yu, J. Inhibition of polysulfide diffusion in lithium–sulfur batteries: Mechanism and improvement strategies. *J. Mater. Chem. A* **2019**, *7*, 12381–12413. [CrossRef]
13. Ogoke, O.; Wu, G.; Wang, X.; Casimir, A.; Ma, L.; Wu, T.; Lu, J. Effective strategies for stabilizing sulfur for advanced lithium–sulfur batteries. *J. Mater. Chem. A* **2017**, *5*, 448–469. [CrossRef]
14. Fu, A.; Wang, C.; Pei, F.; Cui, J.; Fang, X.; Zheng, N. Recent advances in hollow porous carbon materials for lithium-sulfur batteries. *Small* **2019**, *15*, 1804786. [CrossRef]
15. Lee, C.; Kim, I. A hierarchical carbon nanotube-loaded glass-filter composite paper interlayer with outstanding electrolyte uptake properties for high-performance lithium-sulphur batteries. *Nanoscale* **2015**, *7*, 10362–10367. [CrossRef]
16. Eroglu, O.; Kiai, M.S.; Kizil, H. Glass fiber separator coated by boron doped anatase TiO$_2$ for high-rate Li–S battery. *Mater. Res. Bull.* **2020**, *129*, 110917. [CrossRef]
17. Wu, J.; Zeng, H.; Li, X.; Xiang, X.; Liao, Y.; Xue, Z.; Ye, Y.; Xie, X. Ultralight layer-by-layer self-assembled MoS$_2$-polymer modified separator for simultaneously trapping polysulfides and suppressing lithium dendrites. *Adv. Energy Mater.* **2018**, *8*, 1802430. [CrossRef]
18. Li, Z.; Zhang, J.; Lou, X.W. Hollow carbon nanofibers filled with MnO$_2$ nanosheets as efficient sulfur hosts for lithium-sulfur batteries. *Angew. Chem. Int. Ed. Engl.* **2015**, *54*, 12886–12890. [CrossRef]
19. Chen, X.; Huang, Y.; Li, J.; Wang, X.; Zhang, Y.; Guo, Y.; Ding, J.; Wang, L. Bifunctional separator with sandwich structure for high-performance lithium-sulfur batteries. *J. Colloid Interf. Sci.* **2020**, *559*, 13–20. [CrossRef]

20. Ni, L.; Zhao, G.; Wang, Y.; Wu, Z.; Wang, W.; Liao, Y.; Yang, G.; Diao, G.A.-O. Coaxial Carbon/MnO$_2$ Hollow Nanofibers as Sulfur Hosts for High-Performance Lithium-Sulfur Batteries. *Chem.-Asian. J.* **2017**, *12*, 3128–3134. [CrossRef]
21. Thompson, L.T.; Li, J.; Wang, Q.; Wang, Z.; Cao, Y.; Zhu, J.; Lou, Y.; Zhao, Y.; Shi, L.; Yuan, S. Evaporation and in-situ gelation induced porous hybrid film without template enhancing the performance of lithium ion battery separator. *Nat. Commun.* **2021**, *595*, 142–150.
22. Yang, Y.; Huang, C.; Gao, G.; Hu, C.; Luo, L.; Xu, J. Aramid nanofiber/bacterial cellulose composite separators for lithium-ion batteries. *Carbohyd. Polym.* **2020**, *247*, 116702. [CrossRef]
23. Tung, S.O.; Fisher, S.L.; Kotov, N.A. Nanoporous aramid nanofibre separators for nonaqueous redox flow batteries. *Nat. Commun.* **2018**, *9*, 4193. [CrossRef] [PubMed]
24. Yoon, E.; Park, J.W.; Kang, J.K.; Kim, S.; Jung, Y. High-energy-density Li-S batteries with additional elemental sulfur coated on a thin-film separator. *J. Nanosci. Nanotechnol.* **2019**, *19*, 4715–4718. [CrossRef] [PubMed]
25. Babu, D.B.; Giribabu, K. Permselective SPEEK/Nafion Composite-Coated Separator as a Potential Polysulfide Crossover Barrier Layer for Li-S Batteries. *ACS Appl. Mater. Inter.* **2018**, *10*, 19721–19729. [CrossRef]
26. Zhang, Z.; Lai, Y.; Zhang, Z.; Zhang, K.; Li, J. Al$_2$O$_3$-coated porous separator for enhanced electrochemical performance of lithium sulfur batteries. *Electrochim. Acta* **2014**, *129*, 55–61. [CrossRef]

Article

Boosting the Oxygen Evolution Reaction by Controllably Constructing FeNi$_3$/C Nanorods

Xu Yu [1,*], Zhiqiang Pan [1], Zhixin Zhao [1], Yuke Zhou [1], Chengang Pei [1], Yifei Ma [2], Ho Seok Park [3] and Mei Wang [2,*]

1. School of Chemistry and Chemical Engineering, Yangzhou University, Yangzhou 225009, China; w461015600@163.com (Z.P.); zhao1zhi2xin3@163.com (Z.Z.); zykds66@163.com (Y.Z.); chengpyzu@163.com (C.P.)
2. State Key Laboratory of Quantum Optics and Quantum Optics Devices, Institute of Laser Spectroscopy, Collaborative Innovation Center of Extreme Optics, Shanxi University, Taiyuan 030006, China; mayifei@sxu.edu.cn
3. Department of Chemical Engineering, College of Engineering, Sungkyunkwan University, 2066 Seobu-ro, Jangan-gu, Suwon-si 440-746, Gyeonggi-do, Korea; phs0727@skku.edu
* Correspondence: yxypz15@yzu.edu.cn (X.Y.); wangmei@sxu.edu.cn (M.W.)

Abstract: Transition bimetallic alloy-based catalysts are regarded as attractive alternatives for the oxygen evolution reaction (OER), attributed to their competitive economics, high conductivity and intrinsic properties. Herein, we prepared FeNi$_3$/C nanorods with largely improved catalytic OER activity by combining hydrothermal reaction and thermal annealing treatment. The temperature effect on the crystal structure and chemical composition of the FeNi$_3$/C nanorods was revealed, and the enhanced catalytic performance of FeNi$_3$/C with an annealing temperature of 400 °C was confirmed by several electrochemical tests. The outstanding catalytic performance was assigned to the formation of bimetallic alloys/carbon composites. The FeNi$_3$/C nanorods showed an overpotential of 250 mV to afford a current density of 10 mA cm^{-2} and a Tafel slope of 84.9 mV dec^{-1}, which were both smaller than the other control samples and commercial IrO$_2$ catalysts. The fast kinetics and high catalytic stability were also verified by electrochemical impendence spectroscopy and chronoamperometry for 15 h. This study is favorable for the design and construction of bimetallic alloy-based materials as efficient catalysts for the OER.

Keywords: FeNi$_3$ alloy; nanorods; bimetallic; oxygen evolution reaction

Citation: Yu, X.; Pan, Z.; Zhao, Z.; Zhou, Y.; Pei, C.; Ma, Y.; Park, H.S.; Wang, M. Boosting the Oxygen Evolution Reaction by Controllably Constructing FeNi$_3$/C Nanorods. *Nanomaterials* 2022, 12, 2525. https://doi.org/10.3390/nano12152525

Academic Editors: Francesc Viñes Solana and Yuichi Negishi

Received: 12 June 2022
Accepted: 20 July 2022
Published: 22 July 2022

Publisher's Note: MDPI stays neutral with regard to jurisdictional claims in published maps and institutional affiliations.

Copyright: © 2022 by the authors. Licensee MDPI, Basel, Switzerland. This article is an open access article distributed under the terms and conditions of the Creative Commons Attribution (CC BY) license (https://creativecommons.org/licenses/by/4.0/).

1. Introduction

There is a global consensus that producing hydrogen energy via electrochemical water splitting will lighten the burden of consuming energy from fossil fuels and replace unsustainable energy sources [1–5]. The oxygen evolution reaction (OER) is a half-reaction of the electrolysis of water, but the issue of slow reaction kinetics during the complicated four-electron transfer process critically needs to be solved [6,7]. The energy conversion efficiency of catalysts for practical water splitting is affected by the high overpotential and energy consumption of catalysts during the OER process [8,9]. Ruthenium dioxide (RuO$_2$) and iridium dioxide (IrO$_2$), as effective OER catalysts [10–12], can facilitate a combination of OH$^-$ ions in the alkaline electrolyte for the OER. However, the high price and scarcity of resources restrict its wide application in energy conversion systems [13,14]. Therefore, cost-competitive catalysts with high catalytic activity urgently need to be explored.

Recently, earth-abundant and low-cost transition metal (TM)-based catalysts with improved catalytic OER stability have been an effective strategy for water splitting. Transition metal (i.e., Fe, Co, Ni, and Mo) oxides or hydroxides can form hydroxide intermediates during the OER process [15–18], but their high energy barrier and sluggish kinetics are still difficult to overcome due to fact of their poor conductivity [19–22]. Therefore, many efforts have been reported on the study of TM-based derivatives with modification of the surface electronic structure [23,24] including metal phosphides [25,26], sulfides [27,28], and fluorides [29,30]. Bimetal-based catalysts have the merits of enhanced reactivity and abundant active sites by the adjusted electronic structure at the metal/metal interfaces, which are important for improving the electrocatalytic performance [31]. Meanwhile, bimetal-based catalysts can form more oxygen vacancies and reduce the adsorption energy of anions in electrolytes [32]. As demonstrated by many reports, the incorporation of Fe^{3+} into transition metal-based catalysts can significantly increase the reactivity and catalytic activity during the OER process [33,34]. The iron–nickel bimetallic catalyst near the top of a volcano plot shows an excellent catalytic OER performance [35–38]. Importantly, the introduction of a conductive matrix in a bimetal system is favorable for increasing the electronic conductivity and catalytic stability, such as graphene [39], nickel foam [40], and amorphous carbon [41]. A couple of FeNi alloys with carbon supports can significantly increase the electrical conductivity, and the controllable construction of an FeNi/C hybrid can provide fast ion diffusion and enhance the electrocatalytic stability during the OER process.

Herein, we report iron–nickel/carbon ($FeNi_3$/C) nanorods as an effective OER catalyst through hydrothermal and activation approaches. The nanorod morphology can increase the amount of exposed surface and the number of effective catalytic active sites, which are beneficial for increasing the electrochemical activity. The effect of the activation temperature on the crystallinity and catalytic OER behavior of the $FeNi_3$/C nanorods were studied by physical characterization and electrochemical tests. As the optimal temperature was 400 °C, the $FeNi_3$/C nanorods showed excellent OER performance. Only 250 mV of the overpotential was required at 10 mA cm^{-2} with a Tafel slope of 84.9 mV dec^{-1}. The improved electrochemical stability was studied by chronoamperometry, which was indexed to the effect of rough morphology and optimal composition.

2. Experimental Section

2.1. Synthesis of the $FeNi_3$/C Nanorods

A mixture of deionized (DI) water (12 mL) and ethylene glycol (36 mL), as the solvent to dissolve 200 mg of $NiCl_2 \cdot 6H_2O$ and 200 mg of $FeCl_2$, and 200 mg of oxalic acid were subsequently slowly added under continuous ultrasonication. The solution was transferred to a stainless-steel autoclave (100 mL) and maintained at 150 °C for 12 h. As the temperature naturally cooled down, the precipitate was repeatedly washed with DI water/ethanol and dried at 60 °C under vacuum conditions overnight to obtain the FeNi nanorods. The FeNi nanorods were further thermally activated at 400 °C for 2 h with flowing N_2 gas, and the target sample was named $FeNi_3$/C. The $FeNi_3$/C was thermally activated at 300 and 500 °C and labeled as $FeNi_3$/C-300 and $FeNi_3$/C-500. The related catalytic performances of $FeNi_3$/C-300 and $FeNi_3$/C-500 were compared. The synthetic process of the Fe nanorods and the Ni nanorods was the same as for the FeNi nanorods, and the precursor only included a single metal salt, either $FeCl_2$ or $NiCl_2 \cdot 6H_2O$. The samples were thermally activated at 400 °C, and the obtained powders were labeled as Fe/C and Ni/C nanorods for further use.

2.2. Characterization

The crystal structure was characterized by powder X-ray diffraction (XRD) (Bruker D8 Advance powder X-ray diffractometer, Cu Kα1, λ = 1.5405 Å, 40 KV, and 40 mA, Bruker, Saarbrucken, Germany). Scanning electron microscopy (SEM) images were obtained using an S-4800 II, Hitachi (Tokyo, Japan). The morphological structure was confirmed by transmission electron microscopy (TEM, Philips, TECNAI 12, Amsterdam, The Netherlands) and high-resolution transmission electron microscopy (HRTEM) (FEI Tecnai G2 F30 STWIN, 300 kV, FEI, Hillsboro, OR, USA). X-ray photoelectron spectroscopy was measured using a Thermo Science ESCALAB 250Xi (Thermofisher, Waltham, MA, USA).

2.3. Electrochemical Measurements

The electrochemical performance was performed using an electrochemical workstation (CHI 660E, Shanghai, China). The active material loaded on a glassy carbon electrode (GC, 3 mm diameter, 0.07 cm^{-2}), graphite rod, and saturated calomel electrode (SCE) acted as the working, counter, and reference electrodes, respectively. The potentials were calculated to the reversible hydrogen electrode (RHE) by E(RHE) = E(SCE) + 0.0591 × pH + 0.24 V. The catalyst ink was prepared by mixing 5 mg of catalyst, 950 µL of ethanol, and 50 µL of Nafion solution (5 wt.% Nafion^{TM} in lower aliphatic alcohols and water, 15–20% water) under bath sonication. Then, the catalyst ink (10 µL) was dropped onto the GC and naturally dried. All data are presented with IR compensation unless otherwise noted.

The polarization curves were measured as the potential from 1.04 to 1.64 V vs. RHE at 5 mV s^{-1}. Electrochemical impedance spectroscopy (EIS) was measured from 10^6 Hz to 10^{-2} Hz. One thousand CV cycles were measured within the potential ranging from 1.04 to 1.64 V vs. RHE in 1 M KOH at a scan rate of 150 mV s^{-1}, and a linear sweep was measured under a sweep rate of 5 mV s^{-1} after 1000 cycles. Chronoamperometry (CA) was tested at a voltage of 1.48 V for 15 h.

3. Results and Discussion

The iron–nickel alloy with conductive carbon (FeNi$_3$/C) nanorods were synthesized via facile hydrothermal and thermal treatment methods. The high reducibility of ethylene glycol as the solvent could strongly couple the metal ions with oxalic acid, and the content of iron was accurately controlled to adjust the morphology of catalysts. After activation at a high temperature of 400 °C, the carbon ligands decomposed into carbon materials, which can improve the conductivity of the catalysts. Finally, the FeNi$_3$/C nanorods were obtained.

To probe the morphological structure of the FeNi$_3$/C nanorods, scanning electron microscopy (SEM) was carried out. The FeNi nanorods were uniform and had an average length of 1.5 µm, as shown in Figure 1a, which was different from the Ni nanorods, which had irregular lengths, and the Fe nanorods, which had longer lengths of approximately 2 µm (Figure S1a,b). As a comparison to the precursors, the morphology of the FeNi nanorods was adjusted by the electrostatic interaction of metal ions, which is favorable for exposing abundant surface area. After thermal activation, the surfaces of the FeNi$_3$/C nanorods became relatively rough due to the formation of the FeNi$_3$ alloy and the decomposition of the carbon ligands (Figure 1b), which can provide abundant active sites. This morphology was further confirmed by transmission electron microscopy (TEM) images, as shown in Figure 1c. The average thickness of the nanorods was approximately determined to be 90 nm (Figure S2). Two d-spacings of 0.176 and 0.203 nm of the FeNi$_3$/C nanorods corresponded to the (200) and (111) planes of the FeNi$_3$ (Figure 1d). Figure 1e shows the corresponding selected area electron diffraction (SAED). The existence of concerned elements were found using energy-dispersive spectroscopy (EDS) (Figure S3a and Table S1), and the elemental mapping results confirmed that the distribution of the Fe, Ni, C, and O elements in FeNi$_3$/C was uniform (Figure 1f).

The change in the crystal structure of FeNi$_3$/C was characterized by X-ray diffraction (XRD) (Figure 2a). The broadened peak at 25.8° corresponded to the (002) plane of graphitic carbon. The characteristic peaks for the FeNi nanorods were indexed to the existence of NiFe$_2$O$_4$ (JCPDS card No. 54-0964). In comparison, the FeNi$_3$/C nanorods showed strong characteristic peaks at 44.2°, 52.0°, and 75.7°, owing to the (111), (200), and (220) planes of the FeNi$_3$ alloy (JCPDS card No. 38-0419), which is agreement with the TEM results. The disappearance of the diffraction peaks of NiFe$_2$O$_4$ for FeNi$_3$/C was attributed to the decomposition of metal oxides during thermal activation. Meanwhile, the effect of the annealing temperature on the crystallinity of the FeNi$_3$/C nanorods was studied by XRD analysis (Figure S3b), and the average crystal size of the FeNi$_3$/ nanorods was approximately 15.6 nm. As the activation temperature increased, the domain characteristic peaks of the FeNi$_3$/C nanorods became stronger than for FeNi$_3$/C-300, indicating the increased crystallinity due to the formation of FeNi$_3$ alloys. There were no obvious changes in the characteristic peaks between FeNi$_3$/C and FeNi$_3$/C-500, demonstrating that the optimal temperature of 400 °C was high enough to form a stable catalyst.

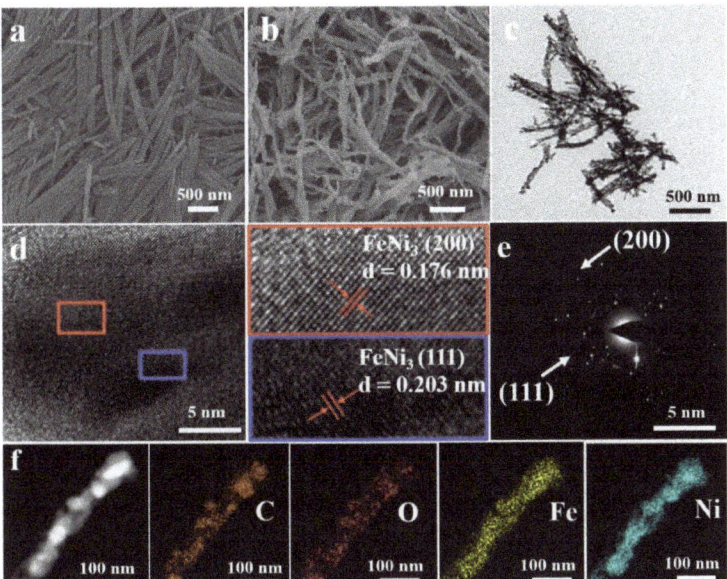

Figure 1. SEM images of (**a**) FeNi and (**b**) FeNi$_3$/C nanorods; (**c**) TEM and (**d**) HR-TEM images of FeNi$_3$/C nanorods; (**e**) SAED pattern; (**f**) STEM and elemental mappings of the FeNi$_3$/C nanorods.

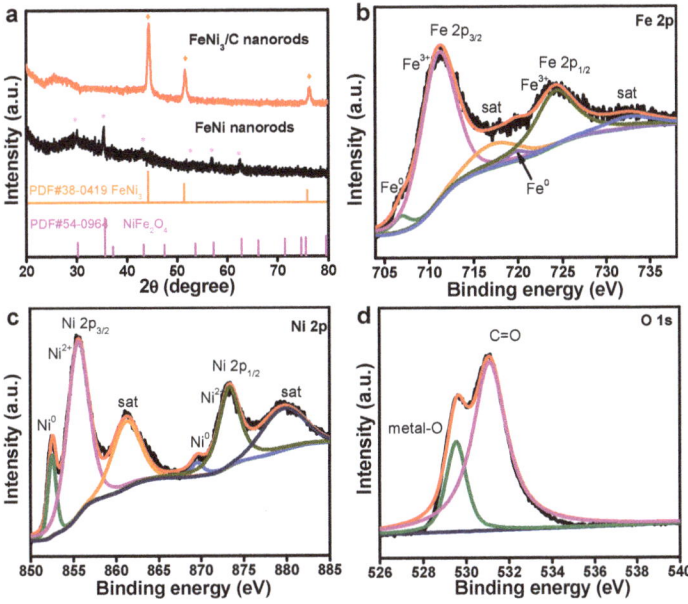

Figure 2. (**a**) XRD spectra of the FeNi and FeNi$_3$/C nanorods (the asterisk and oranges squares represent the diffraction peaks of NiFe$_2$O$_2$ and FeNi$_3$); high-resolution (**b**) Fe 2p, (**c**) Ni 2p, and (**d**) O1s XPS spectra of the FeNi$_3$/C nanorods.

The surface chemical circumstances of the FeNi$_3$/C nanorods were probed by X-ray photoelectron spectroscopy (XPS), and C 1s at 284.8 eV was applied to standardize the binding energy. From the full scan of the XPS spectra, the FeNi$_3$/C nanorods contained 11.6 atom% of O, 22.0 atom% of C, 19.5 atom% of Fe, and 46.9 atom% of Ni elements (Figure S4a). The C 1s spectra showed two dominant peaks at 284.8 eV for a C-C bond and at 288.6 eV for C-O bonds (Figure S4b). Figure 2b shows the deconvoluted Fe 2p spectra, and two distinct peaks were indexed to the spin-orbit coupling of Fe $2p_{1/2}$ and Fe $2p_{3/2}$ accompanying the satellite peaks. The peak was divided into Fe^0 (706.9 and 719.8 eV) and Fe^{3+} peaks (711.3 and 724.8 eV), respectively [42]. The Ni 2p spectra were deconvoluted into Ni^0 (852.5 and 869.7 eV) and Ni^{2+} (855.3 and 873.2 eV) with the related satellite peaks in Figure 2c [43], respectively. An energy difference of 17.8 eV was calculated between Ni $2p_{3/2}$ and Ni $2p_{1/2}$, implying that the Ni^{2+} state was dominant [44]. In addition, two peaks at 852.2 eV and 869.3 eV corresponded to Ni metal. The FeNi$_3$/C nanorods with contents of Fe^{3+} and Ni^{2+} can act as active material, and the electrocatalytic behavior can be dramatically affected by the boosted active sites arising from the conversion of Ni^{2+} to Ni^{3+} during the OER process [45]. The deconvoluted O 1s spectra are shown in Figure 2d, and the two peaks at 529.5 and 532.3 eV corresponded to metal-O and C=O bonds [41]. The existence of metal-O bonds can probably be ascribed to the formation of oxidized states on FeNi$_3$ alloy surfaces during thermal activation, and the internal high-oxygen coordination defects of the nanorods are generally considered as the dominant catalytic sites for increasing the oxidation kinetics and catalytic activity during the OER.

The electrocatalytic OER performance of the FeNi$_3$/C nanorods was initially evaluated by cyclic voltammetry (CV) using a three-electrode configuration, and an aqueous 1 M KOH was the electrolyte with N$_2$ purification. To reflect the effect of the activation temperature on the catalytic OER performance, the polarization curves of the FeNi$_3$/C nanorods were compared (Figure S5a). To receive a current density of 10 mA cm^{-2}, the overpotential of the FeNi$_3$/C nanorods (250 mV) was smaller than 280 mV for FeNi$_3$/C-300 and 290 mV for FeNi$_3$/C-500, and the Tafel slope for the FeNi$_3$/C nanorods (84.9 mV dec^{-1}) was lower than 99.2 and 101.1 mV dec^{-1} for FeNi$_3$/C-300 and FeNi$_3$/C-500 (Figure S5b). As shown in Figure 3a, the FeNi$_3$/C nanorods showed a lower overpotential at 10 mA cm^{-2} than that of Fe/C (370 mV), Ni/C (330 mV), commercial IrO$_2$/C (327 mV), and other reported FeNi-based electrocatalysts for the OER (Table S2). The low overpotential of FeNi$_3$/C implies a high OER activity due to the incorporation of Fe ions with Ni ions [46,47]. In addition, The Tafel slope can reflect the rate-determining step with the related reaction mechanism during the OER process, and Tafel slopes of 120, 60, and 40 mV represent the RSD of one-electron, chemical, and electron–proton reaction steps [48–50]. According to the Tafel slopes shown in Figure 3b, FeNi$_3$/C had a smaller value of 84.9 mV dec^{-1} compared to 102.2 and 121.2 mV dec^{-1} for Fe/C and Ni/C, indicating faster catalytic kinetics for FeNi$_3$/C. Chemical reactions with O$_2$ formation as an intermediate on the catalytic sites was dominant for FeNi$_3$/C, and the impact of the electron transfer process was no longer the primary step for the OER. The electrochemical dynamics and interfacial properties of the electrode were elucidated by electrochemical impedance spectroscopy (Figure 3c), and the calculated resistances are listed in Table S3 and were fit using Nyquist plots with an equivalent circuit in Figure S6. The charge transfer resistances (Rct) were 15.2, 110, and 26 Ω for the FeNi$_3$/C, Fe/C, and Ni/C nanorods, respectively. The smaller Rct value indicates a faster charge transfer behavior as well as higher catalytic activity of FeNi$_3$/C.

A catalyst exposing abundant active sites can show high electrochemical activity. The electrochemical surface area (ECSA) was estimated by CV measurement in a non-Faradic field (Table S4), and the double-layer capacitance (C$_{dl}$) value was calculated by linearly fitting the current density versus scan rates (Figure S7). Specific activity was obtained by normalizing the origin current to the ECSA. The FeNi$_3$/C nanorods with the optimal temperature of 400 °C had a specific activity of 0.24 mA cm^{-2} at the overpotential of 300 mV, which was higher than that of all control samples (Figure S8). The FeNi$_3$/C nanorods had a C$_{dl}$ value of 4.14 mF cm^{-2}, as shown in Figure 3d, which was approximately 9.6 and

3.23 times higher than 0.43 and 1.28 mF cm^{-2} for Fe/C and Ni/C, respectively. This result confirms that FeNi$_3$/C provided an enlarged catalytic active surface for facilitating ion diffusion and promoting the electrochemical reaction.

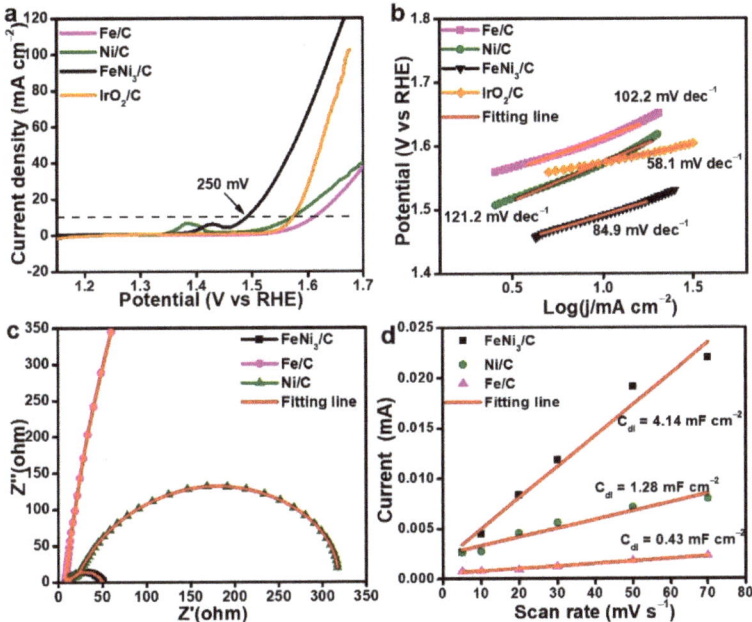

Figure 3. (**a**) Polarization curves of FeNi$_3$/C, IrO$_2$/C, Fe/C, and Ni/C nanorods at 5 mV s^{-1}; (**b**) Tafel slopes of FeNi$_3$/C, Fe/C, Ni/C, and IrO$_2$/C; (**c**) Nyquist plots; (**d**) C$_{dl}$ values of FeNi$_3$/C, Fe/C, and Ni/C nanorods.

The long-term stability of the FeNi$_3$/C nanorods was initially evaluated by performing 1000 CV cycles, as shown in Figure 4a. The initial and 1001st CV curves almost overlap, and the overpotential at 10 mA cm^{-2} was a negligible change. Furthermore, chronoamperometry (CA) was measured at the potential of 1.48 V for 15 h (Figure 4b). The FeNi$_3$/C nanorods exhibited no obvious change in current density at the initial 10 h, and the current density remained at 90% for the next 5 h. The Faraday efficiency of FeNi$_3$/C was measured by comparing the experimental and theoretic amounts of oxygen gas produced during constant voltage electrolysis for 60 min (Figure S9), and the experimental volume was close to the theoretical oxygen volume, indicating that the oxygen evolution efficiency was close to 100%. These results demonstrate the outstanding electrocatalytic OER activity and stability of the FeNi$_3$/C nanorods, attributed to the in situ formation of the FeNi$_3$/C composites.

The morphological change of the FeNi$_3$/C surface after the stability test was characterized by TEM, as shown in Figure 5a. The morphology of FeNi$_3$/C nanorods was maintained, and the slight collapse or fracture phenomena were caused by the partial oxidation of FeNi$_3$ during the catalytic reaction in an alkaline solution. The change in the surface chemistry after the electrocatalytic test was confirmed by XPS. In comparison to the pristine state, the intensity of the metal-O bond after the CA test slightly increased with a shift of 0.4 eV because of the formation of intermediates. For the Ni element, the change in the Ni 2p spectra can be seen in Figure 5c, and the Ni 2p$_{3/2}$ peak downshifted with by a value of 0.1 eV, attributed to the formation of nickel hydroxides or hydroxyl oxides after long-term CA testing. The Fe element showed a similar result, suggesting the formation of electroactive intermediates during the stability test, as shown in Figure 5d.

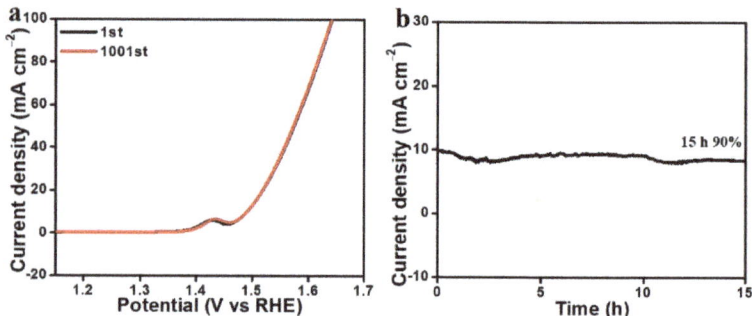

Figure 4. (a) The 1st and 1001st CV curves of FeNi$_3$/C; (b) chronoamperometry test of FeNi$_3$/C at the potential of 1.48 V.

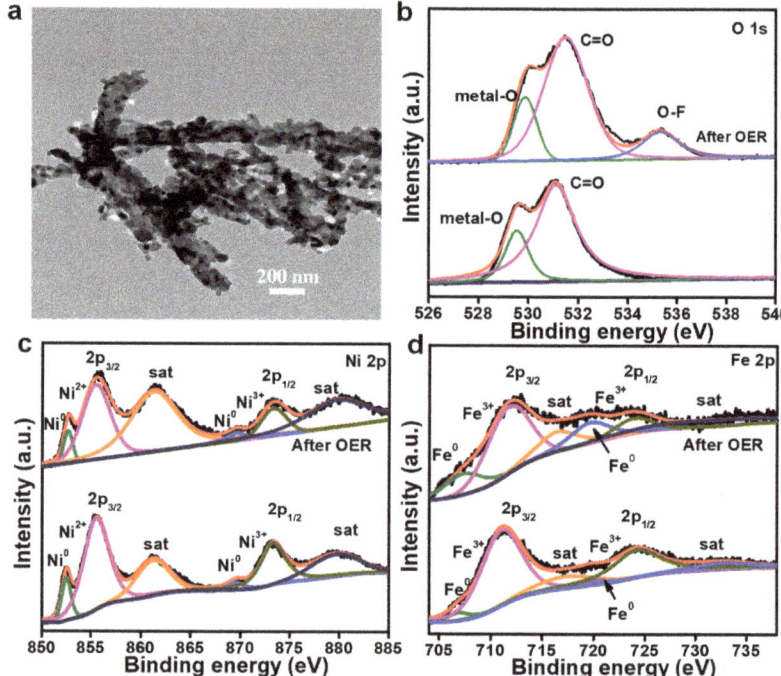

Figure 5. (a) TEM image of FeNi$_3$/C after the stability test for the OER; (b–d) XPS spectrum of O 1s (b), Ni 2p (c), and Fe 2p (d) for the FeNi$_3$/C alloy nanorod after the stability test for the OER.

During the OER process, the high-valence nickel in the catalysts was more conducive to the rapid formation of intermediates (Ni-OH) in the electrolyte, which was further combined with the OH$^-$ to form a nickel oxyhydroxide followed by the removal of oxygen. Therefore, the high content of Ni^{2+} in FeNi$_3$/C was more conducive to the OER, and the adjusted surface electronic structure by incorporation of Fe^{3+} increased the absorbability of OH$^-$, which resulted in the boosted catalytic activity of the catalysts during the OER process.

4. Conclusions

In summary, FeNi$_3$/C nanorods as effective catalysts for the OER were constructed by combining the facile hydrothermal reaction and further thermal annealing treatment. The temperature and compositional effect on the catalyst were discussed. The FeNi$_3$/C nanorods with an annealing temperature of 400 °C showed the best electrocatalytic performance such as a low overpotential of 250 mV at 10 mA cm^{-2}, small Tafel slope of 84.9 mV dec^{-1}, and high catalytic stability after CA testing for 15 h. The improved electrocatalytic behaviors were indexed to the controllable structure and optimal chemical composition by hybridizing bimetallic alloy with carbon. This work provides a strategy for preparing efficient catalysts for the OER by coupling bimetallic alloys with a carbon matrix.

Supplementary Materials: The following supporting information can be downloaded at: https://www.mdpi.com/article/10.3390/nano12152525/s1, Figure S1: SEM images of (a) Ni and (b) Fe nanorods; Figure S2: TEM image of FeNi$_3$/C nanorods; Figure S3; (a) EDS data of FeNi$_3$/C nanorods; (b) XRD patterns of FeNi$_3$/C nanorods with different thermal annealing temperatures; Figure S4: XPS spectrum of FeNi$_3$/C nanorods at (a) full scan and (b) C 1s; Figure S5: (a) Polarization curves and (b) Tafel plots of FeNi$_3$/C nanorods with different thermal annealing temperatures; Figure S6: The equivalent circuit model of the EIS analysis of all samples; Figure S7: CV curves of (a) FeNi$_3$/C nanorods; (b) Ni/C nanorods; (c) Fe/C nanorods at the potential of 1.04 V–1.14 V in 1 M KOH; Figure S8: (a) The specific activity of FeNi$_3$/C at different activation temperatures; (b) the specific activity of FeNi$_3$/C, Ni/C and Fe/C at the overpotential of 300 mV; Figure S9: Faraday efficiency of FeNi$_3$/C for OER; Table S1: The atomic ratio of all elements from EDS; Table S2: The comparison of other FeNi-based OER catalysts in alkaline medium; Table S3: EIS fitting parameters from equivalent circuits for as-prepared catalysts; Table S4: The value of C_{dl} and ECSA for FeNi$_3$/C with different annealing temperature. References [51–60] are cited in the Supplementary Materials.

Author Contributions: Conceptualization, Z.P. and Z.Z.; Formal analysis, Y.Z. and C.P.; Funding acquisition, X.Y., Y.M. and M.W.; Investigation, Y.Z. and C.P.; Methodology, Z.P. and Z.Z.; Project administration, X.Y.; Resources, X.Y.; Writing—original draft, Z.P. and Z.Z.; Writing—review and editing, X.Y., Y.M., H.S.P. and M.W. All authors have read and agreed to the published version of the manuscript.

Funding: This research was funded by the Natural Science Foundation of the Jiangsu Higher Education Institutions of China (18KJB150034); "Six Talent Peaks Project" in Jiangsu Province (XCL-104); "High-End Talent Project" of Yangzhou University; the National Natural Science Foundation of China (51902190); the Key Research and Development Program of Shanxi Province for International Cooperation (201803D421082); the Scientific and Technological Innovation Programs of Higher Education Institutions in Shanxi (2019L0013 and 2019L0018); the Shanxi Scholarship Council of China (2021-004).

Institutional Review Board Statement: Not applicable.

Informed Consent Statement: Not applicable.

Data Availability Statement: The data presented in this study are available on request from the corresponding author.

Acknowledgments: The authors would like to thank the anonymous reviewers who helped to significantly improve the quality of the research article.

Conflicts of Interest: The authors declare no conflict of interest.

References

1. Anantharaj, S.; Ede, S.R.; Karthick, K.; Sam Sankar, S.; Sangeetha, K.; Karthik, P.E.; Kundu, S. Precision and correctness in the evaluation of electrocatalytic water splitting: Revisiting activity parameters with a critical assessment. *Energy Environ. Sci.* **2018**, *11*, 744–771. [CrossRef]
2. Jiang, W.J.; Tang, T.; Zhang, Y.; Hu, J.S. Synergistic Modulation of Non-Precious-Metal Electrocatalysts for Advanced Water Splitting. *Acc. Chem. Res.* **2020**, *53*, 1111–1123. [CrossRef]
3. Guo, Y.; Tang, J.; Qian, H.; Wang, Z.; Yamauchi, Y. One-Pot Synthesis of Zeolitic Imidazolate Framework 67-Derived Hollow Co$_3$S$_4$@MoS$_2$ Heterostructures as Efficient Bifunctional Catalysts. *Chem. Mater.* **2017**, *29*, 5566–5573. [CrossRef]

4. Zhu, H.; Zhang, J.; Yanzhang, R.; Du, M.; Wang, Q.; Gao, G.; Wu, J.; Wu, G.; Zhang, M.; Liu, B.; et al. When cubic cobalt sulfide meets layered molybdenum disulfide: A core-shell system toward synergetic electrocatalytic water splitting. *Adv. Mater.* **2015**, *27*, 4752–4759. [CrossRef] [PubMed]
5. Kwon, I.S.; Debela, T.T.; Kwak, I.H.; Park, Y.C.; Seo, J.; Shim, J.Y.; Yoo, S.J.; Kim, J.G.; Park, J.; Kang, H.S. Ruthenium Nanoparticles on Cobalt-Doped 1T' Phase MoS$_2$ Nanosheets for Overall Water Splitting. *Small* **2020**, *16*, 2000081. [CrossRef]
6. Zhuang, L.; Ge, L.; Yang, Y.; Li, M.; Jia, Y.; Yao, X.; Zhu, Z. Ultrathin iron-cobalt oxide nanosheets with abundant oxygen vacancies for the oxygen evolution reaction. *Adv. Mater.* **2017**, *29*, 1606793. [CrossRef]
7. Lu, B.; Cao, D.; Wang, P.; Wang, G.; Gao, Y. Oxygen evolution reaction on Ni-substituted Co$_3$O$_4$ nanowire array electrodes. *Int. J. Hydrogen Energy* **2011**, *36*, 72–78. [CrossRef]
8. Mom, R.V.; Cheng, J.; Koper, M.T.M.; Sprik, M. Modeling the oxygen evolution reaction on metal oxides: The infuence of unrestricted DFT calculations. *J. Phys. Chem. C* **2014**, *118*, 4095–4102. [CrossRef]
9. Zhang, J.; Zhao, Z.; Xia, Z.; Dai, L. A metal-free bifunctional electrocatalyst for oxygen reduction and oxygen evolution reactions. *Nat. Nanotechnol.* **2015**, *10*, 444–452. [CrossRef]
10. Li, C.; Baek, J.B. Recent Advances in Noble Metal (Pt, Ru, and Ir)-Based Electrocatalysts for Efficient Hydrogen Evolution Reaction. *ACS Omega* **2020**, *5*, 31–40. [CrossRef]
11. Reier, T.; Oezaslan, M.; Strasser, P. Electrocatalytic oxygen evolution reaction (OER) on Ru, Ir, and Pt catalysts: A comparative study of nanoparticles and bulk materials. *ACS Catal.* **2012**, *2*, 1765–1772. [CrossRef]
12. Park, S.; Shao, Y.; Liu, J.; Wang, Y. Oxygen electrocatalysts for water electrolyzers and reversible fuel cells: Status and perspective. *Energy Environ. Sci.* **2012**, *5*, 9331–9344. [CrossRef]
13. Stamenkovic, V.R.; Fowler, B.; Mun, B.S.; Wang, G.; Ross, P.N.; Lucas, C.A.; Marković, N.M. Improved oxygen reduction activity on Pt$_3$Ni(111) via increased surface site availability. *Science* **2007**, *315*, 493–497. [CrossRef] [PubMed]
14. Zheng, Y.; Jiao, Y.; Zhu, Y.; Li, L.H.; Han, Y.; Chen, Y.; Jaroniec, M.; Qiao, S.-Z. High electrocatalytic hydrogen evolution activity of an anomalous ruthenium catalyst. *J. Am. Chem. Soc.* **2016**, *138*, 16174–16181. [CrossRef]
15. Wang, D.; Chen, X.; Evans, D.G.; Yang, W. Well-dispersed Co$_3$O$_4$/Co$_2$MnO$_4$ nanocomposites as a synergistic bifunctional catalyst for oxygen reduction and oxygen evolution reactions. *Nanoscale* **2013**, *5*, 5312–5315. [CrossRef]
16. Meng, Y.; Song, W.; Huang, H.; Ren, Z.; Chen, S.-Y.; Suib, S.L. Structure–property relationship of bifunctional MnO$_2$ nanostructures: Highly efficient, ultra-stable electrochemical water oxidation and oxygen reduction reaction catalysts identified in alkaline media. *J. Am. Chem. Soc.* **2014**, *136*, 11452–11464. [CrossRef]
17. Friebel, D.; Louie, M.W.; Bajdich, M.; Sanwald, K.E.; Cai, Y.; Wise, A.M.; Cheng, M.-J.; Sokaras, D.; Weng, T.-C.; Alonso-Mori, R.; et al. Identification of highly active Fe sites in (Ni,Fe)OOH for electrocatalytic water splitting. *J. Am. Chem. Soc.* **2015**, *137*, 1305–1313. [CrossRef]
18. Song, F.; Hu, X. Exfoliation of layered double hydroxides for enhanced oxygen evolution catalysis. *Nat. Commun.* **2014**, *5*, 4477. [CrossRef]
19. Yang, L.; Chen, L.; Yang, D.; Yu, X.; Xue, H.; Feng, L. NiMn layered double hydroxide nanosheets/NiCo$_2$O$_4$ nanowires with surface rich high valence state metal oxide as an efficient electrocatalyst for oxygen evolution reaction. *J. Power Sources* **2018**, *392*, 23–32. [CrossRef]
20. Chen, G.-F.; Ma, T.Y.; Liu, Z.-Q.; Li, N.; Su, Y.-Z.; Davey, K.; Qiao, S.-Z. Efficient and stable bifunctional electrocatalysts ni/nixmy (M = P, S) for overall water splitting. *Adv. Funct. Mater.* **2016**, *26*, 3314–3323. [CrossRef]
21. Xuan, C.; Wang, J.; Xia, W.; Peng, Z.; Wu, Z.; Lei, W.; Xia, K.; Xin, H.L.; Wang, D. Porous Structured Ni–Fe–P Nanocubes Derived from a Prussian Blue Analogue as an Electrocatalyst for Efficient Overall Water Splitting. *ACS Appl. Mater. Interfaces* **2017**, *9*, 26134–26142. [CrossRef] [PubMed]
22. Huang, C.; Zou, Y.; Ye, Y.-Q.; Ouyang, T.; Xiao, K.; Liu, Z.-Q. Unveiling the active sites of Ni–Fe phosphide/metaphosphate for efficient oxygen evolution under alkaline conditions. *Chem. Commun.* **2019**, *55*, 7687–7690. [CrossRef] [PubMed]
23. Zheng, S.; Guo, X.; Xue, H.; Pan, K.; Liu, C.; Pang, H. Facile one-pot generation of metal oxide/hydroxide@metal–organic framework composites: Highly efficient bifunctional electrocatalysts for overall water splitting. *Chem. Commun.* **2019**, *55*, 10904–10907. [CrossRef] [PubMed]
24. Li, Q.Y.; Zhang, L.; Xu, Y.X.; Li, Q.; Xue, H.; Pang, H. Smart Yolk/Shell ZIF-67@POM Hybrids as Efficient Electrocatalysts for the Oxygen Evolution Reaction. *ACS Sustain. Chem. Eng.* **2019**, *7*, 5027–5033. [CrossRef]
25. Li, D.; Liu, C.; Ma, W.; Xu, S.; Lu, Y.; Wei, W.; Zhu, J.; Jiang, D. Fe-doped NiCoP/Prussian blue analog hollow nanocubes as an efficient electrocatalyst for oxygen evolution reaction. *Electrochim. Acta* **2021**, *367*, 137492. [CrossRef]
26. Wang, F.; Yang, X.; Dong, B.; Yu, X.; Xue, H.; Feng, L. A FeP powder electrocatalyst for the hydrogen evolution reaction. *Electrochem. Commun.* **2018**, *92*, 33–38. [CrossRef]
27. Gao, M.-R.; Cao, X.; Gao, Q.; Xu, Y.-F.; Zheng, Y.-R.; Jiang, J.; Yu, S.-H. Nitrogen-doped graphene supported CoSe$_2$ nanobelt composite catalyst for efficient water oxidation. *ACS Nano* **2014**, *8*, 3970–3978. [CrossRef]
28. Dou, S.; Tao, L.; Huo, J.; Wang, S.; Dai, L. Etched and doped Co$_9$S$_8$/graphene hybrid for oxygen electrocatalysis. *Energy Environ. Sci.* **2016**, *9*, 1320–1326. [CrossRef]
29. Liu, H.; Zha, M.; Liu, Z.; Tian, J.; Hu, G.; Feng, L. Synergistically boosting the oxygen evolution reaction of an Fe-MOF via Ni doping and fluorination. *Chem. Commun.* **2020**, *56*, 7889–7892. [CrossRef]

30. Zha, M.; Pei, C.; Wang, Q.; Hu, G.; Feng, L. Electrochemical oxygen evolution reaction efficiently boosted by selective fluoridation of FeNi$_3$ alloy/oxide hybrid. *J. Energy Chem.* **2020**, *47*, 166–171. [CrossRef]
31. Jung, S.; McCrory, C.C.L.; Ferrer, I.M.; Peters, J.C.; Jaramillo, T.F. Benchmarking nanoparticulate metal oxide electrocatalysts for the alkaline water oxidation reaction. *J. Mater. Chem. A* **2016**, *4*, 3068–3076. [CrossRef]
32. Lyons, M.E.G.; Brandon, M.P. A comparative study of the oxygen evolution reaction on oxidised nickel, cobalt and iron electrodes in base. *J. Electroanal. Chem.* **2010**, *641*, 119–130. [CrossRef]
33. Trotochaud, L.; Young, S.L.; Ranney, J.K.; Boettcher, S.W. Nickel–Iron Oxyhydroxide Oxygen-Evolution Electrocatalysts: The Role of Intentional and Incidental Iron Incorporation. *J. Am. Chem. Soc.* **2014**, *136*, 6744–6753. [CrossRef]
34. Trześniewski, B.J.; Diaz-Morales, O.; Vermaas, D.A.; Longo, A.; Bras, W.; Koper, M.T.M.; Smith, W.A. In situ observation of active oxygen species in Fe-containing Ni-based oxygen evolution catalysts: The effect of pH on electrochemical activity. *J. Am. Chem. Soc.* **2015**, *137*, 15112–15121. [CrossRef] [PubMed]
35. Swierk, J.R.; Klaus, S.; Trotochaud, L.; Bell, A.T.; Tilley, T.D. Electrochemical study of the energetics of the oxygen evolution reaction at nickel iron (Oxy)hydroxide catalysts. *J. Phys. Chem. C* **2015**, *119*, 19022–19029. [CrossRef]
36. Jiang, J.; Zhang, C.; Ai, L. Hierarchical iron nickel oxide architectures derived from metal-organic frameworks as efficient electrocatalysts for oxygen evolution reaction. *Electrochim. Acta* **2016**, *208*, 17–24. [CrossRef]
37. Rossmeisl, J.; Qu, Z.W.; Zhu, H.; Kroes, G.J.; Nørskov, J.K. Electrolysis of water on oxide surfaces. *J. Electroanal. Chem.* **2007**, *607*, 83–89. [CrossRef]
38. Li, Q.; Song, Y.; Xu, R.; Zhang, L.; Gao, Z.; Xia, Z.; Tian, Z.; Wei, N.; Rümmeli, M.H.; Zou, X.; et al. Biotemplating Growth of Nepenthes-like N-Doped Graphene as a Bifunctional Polysulfide Scavenger for Li–S Batteries. *ACS Nano* **2018**, *12*, 10240–10250. [CrossRef]
39. Wang, C.; Li, X.; Li, Q.; Pang, H. Graphene/Co$_3$O$_4$ composites in application of electrochemical energy conversion and storage. *FlatChem* **2019**, *16*, 100107. [CrossRef]
40. Muthurasu, A.; Maruthapandian, V.; Kim, H.Y. Metal-organic framework derived Co$_3$O$_4$/MoS$_2$ heterostructure for efficient bifunctional electrocatalysts for oxygen evolution reaction and hydrogen evolution reaction. *Appl. Catal. B Environ.* **2019**, *248*, 202–210. [CrossRef]
41. Liu, Z.; Yu, X.; Yu, H.; Xue, H.; Feng, L. Nanostructured FeNi$_3$ Incorporated with Carbon Doped with Multiple Nonmetal Elements for the Oxygen Evolution Reaction. *ChemSusChem* **2018**, *11*, 2703–2709. [CrossRef] [PubMed]
42. Liu, L.; Yan, F.; Li, K.; Zhu, C.; Xie, Y.; Zhang, X.; Chen, Y. Ultrasmall FeNi$_3$N particles with an exposed active (110) surface anchored on nitrogen-doped graphene for multifunctional electrocatalysts. *J. Mater. Chem. A* **2019**, *7*, 1083–1091. [CrossRef]
43. Yu, X.; Zhao, Z.; Pei, C. Surface oxidized iron-nickel nanorods anchoring on graphene architectures for oxygen evolution reaction. *Chin. Chem. Lett.* **2021**, *32*, 3579–3583. [CrossRef]
44. Zhang, B.; Xiao, C.; Xie, S.; Liang, J.; Chen, X.; Tang, Y. Iron–Nickel Nitride Nanostructures in Situ Grown on Surface-Redox-Etching Nickel Foam: Efficient and Ultrasustainable Electrocatalysts for Overall Water Splitting. *Chem. Mater.* **2016**, *28*, 6934–6941. [CrossRef]
45. Xue, Y.; Wang, Y.; Liu, H.; Yu, X.; Xue, H.; Feng, L. Electrochemical oxygen evolution reaction catalyzed by a novel nickel–cobalt-fluoride catalyst. *Chem. Commun.* **2018**, *54*, 6204–6207. [CrossRef]
46. Chung, D.Y.; Lopes, P.P.; Farinazzo Bergamo Dias Martins, P.; He, H.; Kawaguchi, T.; Zapol, P.; You, H.; Tripkovic, D.; Strmcnik, D.; Zhu, Y.; et al. Dynamic stability of active sites in hydr(oxy)oxides for the oxygen evolution reaction. *Nat. Energy* **2020**, *5*, 222–230. [CrossRef]
47. Dionigi, F.; Zeng, Z.; Sinev, I.; Merzdorf, T.; Deshpande, S.; Lopez, M.B.; Kunze, S.; Zegkinoglou, I.; Sarodnik, H.; Fan, D.; et al. In-situ structure and catalytic mechanism of NiFe and CoFe layered double hydroxides during oxygen evolution. *Nat. Commun.* **2020**, *11*, 2522. [CrossRef]
48. Suen, N.-T.; Hung, S.-F.; Quan, Q.; Zhang, N.; Xu, Y.-J.; Chen, H.M. Electrocatalysis for the oxygen evolution reaction: Recent development and future perspectives. *Chem. Soc. Rev.* **2017**, *46*, 337–365. [CrossRef] [PubMed]
49. Zhang, J.; Tao, H.B.; Kuang, M.; Yang, H.B.; Cai, W.; Yan, Q.; Mao, Q.; Liu, B. Advances in Thermodynamic-Kinetic Model for Analyzing the Oxygen Evolution Reaction. *ACS Catal.* **2020**, *10*, 8597–8610. [CrossRef]
50. Shinagawa, T.; Garcia-Esparza, A.T.; Takanabe, K. Insight on Tafel slopes from a microkinetic analysis of aqueous electrocatalysis for energy conversion. *Sci. Rep.* **2015**, *5*, 13801. [CrossRef]
51. Wang, J.; Li, K.; Zhong, H.-X.; Xu, D.; Wang, Z.-L.; Jiang, Z.; Wu, Z.-J.; Zhang, X.-B. Synergistic Effect between Met-al–Nitrogen–Carbon Sheets and NiO Nanoparticles for Enhanced Electrochemical Water—Oxidation Performance. *Angew. Chem. Int. Ed.* **2015**, *54*, 10530–10534. [CrossRef] [PubMed]
52. Narendra Kumar, A.V.; Li, Y.; Yu, H.; Yin, S.; Xue, H.; Xu, Y.; Li, X.; Wang, H.; Wang, L. 3D graphene aerogel sup-ported FeNi-P derived from electroactive nickel hexacyanoferrate as efficient oxygen evolution catalyst. *Electrochim. Acta* **2018**, *292*, 107–114. [CrossRef]
53. Li, Y.; Zhao, M.; Zhao, Y.; Song, L.; Zhang, Z. FeNi layered double-hydroxide nanosheets on a 3D carbon network as an efficient electrocatalyst for the oxygen evolution reaction. *Part. Part. Syst. Char.* **2016**, *33*, 158–166. [CrossRef]
54. Zhao, Y.; Chen, S.; Sun, B.; Su, D.; Huang, X.; Liu, H.; Yan, Y.; Sun, K.; Wang, G. Graphene-Co$_3$O$_4$ nanocomposite as electrocatalyst with high performance for oxygen evolution reaction. *Sci. Rep.* **2015**, *5*, 7629.

55. Yang, J.; Zhu, G.; Liu, Y.; Xia, J.; Ji, Z.; Shen, X.; Wu, S. Fe_3O_4-decorated Co9S8 nanoparticles in situ grown on re-duced graphene oxide: A new and efficient electrocatalyst for oxygen evolution reaction. *Adv. Funct. Mater.* **2016**, *6*, 4712–4721. [CrossRef]
56. Elizabeth, I.; Nair, A.K.; Singh, B.P.; Gopukumar, S. Multifunctional Ni-NiO-CNT composite as high performing free standing anode for Li ion batteries and advanced electro catalyst for oxygen evolution reaction. *Electrochim. Acta* **2017**, *230*, 98–105. [CrossRef]
57. Lu, X.; Zhao, C. Highly efficient and robust oxygen evolution catalysts achieved by anchoring nanocrystalline cobalt oxides onto mildly oxidized multiwalled carbon nanotubes. *J. Mater. Chem. A* **2013**, *1*, 12053–12059. [CrossRef]
58. Han, G.-Q.; Liu, Y.-R.; Hu, W.-H.; Dong, B.; Li, X.; Shang, X.; Chai, Y.-M.; Liu, Y.-Q.; Liu, C.-G. Three dimensional nickel oxides/nickel structure by in situ electro-oxidation of nickel foam as robust electrocatalyst for oxygen evolution reaction. *Appl. Surf. Sci.* **2015**, *359*, 172–176. [CrossRef]
59. Pei, C.; Chen, H.; Dong, B.; Yu, X.; Feng, L. Electrochemical oxygen evolution reaction efficiently catalyzed by a novel porous iron-cobalt-fluoride nanocube easily derived from 3-dimensional Prussian blue analogue. *J. Power Sources* **2019**, *424*, 131–137. [CrossRef]
60. Mao, S.; Wen, Z.; Huang, T.; Hou, Y.; Chen, J. High-performance bi-functional electrocatalysts of 3D crumpled graphene–cobalt oxide nanohybrids for oxygen reduction and evolution reactions. *Energ. Environ. Sci.* **2014**, *7*, 609–616. [CrossRef]

Article

Facile Synthesis of 4,4′-biphenyl Dicarboxylic Acid-Based Nickel Metal Organic Frameworks with a Tunable Pore Size towards High-Performance Supercapacitors

Wenlei Zhang [1], Hongwei Yin [1], Zhichao Yu [1], Xiaoxia Jia [1], Jianguo Liang [2], Gang Li [1,3,*], Yan Li [3] and Kaiying Wang [1,4,*]

[1] Institute of Energy Innovation, College of Materials Science and Engineering & College of Information and Computer, Taiyuan University of Technology, Taiyuan 030024, China; zhangwenlei@tyut.edu.cn (W.Z.); yinhongwei@tyut.edu.cn (H.Y.); yuzhichao@tyut.edu.cn (Z.Y.); jiaxiaoxia@tyut.edu.cn (X.J.)
[2] College of Mechanical and Vehicle Engineering, Taiyuan University of Technology, Taiyuan 030024, China; liangjianguo20@tyut.edu.cn
[3] College of Physics and Information Engineering, Minnan Normal University, Zhangzhou 361000, China; liyan_nmsd@163.com
[4] Department of Microsystems-IMS, University of South-Eastern Norway, 3184 Horten, Norway
* Correspondence: ligang02@tyut.edu.cn (G.L.); kaiying.wang@usn.no (K.W.)

Abstract: Metal-organic frameworks (MOFs) have attracted significant research interest for supercapacitor applications due to their high-tunable conductivity and their structure's pore size. In this work, we report a facile one-step hydrothermal method to synthesize nickel-based metal-organic frameworks (MOF) using organic linker 4,4′-biphenyl dicarboxylic acid (BPDC) for high-performance supercapacitors. The pore size of the Ni-BPDC-MOF nanostructure is tuned through different synthesization temperatures. Among them, the sample synthesized at 180 °C exhibits a nanoplate morphology with a specific surface area of 311.99 $m^2 \cdot g^{-1}$, a pore size distribution of 1–40 nm and an average diameter of ~29.2 nm. A high specific capacitance of 488 $F \cdot g^{-1}$ has been obtained at a current density of 1.0 $A \cdot g^{-1}$ in a 3 M KOH aqueous electrolyte. The electrode shows reliable cycling stability, with 85% retention after 2000 cycles. The hydrothermal process Ni-BPDC-MOF may provide a simple and efficient method to synthesize high-performance hybrid MOF composites for future electrochemical energy storage applications.

Keywords: metal organic framework; 4,4′-biphenyl dicarboxylic acid; one-step hydrothermal method; tunable pore size; nanoplate structure; supercapacitors

1. Introduction

The supercapacitor is a promising electrochemical energy storage device and plays an important role of bridging the gap between conventional capacitors and batteries [1,2]. Taking advantage of its long lifetime, high power density and fast charging-discharging rate, the supercapacitor attracts a lot of attention in fields ranging from portable electronic products to vehicle parts [3,4]. From the view of material design, the supercapacitive performance of electrode materials can be improved through a higher conductivity and surface area as well as an appropriate pore size for enhanced efficient ion transmission [5].

Compared with other types of electrode materials such as carbon-based materials, transition metal oxides and conducting polymers, the metal-organic framework (MOF), a porous crystalline material, is emerging as a promising material due to its variable structure, tunable pore size and high surface area [6]. Since the MOF is composed of central metal ions and organic ligands [7], it combines the high specific capacitance of pseudo capacitance and the reliable cycling stability of electric double-layer capacitance at the same time. So far, more than 20,000 different MOFs have been created, and the number is still growing [8]. Although they have relatively high porosity, few MOFs are used as electrode materials,

mainly due to the insufficient pore size for the diffusion of electrolyte ions and the low conductivity [9].

Organic ligands play an important role in the morphology and electrical properties of MOF [10]. Among them, 4,4′-Biphenyl dicarboxylic acid (BPDC), 2, 6-Naphthalene dicarboxylic acid (NDC) and Benzene dicarboxylic acid (BDC) are the most common carboxylic acids used as the organic linkers for synthesizing MOF material. Some researchers have demonstrated that the molecular length of organic linkers influences the pore size and surface area of MOFs [11]. Longer linkers, such as BPDC, can provide MOFs with larger pores and surface areas and thus improve their capacitance properties [12,13]. Since the BPDC-based MOFs are inherently non-conductive and lack thermal and mechanical stability, a conscious method is to fabricate MOF-based composites with common transition metals such as Zn, Cd, Co, Mn, Cu and Ni [14–17], which may integrate the advantages and mitigate the drawbacks of individual components [16].

Herein, we attempt to design a BPDC-based nickel MOF (Ni-BPDC-MOF) with a different surface area, pore size and morphology for high-performance supercapacitor applications. Through a simple hydrothermal reaction between BPDC and nickel nitrate hexahydrate, the typical Ni-BPDC-MOF was prepared under four different hydrothermal temperatures to tune the pore size of its nanostructure. The morphology, surface area, porosity and conductivity of the Ni-BPDC-MOF varied with the hydrothermal temperature, leading to different capacitive properties. The sample fabricated at 180 °C showed a relatively high surface area and pore size and a good thermal stability, with a specific capacitance of 488 $F·g^{-1}$ at 1.0 $A·g^{-1}$ in the 3 M KOH aqueous electrolyte. Further, it exhibited excellent cycling stability, with a capacitance retention of 85% after 2000 cycles.

2. Materials and Methods

2.1. Characterization Techniques and Electrochemical Measurements

The surface morphology of the Ni-BPDC-MOF was studied by field emission scanning electron microscopy (FESEM, SU8010, Hitachi Inc., Tokyo, Japan). Its crystal structure was tested by X-ray diffraction (XRD, D8 Advance, Bruker Inc., Billerica, MA, USA) in a 2θ range of 10–70 degrees with CuKα (α = 1.541 Å) radiation. The chemical analysis was evaluated under Fourier transform infrared spectroscopy (FTIR, NicoletiS10, Thermo Fisher Inc., Waltham, MA, USA) and X-ray photoelectron spectroscopy (XPS, EscaLab250Xi, Thermo Fisher Inc., Waltham, MA, USA). The thermal stability of the Ni-BPDC-MOF was investigated by thermo-gravimetric analysis (TGA, Pyris 1, PerkinElmer Inc., Waltham, MA, USA). To ensure the full decomposition, the sample was heated gradually up to 800 °C at a rate of ~10 °C/min in a nitrogen atmosphere. The Brunauer Emmet Teller (BET) surface area was characterized with the nitrogen adsorption method by a porosity analyzer (QuadraSorb S1, Quantachrome Inc., Graz, Austria).

The supercapacitive performance of the Ni-BPDC-MOF sample was studied by cyclic voltammetry (CV), electrochemical impedance spectra (EIS) and galvanostatic charge-discharge (GCD) tests employing an electrochemical workstation (IM6, Zahner-Elektrik GmbH Inc., Kronach Gundelsdorf, Germany). All of the tests were tested by using a standard three-electrode system, where the fabricated sample, platinum mesh and Hg/HgO were used as the working, counter and reference electrode, respectively. Then, 3.0 M KOH (85.0%, Sinopharm Chemical Reagent Co., Ltd., Shanghai, China) aqueous solution was served as the electrolyte in the electrochemical measurements due to the stability issue of the Ni-BPDC-MOF. For the EIS test, the perturbation potential was set as 5 mV, and the the frequency ranged from 0.01 Hz to 100 KHz.

2.2. Synthesis Methods

The Ni-BPDC-MOF was synthesized through a facile one-step hydrothermal method. In brief, 0.149 g of BPDC (97.0%, Sinopharm Chemical Reagent Co., Ltd., Shanghai, China) and 0.096 g of nickel nitrate hexahydrate (Ni(NO$_3$)$_2$·6H$_2$O, 98.0%, Sinopharm Chemical Reagent Co., Ltd., Shanghai, China) were dissolved in 20 mL N, N-dimethylformamide

(DMF, 99.5%, Sinopharm Chemical Reagent Co., Ltd., Shanghai, China) followed by continuous stirring. The molar ratio between the organic ligands and metal ions was kept as 3:2. Then, 0.032 g of sodium hydroxide (NaOH, 96.0%, Sinopharm Chemical Reagent Co., Ltd., Shanghai, China) was added into the solution. The configured solution was then shifted into a 50 mL Teflon-lined stainless steel autoclave. The hydrothermal process was done in atmosphere environment and kept at 180 °C for 10 h. Some light green precipitate appeared after the hydrothermal process, which was collected by centrifugation, washed in DMF and alcohol and tried in a nitrogen atmosphere. The temperature influence of the hydrothermal process was investigated with four different values: 120 °C, 150 °C, 180 °C and 210 °C, as described in the Supplementary Materials. The four types of Ni-BPDC-MOFs were activated at 120 °C in vacuum for 24 h [18].

For the consolidation of the sample on a substrate, nickel foam (Jiayisheng Electronics Co., Ltd., Kunshan, China) was adopted as the current collector, taking advantage of its three-dimensional structure for full electrolyte penetration, fast ion diffusion and excellent conductivity [19]. Then, 80 wt.% Ni-BPDC-MOF, 10 wt.% carbon black (Sinopharm Chemical Reagent Co., Ltd., Shanghai, China) and 10 wt.% polytetrafluoroethylene (60%, Aladdin Biochemical Technology Co., Ltd., Shanghai, China) were blended together in ethanol. The mixture was painted on one piece of nickel foam with a size of 10.0 × 20.0 mm, dried at 60 °C in a nitrogen atmosphere for 12 h and then compressed at a pressure of 8 MPa for the following electrochemical measurements.

3. Results and Discussion

Figure 1 shows the typical morphology of the as-prepared Ni-BPDC-MOF. It can be seen that the Ni-BPDC-MOF is comprised of a lot of interconnected nanoplates with an approximate feature size of ~500 nm. As shown in Figure 1a, the high concentrated BPDC molecules (12.3 mmol/L) are easily ionized and construct a monolayer liquid membrane [17]. When added into the solution, the nickel ions are coordinated with the carboxyl ions on the top surface of one BPDC membrane and the bottom surface of another BPDC membrane simultaneously, forming the interconnected nanoplate morphology. The high magnification images of Figure 1b,c present a loose connection and large separation between two nanoplates [20], which may also allow for a continuous charge distribution as well as a large available surface to accelerate the electron transport throughout the framework. The SEM graphs by three different hydrothermal temperatures are displayed in Figure S1. With a lower temperature, the size of the nanoplate is small, with some cracks on its surface. The size increases with the increase in temperature. At 180 °C, the nanoplate is large and integrated. However, with higher temperatures, e.g., 210 °C, the structure seems to melt, leading to the smooth plate surface and edge.

Figure 2a shows the XRD patterns of the Ni-BPDC-MOF and the pure BPDC samples. Compared with those of the BPDC, several new diffraction peaks at 12.4°, 15.3°, 18.7° and 30.3° are observed in the case of the Ni-BPDC-MOF. The 2θ positions of these new peaks are similar to those of other reported isostructural Ni-BPDC-MOFs [21,22], suggesting a similar crystalline structure and the high crystallization of our samples. The XRD patterns of the Ni-BPDC-MOF with hydrothermal temperatures of 120 °C and 180 °C are compared in Figure S2. Since no more diffraction peaks appeared, the chemical structures of our samples may remain the same by changing the hydrothermal temperatures. The chemical bonds are further analyzed by FTIR. As shown in Figure 2b, a couple of intense bands at 847 cm^{-1}, 1406 cm^{-1}, 1532 cm^{-1} and 1586 cm^{-1} are observed, which could be assigned to the stretching vibration of the carboxylate groups (COO−) coordinated to the metal center [23]. Among them, the two bands located at 1406 cm^{-1} and 1586 cm^{-1} are assigned to the symmetric and asymmetric stretching modes of the COO− groups, respectively, which are in accordance with Rawool's report [16]. The difference between these two bands indicates that the COO− groups are coordinated to the metal center through the bidentate mode. Three peaks at 679 cm^{-1}, 770 cm^{-1} and 3082 cm^{-1} corresponded to the stretching vibration of the C−H bond, the ring deformation vibration of the benzene ring and the antiplane bending vibration of the aromatic ring in the BPDC, respectively [24]. The peak at 524 cm^{-1} was related to the Ni−O stretching vibration [21]. The XRD and FTIR analyses conclude a proper coordination between the organic ligand BPDC and the nickel ions.

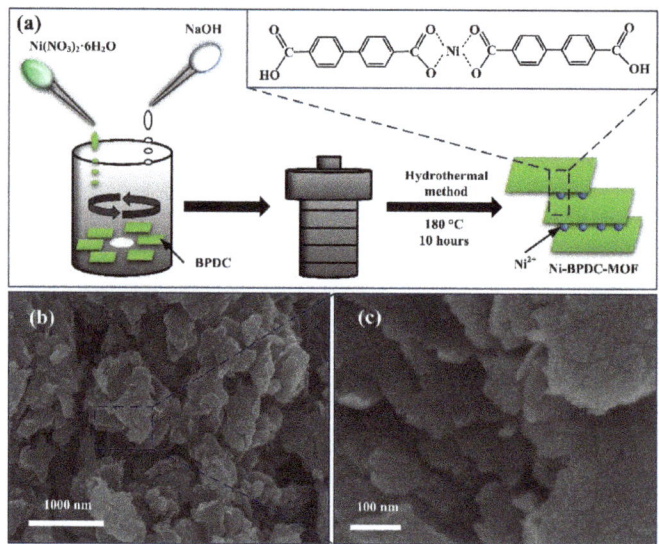

Figure 1. (a) Synthetic scheme for the preparation of Ni-BPDC-MOF nanoplates. (b,c) SEM graphs of the Ni-BPDC-MOF sample with different magnifications.

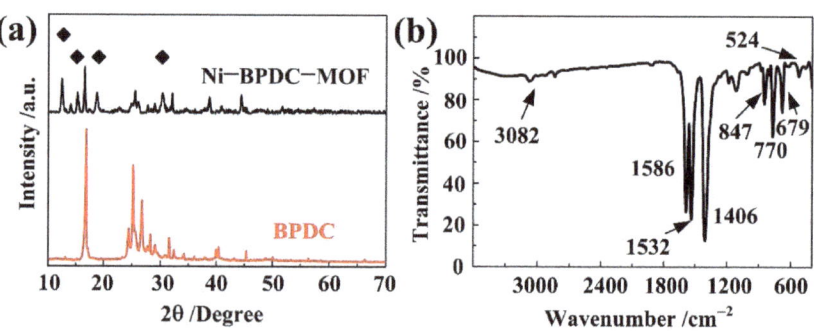

Figure 2. (a) XRD patterns of the Ni-BPDC-MOF and the pure BPDC samples. (b) FTIR spectrum of the Ni-BPDC-MOF sample.

The chemical valence states of the Ni-BPDC-MOF are studied by XPS. The survey spectrum in Figure 3a indicates the existence of C, O and Ni elements. Two major peaks located at 855.8 eV and 873.3 eV are observed in the spectra of Ni 2p (Figure 3b), which correspond to Ni $2p_{1/2}$ and Ni $2p_{3/2}$, respectively [25]. The other two peaks located at 861.3 eV and 879.6 eV are considered as satellite peaks for Ni^{2+} [26]. Figure 3c shows the C 1s of the Ni-BPDC-MOF. The two peaks at 284.8 eV and 288.6 eV represent the sp^2 phase (C=C) in the benzene ring and (CO) bond, respectively [27]. The O 1s spectrum in Figure 3d exhibits one predominant peak at 531.5 eV, which is attributed to the adsorbed −OH on the surface of the Ni-BPDC-MOF [28], while another smaller peak at 536.5 eV is directly related to the hydrocarbon oxides in the BPDC [29].

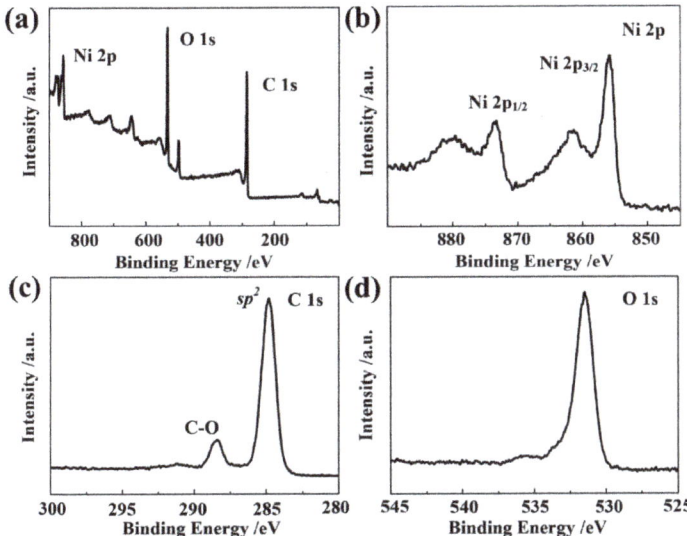

Figure 3. XPS spectrum of the Ni-BPDC-MOF sample. (**a**) Survey spectrum. (**b**–**d**) High resolution XPS spectrum of (**b**) Ni 2p, (**c**) C 1s and (**d**) O 1s.

TGA is performed to evaluate the thermal stability of **the** Ni-BPDC-MOF. As shown in Figure 4a, four mass drops were observed. The first (8.5%) mass drop happened from the RT to 390 °C and was caused by the desorption of physiosorbed water and the chemisorbed organic solvent on the surface of the Ni-BPDC-MOF [30]. The second (21.4%) and the third (34.6%) mass drops happened sharply from 390 °C to 600 °C, with the fastest drop at 428 °C. This can be concluded as the decomposition of the Ni-BPDC-MOF's organic skeleton and the generation of nickel oxide. The final mass loss was observed from 600 °C to 700 °C and is related to the exclusion of CO_2 from the evolution of the carboxylic groups. When the applied temperature is higher than 700 °C, the mass remained stable, and less than 30% of the residual was left. The TGA demonstrates that the Ni-BPDC-MOF is thermally stable below 400 °C in the open air.

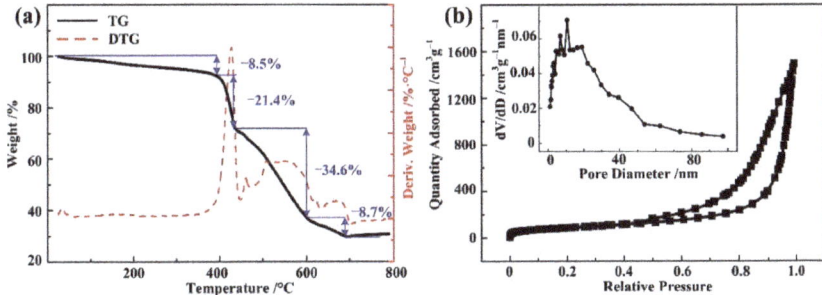

Figure 4. (**a**) TGA spectra of the Ni-BPDC-MOF sample. (**b**) N_2 adsorption–desorption isotherms and the corresponding pore size distribution curve of the Ni-BPDC-MOF sample.

Figure 4b shows the N_2 adsorption–desorption isotherm and the corresponding pore size distribution curve of the Ni-BPDC-MOF. It shows a type-IV curve with a type-H3 hysteresis loop [31], indicating the highly porous structure of the as-prepared sample. The BET specific surface area is 311.99 $m^2 \cdot g^{-1}$. The pore sizes of the Ni-BPDC-MOF are further calculated by the Barrett–Joyner–Halenda (BJH) method. The pore size of the Ni-BPDC-MOF exhibits a bimodal distribution with a diameter of 1–40 nm, while the calculated average value is calculated as 29.2 nm. The average pore size seems to be higher than that of some common MOFs, which is proved by other BPDC-based MOFs with similar structures [20,32]. The relatively larger surface area and pore size may be due to the micropores produced by the long organic linker BPDC and the mesopores caused by the stacked nanoplates. The relatively larger surface area and pore size are likely to lead to the easy diffusion of electrolyte ions, which is beneficial to high charge storage capacity [16].

The adsorption–desorption isotherms and pore size distribution curves of the other samples are shown in Figure S4. It can be seen that the sample synthesized at 180 °C has the largest surface area. The result is in good agreement with the SEM observation. With lower temperatures, the size of the nanoplate is too small and cannot supply enough surface area, while with higher temperatures, the nanoplate is melted and sticks with the other, so the surface area value drops again. A temperature of 180 °C is considered the optimized value for Ni-BPDC-MOF synthesization.

Electrochemical measurements have been carried out to evaluate the supercapacitive performance of the Ni-BPDC-MOF. Figure 5a depicts the CV curves of the Ni-BPDC-MOF sample at three different scan rates from 10 to 100 mV·s^{-1}, with a potential range 0–0.6 V. All of the CV curves displayed a couple of obvious redox peaks, indicating the typical pseudocapacitive property of the samples [33]. With increasing scan rates, the oxidation peak is shifted positively, while the reduction peak is shifted negatively, resulting from the resistance of the electrodes [34]. The pseudocapacitance is generated from faradic redox reactions of the intercalation and deintercalation of the OH$^-$ during electrochemical reactions in the Ni-BPDC-MOF. The anodic and cathodic peaks in the CV curves are considered to be different oxidation states of nickel, corresponding to the following process:

$$\text{Ni(II)}_s + \text{OH}^- \leftrightarrow \text{Ni(II)(OH)}_{ad} + e^- \tag{1}$$

$$\text{Ni(II)(OH)}_{ad} \leftrightarrow \text{Ni(III)(OH)}_{ad} + e^- \tag{2}$$

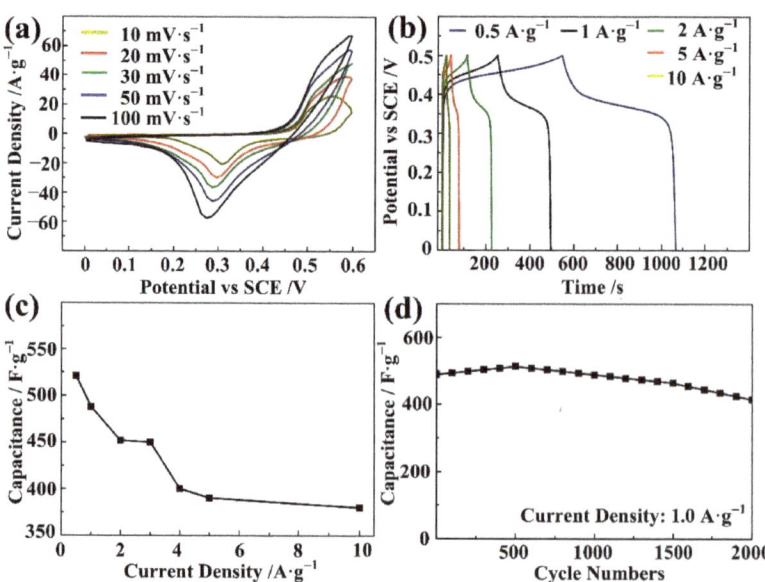

Figure 5. Capacitance performances of the Ni-BPDC-MOF sample. (**a**) CV curves at different scan rates. (**b**) GCD curves at different current densities. (**c**) Specific capacitances at different current densities. (**d**) Cyclic stability.

Figure 5b presents the GCD curves of the Ni-BPDC-MOF with three different current densities. The charge time and discharge time in each GCD curve are almost equal, demonstrating a good capacitance performance and the reversible redox process of the Ni-BPDC-MOF sample. For the discharging part, an obvious potential plateau at 0.43 V is observed. This plateau is interpreted as the electrochemical adsorption–desorption and the faradic redox reaction process on the interface of the electrode and electrolyte [35], which is in accordance with the CV curves. The specific capacitance is calculated from the discharging time of the GCD curve. At the commonly used current density (1.0 A·g^{-1}), the Ni-BPDC-MOF sample delivers a value of 488 F·g^{-1}. Figure 5c concludes the specific capacitances of the Ni-BPDC-MOF with different current densities. With increasing current densities, the calculated capacitance decreases. However, even the current density is changed from 0.5 F·g^{-1} to 10 F·g^{-1}. The capacitance is dropped from 521 F·g^{-1} to 380 F·g^{-1}, with 72.9% of the capacitance retained. The result shows the good rate capability of the Ni-BPDC-MOF. The GCD test was circulated 2000 times at a current density of 1.0 A·g^{-1} to evaluate its cyclic stability. As shown in Figure 5d, more than 85.0% of the specific capacitance is reserved. In the first 500 cycles of the test, the capacitance increased, which may result from the electro-activation of the active material [36]. The value was decreased from 500 to 2000 cycles continuously, owing to the shedding of active material from the nickel foam during the cycling test.

In Figure 6, R_s is the internal resistance of the electrode, including the contact resistance of the active material and substrate, the intrinsic resistance of the active material and the ionic resistance of the electrolyte. R_{ct} and W_s are the charge transfer resistance and Warburg diffusion impedance, which are related to the pseudocapacitive reaction. CPE1 is the double-layer capacitance element and can be neglected in this research. In the low frequency part, the slope of the straight lines in the low frequency is larger than 45°, confirming a smaller W_s and a fast migration of the ions [37]. R_{ct} and R_s are calculated from the radius of the semicircle and the intercept of the real axis in the high frequency range, with relatively low values of 0.49 Ω and 18.55 Ω [38], respectively. The EIS result indicates that the Ni-BPDC-MOF has a good ability to deliver charges on the interface of the electrode and electrolyte and thus led to a positive effect on its specific capacitance. The relationship between the hydrothermal temperature and the supercapacitive performance of the Ni-BPDC-MOF is displayed in Figure S4 and Table S1. All of the Ni-BPDC-MOFs exhibit a typical seudocapacitive property, and the sample synthesized at 180 °C had the highest capacitance.

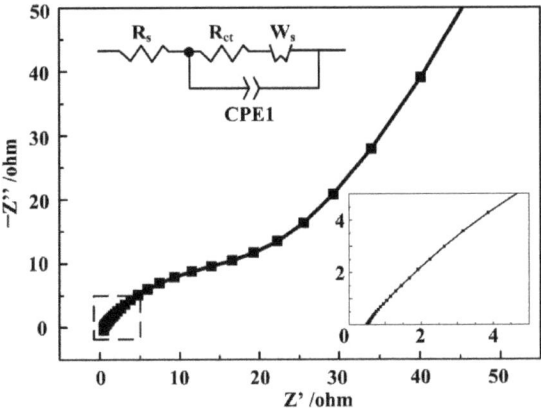

Figure 6. Nyquist plots of the Ni-BPDC-MOF sample.

Table 1 compares recent reports on the porosity and supercapacitive performance of BPDC-based MOF materials. The specific capacitance is superior to that found by most other researchers in the absence of the highest surface area value. For the research on

BPDC-based porous carbon nanoparticles (C-BPDC) [15,30], the specific surface area is even 2~3 times higher than that found in our report. However, because of their small pore size, the electrode materials may be not well-soaked in the electrolyte, leading to a relatively low specific capacitance. For the BPDC-based metal nanoparticles (Al-BPDC, Ni-BPDC) [16,17,39], the surface area and pore size may reach a sufficient amount for the better intercalation of electrolyte ions. Still, our report has the lowest R_{ct} value, which might be related to our unique structure: the interconnected network of the Ni-BPDC nanoplates results in a continuous charge distribution, leading to a faster electron transport throughout the whole framework and a lower R_{ct} value [16].

Table 1. Comparison of the present work with previously reported BPDC-based MOFs for porosity and supercapacitive performance.

Sample	Surface Area ($m^2 \cdot g^{-1}$)	Average Diameter (nm)	Electrolyte	Scan Rate ($mV \cdot s^{-1}$)	Current Density ($A \cdot g^{-1}$)	R_{ct} (Ω)	Capacitance ($F \cdot g^{-1}$)	Ref.
C-BPDC	1137	4.2	6M KOH	1	—	0.34	170	[40]
Al-BPDC	415.2	18.5	6M KOH	—	0.25	3.25	119	[39]
C-BPDC	843	0.53	6M KOH	—	1	1.23	256	[41]
Ni-BPDC	347	32.7	6M KOH	—	1	1.10	328	[16]
Ni-BPDC	—	—	6M KOH	—	1	0.65	432	[17]
Ni-BPDC	311.99	29.16	3M KOH	—	1	0.49	488	This Work

4. Conclusions

In conclusion, we demonstrate a facile one-step hydrothermal method for synthesizing a BPDC-based nickel MOF with an optimized reaction temperature of 180 °C. A chemical characterization confirmed a good coordination between the organic ligand BPDC and the nickel ions. The as-synthesized Ni-BPDC-MOF presents a nanoplate structure with a relatively high specific surface area of 311.99 $m^2 \cdot g^{-1}$, a pore size distribution of 1–20 nm and an average diameter of 29.2 nm. As a result, the capacitance properties of the Ni-BPDC-MOF exhibited a great performance, with a specific capacitance of 488 $F \cdot g^{-1}$ at a current density of 1.0 $A \cdot g^{-1}$. The capacitance retention is kept at 85% of its initial value after 2000 cycles. This study demonstrates a simple synthesizing method for high-performance BPDC-based MOF material, which may become a promising candidate for electrochemical capacitive energy storage.

Supplementary Materials: The following supporting information can be downloaded at: https://www.mdpi.com/article/10.3390/nano12122062/s1, Figure S1: SEM graphs of Ni-BPDC-MOF samples with different hydrothermal temperature; Figure S2: XRD patterns of Ni-BPDC-MOF with hydrothermal temperatures of 120 °C and 180 °C; Figure S3: N2 adsorption/desorption isotherms and corresponding pore size distribution curve of Ni-BPDC-MOF samples with different hydrothermal temperature; Figure S4: Capacitance performances of Ni-BPDC-MOF samples with different hydrothermal temperature; Table S1: The comparison for porosity and supercapacitive performance of Ni-BPDC-MOF samples with different hydrothermal temperature.

Author Contributions: Conceptualization, G.L. and K.W.; methodology, Y.L.; validation, Z.Y., G.L. and K.W.; investigation, X.J.; data curation, Y.L.; writing—original draft preparation, H.Y.; writing—review and editing, W.Z. and J.L.; visualization, W.Z.; supervision, G.L.; funding acquisition, W.Z., J.L. and G.L. All authors have read and agreed to the published version of the manuscript.

Funding: This research was supported by the National Natural Science Foundation of China (52005363, 51622507, 52075361 and 61471255), the Natural Science Foundation of Shanxi Province, China (2016011040), and the Scientific and Technological Innovation Programs of Higher Education Institutions in Shanxi Province, China (2016138).

Data Availability Statement: The data presented in this study are available on request from the corresponding author.

Conflicts of Interest: The authors declare no conflict of interest.

References

1. Wang, L.; Han, Y.; Feng, X.; Zhou, J.; Qi, P.; Wang, B. Metal-organic frameworks for energy storage: Batteries and supercapacitors. *Coordin. Chem. Rev.* **2016**, *307*, 361–381. [CrossRef]
2. Morozan, A.; Jaouen, F. Metal organic frameworks for electrochemical applications. *Energy Environ. Sci.* **2012**, *5*, 9269–9290. [CrossRef]
3. Naoi, K.; Naoi, W.; Aoyagi, S.; Miyamoto, J.; Kamino, T. New generation "nanohybrid supercapacitor". *Accounts Chem. Res.* **2014**, *46*, 1075–1083. [CrossRef] [PubMed]
4. Wu, C.; Feng, F.; Xie, Y. Design of vanadium oxide structures with controllable electrical properties for energy applications. *Chem. Soc. Rev.* **2013**, *42*, 5157–5183. [CrossRef] [PubMed]
5. Li, Q.; Jiang, Y.; Zhu, J.; Gan, X.; Qin, F.; Tang, T.; Luo, W.; Guo, N.; Liu, Z.; Wang, L.; et al. Ultrafast pore-tailoring of dense microporous carbon for high volumetric performance asupercapacitors in organic electrolyte. *Carbon* **2022**, *191*, 19–27. [CrossRef]
6. Furukawa, H.; Cordova, K.; O'Keeffe, M.; Yaghi, O. The chemistry and applications of metal-organic frameworks. *Science* **2013**, *341*, 1230444. [CrossRef]
7. Ke, F.; Wu, Y.; Deng, H. Metal-organic frameworks for lithium ion batteries and supercapacitors. *J. Solid. State. Electr.* **2015**, *223*, 109–121. [CrossRef]
8. Li, Y.; Xu, Y.; Liu, Y.; Pang, H. Exposing {001} crystal plane on hexagonal Ni-MOF with surface-grown cross-linked mesh-structures for electrochemical energy storage. *Small* **2019**, *15*, 1902463. [CrossRef]
9. Yan, Y.; Gu, P.; Zheng, S.; Zheng, M.; Pang, H.; Xue, H. Facile synthesis of an accordion-like Ni-MOF superstructure for high-performance flexible supercapacitors. *J. Mater. Chem. A* **2016**, *4*, 19078–19085. [CrossRef]
10. Maina, J.W.; Gonzalo, C.P.; Kong, L.; Schutz, J.; Hill, M.; Dumee, L.F. Metal organic framework based catalysts for CO_2 conversion. *Mater. Horiz.* **2017**, *4*, 345–361. [CrossRef]
11. Li, L.; Jung, H.S.; Lee, J.W.; Kang, Y.T. Review on applications of metal-organic frameworks for CO_2 capture and the performance enhancement mechanisms. *Renew. Sust. Energ. Rev.* **2022**, *162*, 112441. [CrossRef]
12. Furukawa, H.; Nakeum, K.; Yong, G.; Aratani, N.; Choi, S.; Yazaydin, A.; Snurr, R.; O'Keeffe, M.; Yaghi, O. Ultrahigh porosity in metal-organic frameworks. *Science* **2010**, *329*, 424–428. [CrossRef] [PubMed]
13. Park, J.; Lim, W.; Oh, T.; Suh, M. A highly porous metal–organic framework: Structural transformations of a guest-free MOF depending on activation method and temperature. *Chem. Eur. J.* **2011**, *17*, 7251–7260. [CrossRef] [PubMed]
14. Yue, M.; Yu, C.; Duan, H.; Yang, B.; Meng, X.; Li, Z. Six isomorphous window-beam MOFs: Explore the effects of metal ions on MOF-derived carbon for supercapacitors. *Chem. Eur. J.* **2018**, *24*, 16160–16169. [CrossRef] [PubMed]
15. Liu, M.; Zhao, F.; Zhu, D.; Duan, H.; Lv, Y.; Li, L.; Gan, L. Ultramicroporous carbon nanoparticles derived from metaleorganic framework nanoparticles for high-performance supercapacitors. *Mater. Chem. Phys.* **2018**, *211*, 234–241. [CrossRef]
16. Rawool, C.; Karna, S.; Srivastava, A. Enhancing the supercapacitive performance of Nickel based metal organic framework-carbon nanofibers composite by changing the ligands. *Electrochim. Acta* **2019**, *294*, 345–356. [CrossRef]
17. He, F.; Yang, N.; Li, K.; Wang, X.; Cong, S.; Zhang, L.; Xiong, S.; Zhou, A. Hydrothermal synthesis of Ni-based metal organic frameworks/graphene oxide composites as supercapacitor electrode materials. *J. Mater. Res.* **2020**, *35*, 1439–1450. [CrossRef]
18. Rahmanifar, M.S.; Hesari, H.; Noori, A.; Masoomi, M.Y.; Morsali, A.; Mousavi, M.F. A dual Ni/Co-MOF-reduced graphene oxide nanocomposite as a high performance supercapacitor electrode material. *Electrochim. Acta* **2018**, *275*, 76–86. [CrossRef]
19. Gu, W.; Wang, Y.; Lu, R.; Guan, L.; Peng, X.; Sha, J. Anodic electrodeposition of a porous nickel oxide–hydroxide film on passivated nickel foam for supercapacitors. *J. Mater. Chem. A* **2014**, *2*, 7161–7164. [CrossRef]
20. Lee, S.; Shinde, D.; Kim, E.; Lee, W.; Oh, I.; Shrestha, N.; Lee, J.; Han, S. Supercapacitive property of metal–organic-frameworks with different pore dimensions and morphology. *Micropor. Mesopor. Mater.* **2013**, *171*, 53–57. [CrossRef]
21. Wang, X.; Li, Q.; Yang, N.; He, F.; Chu, J.; Gong, M.; Wu, B.; Zhang, R.; Xiong, S. Hydrothermal synthesis of NiCo-based bimetal-organic frameworks as electrode materials for supercapacitors. *J. Solid State Chem.* **2019**, *270*, 370–378. [CrossRef]
22. Wang, X.; Yang, N.; Li, Q.; He, F.; Yang, Y.; Cong, S.; Li, K.; Xiong, S.; Zhou, A. Hydrothermal synthesis of humate-layer-based bimetal organic framework composites as high rate-capability and enery-density electrode materials for supercapacitors. *ChemistrySelect* **2020**, *5*, 2794–2804. [CrossRef]
23. Xu, J.; Yang, C.; Xue, Y.; Wang, C.; Cao, J.; Chen, Z. Facile synthesis of novel metal-organic nickel hydroxide nanorods for high performance supercapacitor. *Electrochim. Acta* **2016**, *211*, 595–602. [CrossRef]
24. Inoue, S.; Fujihara, S. Liquid-liquid biphasic synthesis of layered Zinc hydroxides intercalated with long-chain carboxylate ions and their conversion into ZnO nanostructures. *Inorg. Chem.* **2011**, *50*, 3605–3612. [CrossRef] [PubMed]
25. Lee, J.; Ahn, T.; Soundararajan, D.; Ko, J.; Kim, J. Non-aqueous approach to the preparation of reduced graphene oxide/α-Ni(OH)$_2$ hybrid composites and their high capacitance behavior. *Chem. Commun.* **2011**, *22*, 6305–6307. [CrossRef]
26. Yan, H.; Bai, J.; Wang, J.; Zhang, X.; Wang, B.; Liu, Q.; Liu, L. Graphene homogeneously anchored with Ni(OH)$_2$ nanoparticles as advanced supercapacitor electrodes. *CrystEngComm* **2013**, *15*, 10007–10015. [CrossRef]
27. Zhang, W.; Hirai, Y.; Tsuchiya, T.; Osamu, T. Effect of substrate bias voltage on tensile properties of single crystal silicon microstructure fully coated with plasma CVD diamond-like carbon film. *Appl. Surf. Sci.* **2018**, *443*, 48–54. [CrossRef]

28. Wang, X.; Sun, W.; Dang, K.; Zhang, Q.; Shen, Z.; Zhan, S. Conjugated π electrons of MOFs drive charge separation at heterostructures interface for enhanced photoelectrochemical water oxidation. *Small* **2021**, *17*, 2100367. [CrossRef]
29. Waki, I.; Fujioka, H.; Ono, K.; Oshima, M.; Miki, H.; Fukizawa, A. The effect of surface cleaning by wet treatments and ultra high vacuum annealing for Ohmic contact formation of P-type GaN. *Jpn. J. Appl. Phys.* **2000**, *39*, 4451–4455. [CrossRef]
30. Salam, H.; Nassar, H.N.; Khidr, A.; Zaki, T. Antimicrobial activities of green synthesized Ag nanoparticles@Ni-MOF nanosheets. *J. Inorg. Organomet. P.* **2018**, *28*, 2791–2798. [CrossRef]
31. Li, G.; Hou, J.J.; Zhang, W.; Li, P.; Liu, G.; Wang, Y.; Wang, K. Graphene-bridged WO_3/MoS_2 Z-scheme photocatalyst for enhanced photodegradation under visible light irradiation. *Mater. Chem. Phys.* **2020**, *246*, 122827. [CrossRef]
32. Du, R.; Wu, Y.; Yang, Y.; Zhai, T.; Zhou, T.; Shang, Q.; Zhu, L.; Shang, C.; Guo, Z. Porosity engineering of MOF-based materials for electrochemical energy storage. *Adv. Energy Mater.* **2021**, *11*, 2100154. [CrossRef]
33. Nie, Y.; Wang, X.; Chen, X.; Zhang, Z. Synergistic effect of novel redox additives of p-nitroaniline and dimethylglyoxime for highly improving the supercapacitor performances. *Phys. Chem. Chem. Phys.* **2016**, *18*, 2718–2729. [CrossRef] [PubMed]
34. Yan, J.; Wei, T.; Qiao, W.; Shao, B.; Zhao, Q.; Zhang, L.; Fan, Z. Rapid microwave-assisted synthesis of graphene nanosheet/Co_3O_4 composite for supercapacitors. *Electrochim. Acta* **2010**, *55*, 6973–6978. [CrossRef]
35. Lu, Z.; Chang, Z.; Liu, J.; Sun, X. Stable ultrahigh specific capacitance of NiO nanorod arrays. *Nano Res.* **2011**, *4*, 658–665. [CrossRef]
36. Zeng, L.; Xiao, F.; Wang, J.; Gao, S.; Ding, X.; Wei, M. ZnV_2O_4-CMK nanocomposite as an anode material for rechargeable lithium-ion batteries. *J. Mater. Chem.* **2012**, *22*, 14284–14288. [CrossRef]
37. Hao, P.; Zhao, Z.; Tian, J.; Li, H.; Sang, Y.; Yu, G.; Cai, H.; Liu, H.; Wong, C.P.; Umar, A. Hierarchical porous carbon aerogel derived from bagasse for high performance supercapacitor electrode. *Nanoscale* **2014**, *6*, 12120. [CrossRef]
38. Wang, J.; Zhong, Q.; Xiong, Y.; Cheng, D.; Zeng, Y.; Bu, Y. Fabrication of 3D Co-doped Ni-based MOF hierarchical micro-flowers as a high-performance electrode material for supercapacitors. *Appl. Surf. Sci.* **2019**, *483*, 1158–1165. [CrossRef]
39. Liu, Y.; Xu, J.; Liu, S. Porous carbon nanosheets derived from Al-based MOFs for supercapacitors. *Micropor. Mesopor. Mat.* **2016**, *236*, 94–99. [CrossRef]
40. Yang, Y.; Hao, S.; Zhao, H.; Wang, Y.; Zhang, X. Hierarchically porous carbons derived from nonporous metal-organic frameworks: Synthesis and influence of struts. *Electrochim. Acta* **2015**, *180*, 651–657. [CrossRef]
41. Yin, W.; Zhang, G.; Wang, X.; Pang, H. One–dimensional metal–organic frameworks for electrochemical applications. *Adv. Colloid Interface Sci.* **2021**, *298*, 102562. [CrossRef] [PubMed]

Review

Carbon Tube-Based Cathode for Li-CO$_2$ Batteries: A Review

Deyu Mao [1,†], Zirui He [1,†], Wanni Lu [1] and Qiancheng Zhu [1,2,*]

[1] School of Mechanical and Automotive Engineering, Guangxi University of Science and Technology, Liuzhou 545006, China; maodeyu2012@126.com (D.M.); hzr13865666819@163.com (Z.H.); wannilu@126.com (W.L.)
[2] Faurecia (Liuzhou) Automotive Interior Systems Co., Ltd., Liuzhou 545000, China
* Correspondence: zhuqc@gxust.edu.cn
† These authors contributed equally to this work.

Abstract: Metal–air batteries are considered the research, development, and application direction of electrochemical devices in the future because of their high theoretical energy density. Among them, lithium–carbon dioxide (Li–CO$_2$) batteries can capture, fix, and transform the greenhouse gas carbon dioxide while storing energy efficiently, which is an effective technique to achieve "carbon neutrality". However, the current research on this battery system is still in the initial stage, the selection of key materials such as electrodes and electrolytes still need to be optimized, and the actual reaction path needs to be studied. Carbon tube-based composites have been widely used in this energy storage system due to their excellent electrical conductivity and ability to construct unique spatial structures containing various catalyst loads. In this review, the basic principle of Li–CO$_2$ batteries and the research progress of carbon tube-based composite cathode materials were introduced, the preparation and evaluation strategies together with the existing problems were described, and the future development direction of carbon tube-based materials in Li–CO$_2$ batteries was proposed.

Keywords: carbon tube-based cathode; Li–CO$_2$ battery; reaction mechanism; performance improvement; carbon neutrality

1. Introduction

Global warming caused by greenhouse gases is an essential factor affecting the current environmental deterioration. The generation of greenhouse gases is inevitable based on the biomass (coal, oil, natural gas, etc.) combustion energy conversion method. Among them, carbon dioxide is the most important greenhouse gas, standing as the core issue that needs to be addressed to realize the currently advocated "low carbon environment" [1]. The solution strategy from "carbon peak" to "carbon neutralization" is divided into two aspects: one is the need to reduce carbon emissions, that is, to minimize energy storage and transformation based on "C", and to gradually promote all types of renewable alternative energy in all walks of life; the second is the treatment of existing carbon dioxide with conventionally applied methods including chemical conversion, photocatalytic reduction, electrochemical reduction, and biological conversion [2]. The conversion efficiency of these methods has yet to be improved, while the biggest limitation is that the direct conversion of C in CO$_2$ inevitably requires additional energy (resulting in extra "carbon emissions"), as C in CO$_2$ is in the highest oxidation state [3]. The products obtained by these methods are carbon monoxide, methane, ethylene, formic acid, methanol, etc. The gaseous or liquid products are involved in compression, packaging, storage, transportation, and other steps before they are used as energy storage materials, which is bound to cause further energy loss [4]. Thus, the conventional CO$_2$ conversion method is also a "high carbon" process.

In recent years, the research on lithium–air batteries has made significant progress, especially the developments focusing on the optimal catalyst selection and the structure of the carbon matrix composite cathode design [5–8]. Because of their high theoretical specific capacity (their theoretical specific capacity is 5–10 times that of lithium-ion batteries) [5,9,10], they are considered the ultimate devices for the energy storage of vehicle power batteries in the future. In the study of lithium–oxygen batteries, the effects of water vapor and CO_2 have to be carefully considered. Research shows that the battery capacity under an O_2/CO_2 mixture is three times that of pure oxygen [11], but the stability decreases significantly. The study of the battery reaction process in a carbon dioxide atmosphere is an essential intermediate link to realize the real application of metal–air batteries in the future [6,12]. With the expansion of this research, Li–CO_2 batteries have gradually developed into an independent research direction because this system can achieve potential applications in particular fields such as Mars (96% of carbon dioxide in the atmosphere with a low temperature) detection [13] and energy storage for submarines. In recent years, the number of related research papers published (Web of Science statistics) has increased year by year (Figure 1a), and the distribution of disciplines is shown in Figure 1b. Previous studies have shown that lithium–carbon dioxide batteries based on carbon-based cathode catalysis can achieve a stable cycle, and it is believed that the charging and discharging process is based on the following reaction: $4Li^+ + 3CO_2 + 4e^- \rightarrow 2Li_2CO_3 + C$ ($E^0 = 2.80$ V versus Li/Li^+) [14–16]. This reaction has attracted wide attention in the fields of energy and the environment because it involves the fixation and transformation of CO_2 in the electrochemical energy storage process. With the deepening of this research, the reversibility of the battery reaction has also sparked a controversial concept. At present, significant progress has been made in the study of the performance improvement of battery systems, such as the number of rechargeable cycles and the reduction in the overpotential. However, the research on the controllable preparation of optimized electrode materials and the corresponding reaction mechanism is still in its infancy.

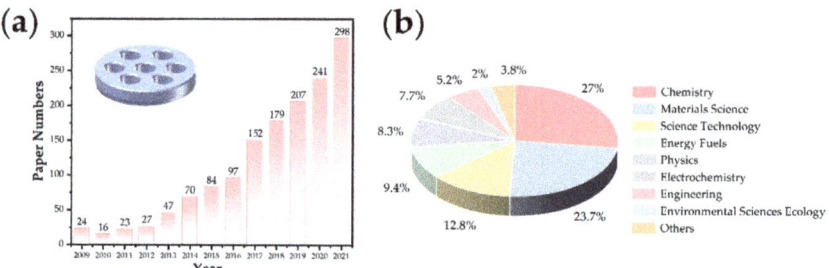

Figure 1. (a) The research progress and (b) distribution of disciplines of Li–CO_2 batteries.

This review introduces the primary mechanism of lithium–carbon dioxide batteries and the latest progress in the application of carbon tube-based materials in battery systems, including the strategy and application of carbon nanotubes (fibers) combined with noble metals, molybdenum-based materials, other metal-based materials, and heteroatoms. This paper focused on the optimum selection and structure construction of carbon tube matrix composites and the improvement and enhancement of the performance of lithium–carbon dioxide batteries. Combined with innovative research methods, the development direction of carbon tube-based Li–CO_2 batteries was proposed.

2. Structure and Reaction Mechanism of a Lithium–Carbon Dioxide Battery
2.1. The Structure of a Li–CO_2 Battery

A typical type of Li–CO_2 battery consists of a porous cathode, electrolyte (liquid, solid), and lithium metal anode [17]. The basic structure is shown in Figure 2, in the form of coin cells from different points of view.

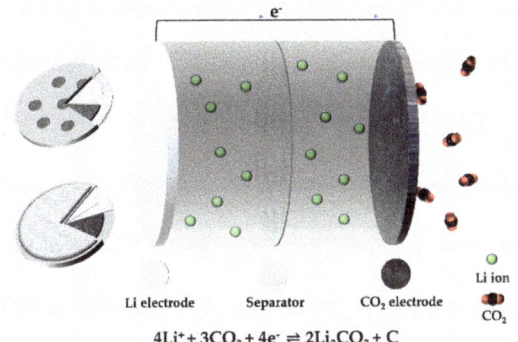

$4Li^+ + 3CO_2 + 4e^- \rightleftharpoons 2Li_2CO_3 + C$

Figure 2. The structure of a Li–CO$_2$ battery.

2.2. The Mechanism of a Li–CO$_2$ Battery

The reaction process of Li–CO$_2$ batteries is closely related to the electrode, electrolyte, and atmosphere environment. Studies have shown that lithium–carbon dioxide batteries cannot discharge in a pure CO$_2$ atmosphere, and there must be a small amount of oxygen involved in the catalysis [11]. A much more critical problem is the lack of research evidence on the generation and decomposition process of discharge product C, and the lack of direct and powerful characterization test data. Zhou's group reported in Joule that lithium–carbon dioxide batteries are rechargeable. Still, their charge and discharge processes are irreversible: lithium carbonate generated during discharge can be decomposed, but the generated carbon will not be decomposed but enriched on the electrode [18]. That is, when charging, the battery reaction is $2Li_2CO_3 \rightarrow 2CO_2 + O_2 + 4Li^+ + 4e^-$ (E^0 = 3.82 V versus Li/Li$^+$). Based on this principle, an energy storage device can be designed, which can not only reduce the emission of CO$_2$ but also use CO$_2$ as the energy storage carrier. The gaseous CO$_2$ is fixed into the solid C; that is, during the charging and discharging process, high specific energy storage and greenhouse gas treatment can be realized at the same time. The energy efficiency reaches 73.3% without pollution, which is of great significance to the solution of energy and environmental problems (Figure 3). Therefore, it is urgent to carry out proper mechanism research to confirm the reaction process of the system [19,20].

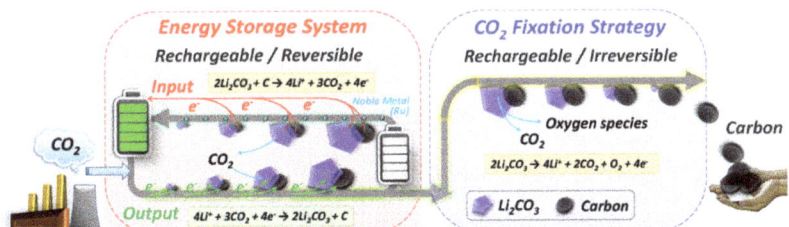

Figure 3. Schematic for the achievement of an energy storage system (reversible process) and a CO$_2$ fixation strategy (irreversible process) via Li–CO$_2$ electrochemistry technology. (Reprinted/adapted with permission from [18]. Copyright 2017 Elsevier).

2.3. The Application of DEMS in Electrode Interface Reaction

The macroscopic property of the electrode catalytic material interface lies in the chemical reaction. Since metal–air batteries involve the gas consumption and emission at the surface interface of catalytic materials, in recent years, the application of differential electrochemical mass spectrometry (DEMS) analysis based on gas detection in the field of lithium–oxygen batteries have realized the continuous measurement of gas, and in situ online analysis of the catalytic cathode interface reaction and possible side reactions [21,22].

The components of DEMS are shown in Figure 4a. In 2006, Bruce et al. added Li_2O_2 to the cathode of a lithium–air battery [23]. With the help of DEMS analysis, the results proved that O_2 could be generated by oxidation during the charging process. The reaction mechanism of lithium–oxygen batteries was directly confirmed by experimental data. McCloskey et al. studied the formation and decomposition process of Li_2O_2 [21]. Peng et al. used nanoporous gold as a simulated cathode instead of a carbon-based cathode material [24]. The battery test results based on the DMSO electrolyte showed that no CO_2 was detected. Based on the test of carbon-based materials, there will be apparent CO_2 emission, indicating that lithium–air batteries with a carbon-based cathode have the possibility of decomposition in the use process, and the safety factor must be considered. The effect of nanocatalysts added to the cathode materials can also be evaluated by the DEMS test and the reaction measurement calculation, such as TiC and Mo_2C, so as to speculate on the actual effect of the catalyst and analyze the stability of the electrode material [25]. The charge–discharge e^-/reaction gas ratio can be calculated by Faraday's law so as to speculate on the reaction path (Figure 4b). Another advantage of mass spectrometry is that it can be combined with isotope calibration methods, such as an isotope reaction gas or electrolyte, to track the intermediate products in the reaction process, and to effectively analyze the ion migration process and catalytic mechanism of the interface by in situ chemical testing methods [26].

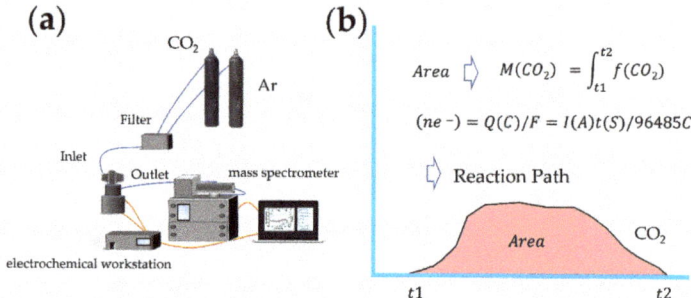

Figure 4. (a) Schematic for a DEMS system; (b) the process of predicting the reaction mechanism.

3. Carbon Tube-Based Cathode for Li–CO$_2$ Battery

Carbon nanotubes are unique 1D materials, consisting of hexagonal carbon atoms to form a single layer to dozens of layers of coaxial circular tubes. With special mechanical, thermal, and electrical properties, they can be used in special applications in the field of engineering materials. Carbon nanotubes have good electronic conductivity and form a unique 3D overlapping space structure, providing sufficient space for the deposition of discharge products [27]. In 2015, Zhou's group applied carbon nanotubes to a Li–CO$_2$ battery cathode, and the battery charge and discharge cycles were realized [28]. However, the catalytic activity of pure carbon materials in the charge–discharge reaction of the Li–CO$_2$ battery was limited, and the battery exhibited a high charge–discharge overpotential and a poor cycle life. Carbon tube matrix composites can effectively encapsulate and load various catalyst nanoparticles so as to achieve the expected carbon dioxide reduction/evolution reaction (CO$_2$-RR/CO$_2$-ER) catalytic performance and increase the reactive sites in Li–CO$_2$ batteries.

3.1. Carbon Tube–Noble Metal-Based Composites

Noble metals such as gold, silver, and the platinum family (ruthenium, rhodium, palladium, osmium, iridium, platinum) have unique activity in catalytic reactions [29,30]. They are used as catalysts for oxygen reduction (ORR) and oxygen precipitation (OER) reactions in the initial stage of lithium–air battery research and are designed as air electrodes with various unique spatial structures [31].

Kong et al. dispersed Au nanoparticles uniformly on the surface of a carbon tube (AuNPs/CNTs) to achieve 46 cycles of a Li–CO_2 battery under the condition of a limited capacity of 100 mAh g^{-1} [32]. In the discharge stage, the voltage platform is obviously higher than that of the carbon nanotube electrode, indicating that it has a catalytic effect in the CO_2 reduction process. In the charging stage, the voltage platform and carbon nanotube electrode are basically the same, reaching more than 4.5 V, indicating that the catalyst effect of the Au electrode in the CO_2 evolution process is limited.

In previous work, ruthenium-based materials were considered efficient catalysts for lithium carbonate decomposition [33,34]. Chen et al. anchored Ru nanoparticles on carbon tubes, which could effectively improve the conductivity of the material matrix, and the porous skeleton formed by cross-linking could promote the diffusion and transmission of CO_2 and the electrolyte [35]. By comparing the attenuation of the battery system at a low specific capacity of 100 mAh g^{-1} and a high specific capacity of 500 mAh g^{-1}, the researchers believed that the dominant role is not the passivation of the cathode, but the effect of lithium dendrites on the anode. After replacing the lithium metal sheet with a C/Li anode deposited on the surface by sputtering, the cycle stability of the battery can be improved by more than three times. Using ruthenium chloride and carbon tubes as raw materials, ruthenium nanoparticles can be attached to the surface of carbon tubes by the reflux method (Figure 5a,b). The obtained composite material can be used as the cathode of lithium–carbon dioxide batteries and sodium–carbon dioxide batteries to realize more than 100 cycles [36]. Li et al. combined a covalent organic framework compound (COF) with Ru-coated carbon nanotubes to explore the adsorption capacity of the material for CO_2 [37]. In the Li–CO_2 battery, the specific capacity reached 27348 mAh g^{-1}. The good performance was attributed to the 1D channel of the material and the functionality of the COF, which facilitate the capture of CO_2 and the rapid transmission of lithium ions. These synergistic effects promote the rapid formation/decomposition of Li_2CO_3 during the discharge–charge process.

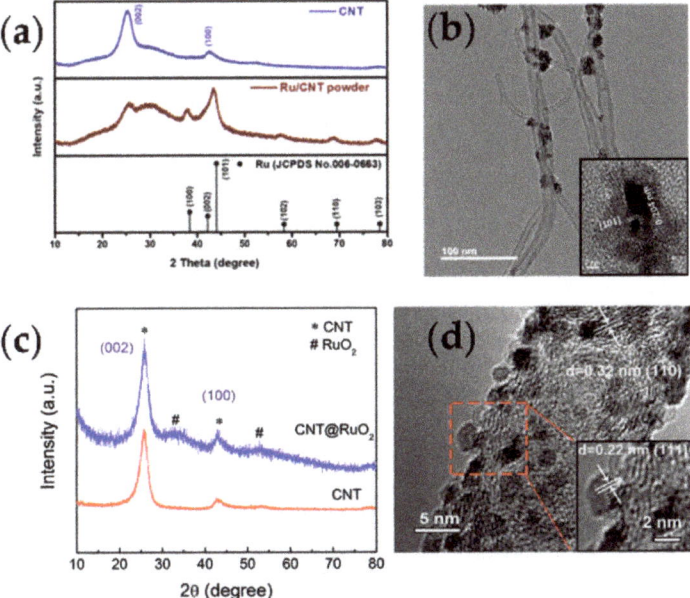

Figure 5. (a) XRD patterns and (b) TEM and high-resolution TEM images of Ru/CNT (reprinted/adapted with permission from [36]; Copyright 2021 American Chemical Society); (c) XRD patterns and (d) HRTEM image of CNT@RuO_2 (reprinted/adapted with permission from [38]; Copyright 2019 American Chemical Society).

Using ruthenium chloride and carbon nanotubes as raw materials, through a simple solution reaction and vacuum sintering at 200 °C, the structure of RuO_2 particles attached to the surface of carbon nanotubes can be obtained (Figure 5c,d) [38]. This type of material was first used in Li–CO_2 batteries. Under the condition of a limited specific capacity of 500 mAh g^{-1}, the initial cycle was 30 cycles, and the charging voltage was lower than 4.0 V. Based on the analysis of XPS, XRD, and SEM, most of the discharge products of lithium carbonate can be effectively decomposed. The catalytic activity of the electrode was analyzed by charging decomposition with pre-filled lithium carbonate. The coulombic efficiency of the carbon tube electrode was 56%, and the charging voltage was close to 4.5 V. The coulombic efficiency of the carbon tube and ruthenium oxide composite electrode was 93%, and the charging voltage was controlled at 3.9 V.

In the practical application of noble metal-based electrode materials, the following issues should be considered: raw material cost, synthesis technology, and performance optimization. Alloying is an effective strategy to reduce costs and optimize the composition. Jin et al. attached Ru and Cu nanoalloys on the surface of carbon nanofibers. The optimized composition and electronic effect can promote the formation and decomposition of lithium carbonate during discharging and charging [39].

3.2. Carbon Tube–Molybdenum-Based Composites

Mo-based materials, due to the multiple valence states of Mo, its ease in the formation of composites, its high electronic conductivity, its high electrocatalytic activity, and the low cost of raw materials, have achieved excellent electrochemical performance in the study of Li–O_2 batteries [40,41]. In Li–CO_2 batteries, Mo-based materials also exhibited excellent properties, such as: improved catalytic activity, reduced charging overpotential, and long cycle stability [42].

Molybdenum carbide (Mo_2C) has been widely studied due to its excellent catalytic properties, similar to VIII metals. It has attracted extensive attention in methane reconstruction, water–gas transfer reactions, hydrogen evolution reactions, and CO_2 reduction reactions. Compared with the metal Mo, the high activity of Mo_2C originates from the electronic properties introduced by carbon, which affects the reaction activity of the Mo-C bond energy and adsorbate. Chen's research group synthesized a composite of Mo_2C and carbon nanotubes and used it for the test of Li–CO_2 batteries [43]. The battery energy efficiency reached 77% and could be recycled for 40 cycles. The mechanism analysis showed that Mo_2C can stabilize the intermediate product of CO_2 in the reduction process during discharging, thereby inhibiting the formation of insulating lithium carbonate. The amorphous discharge product $Li_2C_2O_4$-Mo_2C can be decomposed at a 3.5 V charge voltage. First-principles calculation can further analyze the role of Mo_2C electrodes in Li–CO_2 batteries. Yang et al. systematically studied the Gibbs free energy changes of $Li_2C_2O_4$ and Li_2CO_3 nucleation intermediates and theoretically proved that $Li_2C_2O_4$ can stably become the final discharge product without forming Li_2CO_3 [44]. The overpotential was analyzed by an electrochemical free-energy level diagram, and the catalytic activity of the catalyst during charging and discharging was evaluated. The electron transfer between the intermediate product and the Mo_2C catalyst plays a crucial role in the stability of the discharge product and the electrochemical mechanism of the battery. Xia's group designed a water-in-salt electrolyte (LiTFSI/H_2O 21.0 mol/1 kg) and studied the reaction mechanism of a Li–CO_2 battery composed of CNT and Mo_2C/CNT cathodes [45]. Through a variety of in situ/non-in situ and qualitative/quantitative characterization analyses, the electrode based on Mo_2C/CNT can realize the reversible conversion between CO_2 and $Li_2C_2O_4$ at a low charge–discharge overpotential. In contrast, the CNT-based electrode needs to form and decompose Li_2CO_3 at a high overpotential. The electrochemistry mechanism is schematically shown in Figure 6.

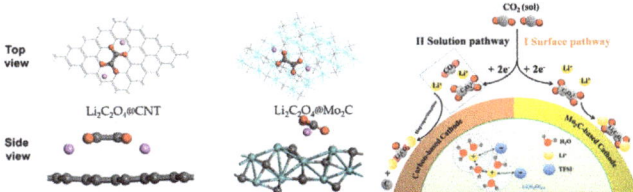

Figure 6. Schematic of the electrochemistry mechanism of CO_2 reduction in WIS-based Li–CO_2 batteries with various cathodes (reprinted/adapted with permission from [45]; Copyright 2021 American Chemical Society).

MoC has also been demonstrated as a highly efficient catalyst in lithium–air batteries [46]. Zhu et al. applied a modified flowing catalyst chemical vapor deposition (FCCVD) method by using dicyandiamide as a carbon and nitrogen source [47]. Nitrogen-doped carbon nanotubes were grown on the surface of the nickel foam skeleton, and MoC nanoparticles were embedded in the carbon nanotubes. The obtained self-supporting structural materials can be directly used for Li–O_2 batteries, Li–CO_2 batteries, and Li–air batteries to avoid the addition of an organic binder. The formation and decomposition of the discharge product lithium carbonate on the electrode surface were verified by XRD, Raman, XPS, FTIR, and SEM analysis. Combined with the DEMS test, the possible reaction mechanism in the charging and discharging process was explained.

Chen et al. prepared an array structure of MoO_3-coated CNTs grown on the surface of a nickel foam [48]. The cross-linking structure of carbon tubes provided abundant channels for electron transport. The large specific surface area provided sufficient ion embedding/desorption sites. Based on the surface self-restriction and self-saturation adsorption, the MoO_3 layer deposited on the outer layer of the carbon tube had good 3D consistency. The Li–CO_2 battery assembled with the material had a discharge capacity of 121.06 mAh cm^{-2}, a charge voltage of less than 3.8 V, and a cycle number of 300 cycles.

Transition metal dichalcogenides have been widely investigated in various electrochemical reactions, including Li–CO_2 batteries [34,49]. The composites of MoS_2 and CNTs also realized a stable cycle in Li–CO_2 batteries [50]. Chen et al. prepared a MoS_2/CNT electrode with a discharge capacity of 8551 mAh g^{-1}, coulombic efficiency of 96.7% and charge voltage of 3.98 V. The DFT calculation results showing the adsorption sites of Li, CO_2, and Li_2CO_3 on MoS_2 materials were analyzed and verified by Raman and X-ray absorption fine structure spectra. In addition, $MoSe_2$@CNT composites were designed, synthesized, and proven to be beneficial to the nucleation and growth of Li_2O_2 in Li–O_2 batteries [51]. The batteries had a specific capacity of 32,000 mAh g^{-1} and a cycle life of 270 at 500 mA g^{-1}.

3.3. Carbon Tube–Other Metal-Based Composites

Transition metals are widely used in various catalytic fields due to their rich con-tent, low cost, and good catalytic activity. Copper was demonstrated effective in the adsorption and activation of CO_2 molecules [52–54]. Xu et al. prepared copper polyphthalocyanine-carbon nanotubes composites (CuPPc-CNTs) by solvothermal in-situ polymerization method. The obtained Li–CO_2 battery show high discharge capacity [55].

Li et al. used an FCCVD method to grow nitrogen-doped carbon tubes on a Ti substrate [56]. The electrode with a self-supporting structure had 45 cycles in a Li–CO_2 battery. The electrode could also assemble a flexible fibrous battery with a semi-solid electrolyte.

Kim et al. used hemoglobin biomolecules as raw materials to analyze the mechanism of Fe nanoparticles embedded in nitrogen-doped carbon nanotubes through capillary action under different heat treatment conditions [57]. Non-in situ characterization proved that the material was not $Li_2C_2O_4$ based on the formation and decomposition of Li_2CO_3 during the battery cycle. It is believed that the material has high catalytic activity in the battery reaction due to the formation of an Fe-O-C bond.

Thoka et al. compared the performance of CNTs, Co_3O_4@CNT and $ZnCo_2O_4$@CNT in Li–CO_2 batteries [58]. The $ZnCo_2O_4$@CNT composite with a spinel structure significantly decreased the overpotential while improving the cycle number performance of the battery.

Zhang et al. deposited NiO nanosheets on the CNT surface using a hydrothermal method. In the first full charge–discharge cycle, the coulombic efficiency reached 97.8%, and the efficiency after five cycles was 91.7% [59]. Inspired by the design of high-nickel and low-cobalt electrode materials, Xiao et al. modified $Co_{0.1}Ni_{0.9}O_x$ nanoparticles on the surface of carbon nanotubes [60]. Through various characterization techniques, the effect of doping Co on the electrochemical performance was systematically studied. The $Co_{0.1}Ni_{0.9}O_x$/CNT electrode achieved 50 cycles without apparent attenuation, which is twice more than that of the NiO/CNT and CNT electrodes. The improvement of the catalytic performance was attributed to the advancement in the P-type electronic conductivity by doping Co^{2+} in the NiO lattice. The Co^{2+}-doped electrode surface formed polymer-like discharge products, which was considered to build a better reaction interface.

Lei et al. prepared an α-MnO_2/CNT electrode, which could increase the number of surface-active sites. A total of 50 cycles could be realized under the limited capacity, and 6 cycles could be realized under the full charge and discharge conditions [61]. Liu et al. prepared layered sodium manganese hydroxide δ-MnO_2-coated carbon nanotubes [62]. Porous lamellar MnO_2 had abundant oxygen vacancy and was evenly distributed on the surface of the carbon tubes. The 3D cross-linked pore network structure formed between the layered MnO_2 and CNTs could effectively promote the permeation of the electrode solution and the diffusion and transmission of CO_2, providing a sufficient Li^+/electron transport path and catalytic activity sites for the CO_2 precipitation reaction. The adsorption energies of CO_2 and Li_2CO_3 in the absence and presence of oxygen vacancies were calculated by DFT, indicating that CO_2 is easier to adsorb at sites containing oxygen vacancies.

Zhang et al. embedded tungsten carbide (W_2C) nanoparticles into the CNT wall to achieve an ultra-low charging voltage of 3.2 V in lithium–carbon dioxide batteries [63]. The ultra-low polarization originates from the electron-rich effect of W atoms in the W-O bond, breaking the stable triangular structure in CO_3^{2-}. The discharge product formed by the W_2C-CNT catalytic electrode was amorphous, which is conducive to decomposition at a low voltage, and the cycle energy efficiency of the battery could reach 90.1%. Combined with the EXAFS spectrum and theoretical calculation results, it was considered that the interaction between lithium carbonate and CNTs was physical, while the interaction between lithium carbonate and W_2C-CNTs was chemical (Figure 7). The problem with the electrode material is that the charging voltage began to increase significantly after 30 cycles.

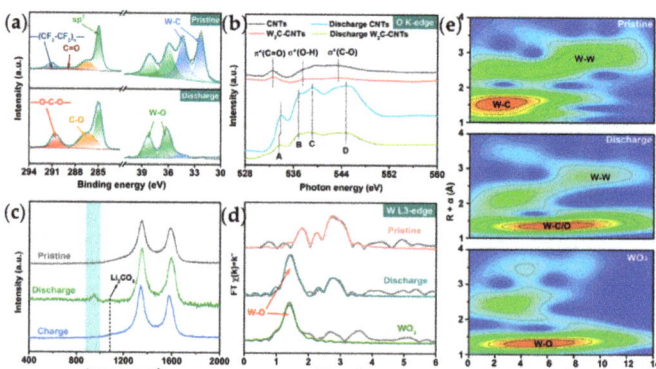

Figure 7. Investigation of the reaction mechanism of Li–CO_2 batteries with W_2C-CNTs. (**a**) XPS spectra, (**b**) O K-edge XANES spectra, (**c**) Raman spectra, (**d**) wavelet transform EXAFS spectra, and (**e**) Fourier transform EXAFS spectra of the cathode (reprinted/adapted with permission from [63]; Copyright 2021 American Chemical Society).

3.4. Heteroatom-Doped Carbon Tube-Based Composites

Due to their low catalytic activity, pure carbon tubes have been proven not suitable for direct use in lithium–carbon dioxide batteries, and a proper adsorbent should be designed [64,65]. Heteroatom doping is an effective strategy. For example, the introduction of nitrogen atoms can enhance the adsorption of CO_2 on the surface through Lewis's acid–base interaction [47]. Li et al. prepared bamboo-like nitrogen-doped carbon tubes using an FCCVD method (Figure 8a) [66]. Studies have shown that nitrogen-doped carbon tube materials can benefit from pyridine nitrogen on the surface, forming abundant defect active sites. Li et al. believed that carbon-based materials were not suitable for direct use in lithium–carbon dioxide batteries without binders, and the introduction of binders would not only cause the loss of active sites on the electrode but also lead to heterogeneous dispersion, resulting in attenuation of the catalytic activity [67,68]. In addition, the contact interface between the catalyst and lithium carbonate will also decrease, resulting in incomplete decomposition, and gradually accumulating and blocking the channels of CO_2 and Li^+ transport, thereby further reducing the active sites. High-capacity Li–CO_2 batteries have become candidates for a flexible power supply, having potential applications in bracelets or wearable electronic devices. On this basis, a vertical array of nitrogen-doped carbon tubes on a titanium mesh was designed, which was used as a self-supporting electrode to achieve an ultra-long cycle, high performance and a flexible lithium–carbon dioxide battery with a gel polymer electrolyte. Two layers of nitrogen-doped graphene-coated CNTs were prepared by Dai's group using anodic aluminum oxide (AAO) as a template (Figure 8b) [69]. They can also be used as a binder-free 3D cathode, as well as a metal-free structure. Song et al. prepared N-S co-doped carbon tubes and confirmed N-S co-doping by high-resolution XPS [70]. The battery test analyzed the high catalytic activity of the composite compared with the CNTs, and a specific capacity of 23,560 mAh g^{-1} could be achieved in the quasi-solid flexible lithium–carbon dioxide battery for 110 days, while the effect of N-S doping remains to be further studied.

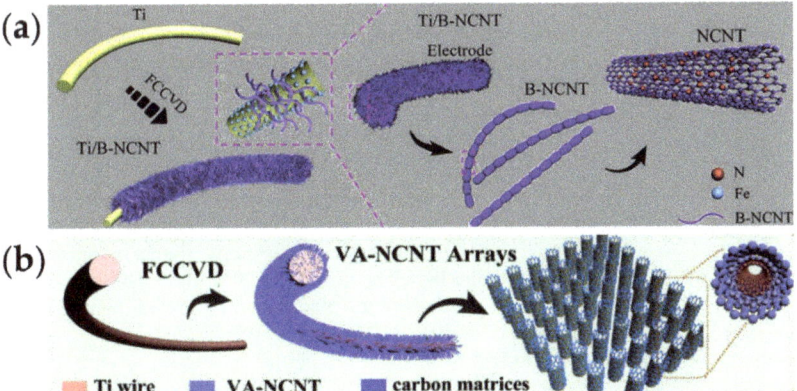

Figure 8. (**a**) Schematic illustration of preparation procedures for B-NCNT electrodes using an FCCVD method (reprinted/adapted with permission from [66]; Copyright 2019 Wiley); (**b**) schematic illustration of the synthesis procedure of VA-NCNT arrays on a Ti wire via an FCCVD method (reprinted/adapted with permission from [69]; Copyright 2020 American Chemical Society).

In summary for this section, the current research progress of CNT-based cathodes for Li–CO_2 batteries is compared in Table 1. The discharge capacity with the corresponding current density, the cycle performances, and the discharge–charge voltage platform are listed in detail.

Table 1. Comparisons of the performances of CNT-based cathodes for Li–CO$_2$ batteries.

Cathode	Discharge Capacity/ Current Density	Cycle Performance (Cutoff Specific Capacity/Current Density)	Discharge–Charge Voltage Platform	Year	Ref.
Mo$_2$C/CNTs	1150 µAh/20 µA	40 (100 µAh/20 µA)	2.65/3.35 V	2017	[43]
MoC/N-CNTs	8227 mAh g^{-1}/100 mA g^{-1}	90 (1000 mAh g^{-1}/1000 mA g^{-1})	2.75/3.79 V	2017	[47]
NiO-CNTs	9000 mAh g^{-1}/100 mA g^{-1}	42 (1000 mAh g^{-1}/50 mA g^{-1})	2.75/4.00 V	2018	[59]
COF-Ru@CNT	27,348 mAh g^{-1}/200 mA g^{-1}	200 (1000 mAh g^{-1}/1000 mA g^{-1})	2.53/4.27 V	2019	[37]
CNTs@RuO$_2$	2187 mAh g^{-1}/50 mA g^{-1}	55 (500 mAh g^{-1}/50 mA g^{-1})	2.48/3.90 V	2019	[38]
N-CNTs@Ti	9292.3 mAh g^{-1}/50 mA g^{-1}	45 (1000 mAh g^{-1}/250 mA g^{-1})	2.60/4.18 V	2019	[56]
MnO$_2$/CNTs	7134 mAh g^{-1}/50 mA g^{-1}	50 (1000 mAh g^{-1}/100 mA g^{-1})	2.62/3.95 V	2019	[61]
N-CNTs	23,328 mAh g^{-1}/50 mA g^{-1}	360 (1000 mAh g^{-1}/1000 mA g^{-1})	2.72/3.98 V	2019	[66]
Ru/CNTs	2882 mAh g^{-1}/100 mA g^{-1}	268 (100 mAh g^{-1}/100 mA g^{-1})	2.56/4.01 V	2020	[35]
ZnCo$_2$O$_4$@CNTs	4275 mAh g^{-1}/100 mA g^{-1}	230 (500 mAh g^{-1}/100 mA g^{-1})	2.52/4.22 V	2020	[58]
Co$_3$O$_4$@CNTs	2473 mAh g^{-1}/100 mA g^{-1}	43 (500 mAh g^{-1}/100 mA g^{-1})	2.45/4.38 V	2020	[58]
3D NCNTs/G	17,534 mAh g^{-1}/50 mA g^{-1}	185 (1000 mAh g^{-1}/100 mA g^{-1})	2.77/3.90 V	2020	[69]
N,S-CNTs	23,560 mAh g^{-1}/200 mA g^{-1}	538 (500 mAh g^{-1}/200 mA g^{-1})	2.63/4.52 V	2020	[70]
Ru/CNTs	4541 mAh g^{-1}/100 mA g^{-1}	45 (500 mAh g^{-1}/100 mA g^{-1})	2.76/4.24 V	2021	[31]
AuNPs/CNTs	6399 mAh g^{-1}/100 mA g^{-1}	46 (1000 mAh g^{-1}/200 mA g^{-1})	2.73/4.30 V	2021	[32]
Ru/CNTs	23,102 mAh g^{-1}/100 mA g^{-1}	100 (500 mAh g^{-1}/100 mA g^{-1})	2.60/4.09 V	2021	[36]
Mo$_2$C/CNTs	0.5 mAh/0.05 mA	20 (1000 mAh g^{-1}/100 mA g^{-1})	2.74/3.41 V	2021	[45]
MoO$_3$@CNTs	30.25 mAh cm^{-2}/0.05 mA cm^{-2}	300 (1 mAh cm^{-2}/0.05 mA cm^{-2})	2.68 /4.03 V	2021	[48]
MoS$_2$/CNTs	8551 mAh g^{-1}/100 mA g^{-1}	140 (500 mAh g^{-1}/100 mA g^{-1})	2.70/3.94 V	2021	[50]
Fe/CNTs	3898 mAh g^{-1}/100 mA g^{-1}	30 (600 mAh g^{-1}/100 mA g^{-1})	2.62/4.24 V	2021	[57]
Co$_{0.1}$Ni$_{0.9}$O$_x$/CNT	5871.4 mAh g^{-1}/100 mA g^{-1}	50 (500 mAh g^{-1}/100 mA g^{-1})	2.55/3.94 V	2021	[60]
CNT@MnO$_2$	-	50 (1000 mAh g^{-1}/200 mA g^{-1})	2.64/4.19 V	2021	[62]
W$_2$C-CNTs	10,632 mAh g^{-1}/100 mA g^{-1}	75 (500 mAh g^{-1}/200 mA g^{-1})	2.81/3.20 V	2021	[63]
N-CNTs	18,652 mAh g^{-1}/100 mA g^{-1}	120 (1000 mAh g^{-1}/250 mA g^{-1})	2.51/4.25 V	2021	[67]
CuPPc-CNTs	18,652.7 mAh g^{-1}/100 mA g^{-1}	160 (1000 mAh g^{-1}/200 mA g^{-1})	2.87/4.32 V	2022	[55]
MWCNTs	5255 mAh g^{-1}/60 mA g^{-1}	50 (600 mAh g^{-1}/60 mA g^{-1})	2.75/4.31 V	2022	[71]
Holey CNTs	17,500 mAh g^{-1}/500 mA g^{-1}	150 (500 mAh g^{-1}/100 mA g^{-1})	2.75/4.31 V	2022	[72]

4. Prospect

The actual process of a Li–CO$_2$ or future genuine Li–air battery is complex [73]. Based on the existing structure and system, a large number of creative works have been carried out, including the battery reaction process assisted by an external light field, the replacement of a liquid electrolyte with a solid one, and other types of metal–carbon dioxide batteries.

4.1. Light Field Assistance

Peng's group designed a carbon nanotube framework with an internal connection structure and coated carbon nitride (C$_3$N$_4$) on the surface to form a heterojunction photocathode for a quasi-solid-state Li–CO$_2$ battery [74]. The composite was prepared at 600 °C to form carbon nitride with more defect structures, which enhanced the absorption of ultraviolet light and produced abundant photoelectrons and vacancies. This further led to the red shift of the absorption edge and the improvement of the carrier separation efficiency in carbon nitride. The accelerated transport of the charge between the carbon nitride and carbon tube structure provides more photoelectron/vacancy migration and promotes the reduction/evolution reaction of carbon dioxide rather than recombination. This process also increases the utilization of light energy. Carbon dioxide was reduced to lithium carbonate/carbon by photoelectrons at 3.24 V (higher than the thermodynamic equilibrium voltage, 2.80 V). In the charging process, the voltage of the oxidation process was only 3.28 V, so the energy efficiency of the charging and discharging cycle process was 98.8%, while the energy efficiency after 100 cycles was 86.1%. This electrode material can also adapt to the application of flexible wearable electronic equipment in the future. Based on the photoelectric effect, the specific incident light can excite a large number of high-energy photoelectrons and holes with high redox activity on semiconductor materials, which is expected to promote the reduction/evolution of CO$_2$ (Figure 9). However, due to the unfavorable charge transfer/separation of typical semiconductor photocatalysts, most photogenerated carriers often recombine before participating in the target battery reaction. The charge transfer/separation of the semiconductor photocatalyst is unfavorable, and most photogenerated carriers tend to recombine before participating in the target battery reaction. Only a tiny fraction of photogenerated carriers can migrate to the surface of semiconductors to promote the CO_2 reduction/evolution reaction. In order to effectively use light to promote the kinetic process of a CO_2 cathode, it is necessary but challenging to encourage the transfer/separation of photogenerated carriers in a semiconductor photocatalyst. Peng's team designed a synergistic two-field auxiliary cathode to solve the inherent limitation of semiconductor photocatalysts by combining plasma metal nanoparticles (such as gold and silver) with semiconductor photocatalytic materials [75]. Under the incident light, the free electrons in the plasma metal nanoparticles can construct a locally enhanced electric field, exerting the opposite force on the electrons and holes to suppress the recombination of carriers. Silver (Ag) nanoparticles were electrodeposited on an anodic oxidation TiO_2 nanotube array as a dual-field auxiliary cathode to promote the CO_2 reduction/evolution reaction. Under the action of a light field, a large number of photogenerated electron holes are generated on TiO_2, and the enhanced electric field around Ag nanoparticles promotes the separation/transfer of photogenerated carriers, thus making better use of carriers for the CO_2 reduction/evolution reaction. The dual-field-assisted Li–CO_2 battery had an ultra-low charging voltage (2.86 V at 0.10 mA cm^{-2}), and the efficiency was 86.9% after 100 cycles, achieving an area capacity of 31.11 mAh cm^{-2}.

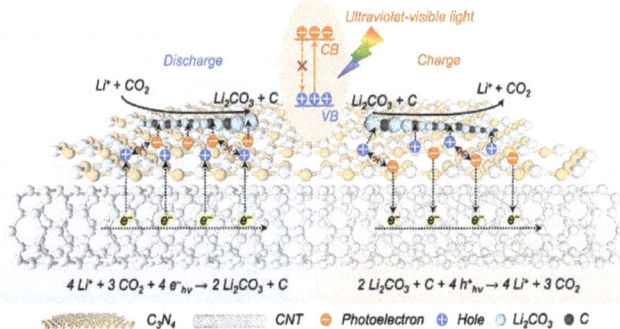

Figure 9. Depiction of the light-assisted discharge–charge processes in a CNT@C_3N_4 heterostructured photocathode (reprinted/adapted with permission from [74]; Copyright 2022 Wiley).

4.2. Solid State

In recent years, the research on Li–CO_2 batteries has been mainly based on the non-proton solvent system, namely, the organic electrolyte system [76,77]. Compared with carbonate electrolytes used in lithium-ion batteries, ether or sulfone electrolytes have wider electrochemical windows and better stability. However, with lithium metal as a negative electrode, it is inevitable that the problems of lithium dendrite growth and a short circuit caused by the penetrating diaphragm will be faced [78,79]. At the same time, as an open system, the battery has problems such as volatilization, leakage of liquid electrolytes, and dissolution of the atmosphere in the air into the corrosion of the lithium sheet, which seriously restrict the improvement of battery performance in terms of stability and a long cycle life. Solid electrolytes have the following advantages:

1. They avoid electrolyte volatilization, which is not flammable;
2. They can inhibit the growth of lithium dendrites with higher safety;
3. They are not prone to inducing side reactions and have a better stability;
4. They can effectively prevent water vapor in the air and reduce the corrosion of lithium anodes.

However, there are still many bottlenecks to be considered in the development and application of solid-state Li–CO_2 batteries, including the composition and structure design of the electrolyte, the contact between the electrolyte and electrode solid interface, the ionic and electronic conductivity and the cost [80]. In existing solid electrolytes, sulfur-based and garnet-based electrolytes are sensitive to water vapor and need to be prepared in an inert atmosphere, which is not suitable for metal–air batteries. Sodium-ion fast-charged $Li_{1+x}Al_xGe_{2-x}(PO_4)_3$ (LAGP) and $Li_{1+x}Al_xTi_{2-x}(PO_4)_3$ (LATP) ceramics can alleviate the grain-limited boundary transport of lithium ions and stabilize in a carbon dioxide atmosphere [81]. Using a commercial LAGP wafer as the electrolyte, a Ru/CNT cathode and a lithium anode, Liu et al. assembled a Li–CO_2 battery and achieved 45 cycles at a limited capacity of 500 mAh g^{-1} [31].

Wang's group prepared a gel polymer electrolyte with a compact structure and good thermal stability by conventional UV curing technology and assembled rechargeable Li–CO_2 batteries based on carbon tube gas electrodes [82]. Under the condition of a limited specific capacity of 1000 mAh g^{-1}, the battery could cycle for 60 cycles. Compared with the liquid electrolyte battery, the discharge product had a granular form with poor crystallinity. In the liquid electrolyte battery, the discharge product created a continuous polymer coating structure, which is not conducive to the transmission of ions, electrons, and gases, thus affecting the cycle stability of the battery. The charging process of the battery was analyzed by in situ differential electrochemical mass spectrometry, but the reaction mechanism needs to be further verified by the isotope method.

The selection of both the electrode and electrolyte needs to be considered in flexible batteries [67]. Li–CO_2 batteries with a liquid electrolyte inevitably have a leakage problem, which cannot meet the requirements of flexible batteries with high safety. Based on a lithium sheet anode and a polymethacrylate/polyethylene glycol-lithium perchlorate-silica composite polymer electrolyte combined with a carbon nanotube cathode, Chen's team constructed a flexible Li–CO_2 battery without a binder and liquid electrolyte [83]. At 55 °C, the ionic conductivity of this organic–inorganic composite electrolyte reached 0.0714 mS cm^{-1}, while the composite of the electrode and electrolyte provided more surface-active sites, reduced the interfacial resistance, and avoided the side reactions caused by the binder. The assembled Li–CO_2 battery effectively cycled for 100 cycles with a specific capacity of 1000 mAh g^{-1}. The assembled soft-packed battery achieved a capacity of 993.3 mAh, reached an energy density of 521 Wh kg^{-1} and cycled for 220 h under the bending condition of 0–360° (Figure 10). The formation and reaction process of lithium carbonate were characterized by in situ Raman technology, and the stability of the composite electrolyte/electrode material was verified.

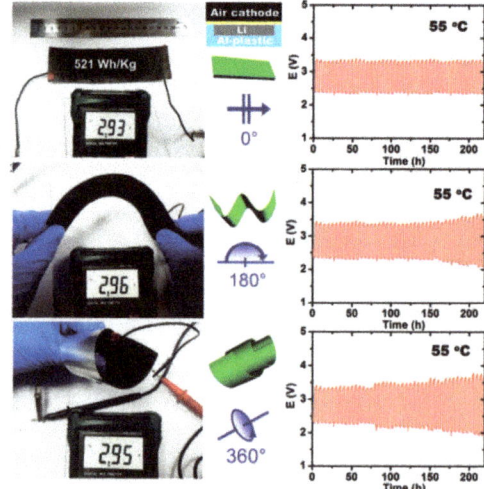

Figure 10. The bending and twisting properties and corresponding cycling numbers of Li–CO_2 batteries (reprinted/adapted with permission from [83]; Copyright 2017 Wiley).

4.3. Other Metal–CO$_2$ Batteries

Due to the shortage of lithium metal resources and the complicated and expensive production process, it has become a hot trend to derive metal–carbon dioxide batteries from other metals (such as sodium or potassium as compared in Table 2) with a higher content in the Earth's crust. Sodium is similar to lithium in terms of physical and chemical properties. The theoretical energy density of sodium–carbon dioxide batteries is 1013 Wh kg^{-1}, and the reaction is based on 4Na + 3CO$_2$→2Na$_2$CO$_3$ + C [84–87]. The discharge voltage of this type of battery is 2.35 V, which is lower than that of a Li–CO$_2$ battery. Still, the advantage is that the charging voltage is reduced accordingly, which has better safety [36]. Because metal sodium is more active and the atomic radius is larger, it will cause higher polarization and more security problems. At present, sodium–carbon dioxide (Na–CO$_2$) primary batteries and rechargeable batteries have great security risks, mainly due to the problems with the liquid electrolyte and pure sodium anode, including the leakage and volatilization of the liquid electrolyte in sodium salt/organic solvent Na–CO$_2$ batteries (especially under high-temperature working conditions). During the battery cycle, dendrite formation or surface cracking will also occur, leading to a short circuit. Therefore, it is crucial to find alternatives to liquid electrolytes and pure sodium anodes [88]. In potassium–carbon dioxide (K–CO$_2$) batteries, using an alloy instead of a pure metal anode will be safer and more stable and will avoid the generation of dendrites, which is an effective strategy. A K–CO$_2$ battery was assembled with potassium-tin alloy as the negative electrode and carbon tubes containing carboxyl functional groups as the positive electrode. Since the -COO$^-$ ions in the carboxyl group were similar to those in potassium carbonate, the C=O bond in potassium carbonate could be weakened, thus promoting the decomposition of potassium carbonate [89]. The specific capacity of the battery reached 3681 mAh g^{-1} at the current density of 100 mA g^{-1}; under the condition of a limited capacity, 400 cycles could be realized.

Table 2. Comparisons of Metal-CO$_2$ batteries.

Metal-CO$_2$ Battery	Earth's Crust	Theoretical Potential	Theoretical Energy Density
Li	0.0017 wt%	2.80 V	1876 Wh Kg^{-1}
Na	2.3 wt%	2.35 V	1130 Wh Kg^{-1}
K	1.5 wt%	2.48 V	921 Wh Kg^{-1}

Theoretical energy densities were calculated based on 4M + 3CO$_2$→2M$_2$CO$_3$ + C (M: Li, Na, K).

5. Conclusions

This review introduced the basic structure and reaction mechanism of Li–CO$_2$ batteries, focusing on the application of carbon nanotube-based composites in Li–CO$_2$ batteries, from material synthesis to performance analysis. New development directions of optical field assistance, the solid state, and other metal–CO$_2$ batteries were also introduced as prospects. Although many breakthroughs have been made, Li–CO$_2$ batteries still have many problems to be solved, including the following:

- Mechanism problems in actual material systems. Through mass spectrometry or other in situ characterization techniques, the dynamic process of the battery system is usually explored by using simulated batteries, which may not effectively correspond to the actual battery system, especially in terms of the battery performance under different specific electrode material systems. For several steps of intermediate product formation, lithium carbonate or Li$_2$C$_2$O$_4$ needs to be combined with the specific actual material system, and even with the different phase structures and tri-configuration systems of the same material, which is complex work.
- Key material selection and structural design issues. The selection of catalytic electrode materials needs to comprehensively consider factors such as the source, cost, preparation process, catalytic performance, and stability. The selection of the electrolyte is primarily concerned with the safety of the voltage window range, as well as the

cost. The cost of existing liquid and solid electrolyte systems is much higher than that of lithium-ion batteries. In addition to the electrode and electrolyte, the type and packaging process of the separator and battery casing will affect the final performance of the battery. That means each component plays a vital role in the final performance and application of Li–CO_2 batteries.

- Characterization method. Non-in situ characterization methods can only provide complementary reference information for the disassembled battery. Effectively combining in situ infrared, Raman, scanning, and transmission electron microscopy with the battery system is the key to the study. In addition, with the combination of experimental characterization and theoretical calculation, the experimental data should provide more guidance to the calculation model, rather than a simple simulation, such as the reaction path, product type, formation, and decomposition of energy, reactant, product adsorption energy, Gibbs free energy change, electron, ion migration rate of thermodynamics, and kinetics analysis.
- Practical application problems. As a high-energy-density energy storage device and carbon dioxide treatment device, the actual reaction process, performance, and effect of the battery at the amplification scale should be considered in the actual application process.

To sum up, Li–CO_2 batteries are a very promising energy storage device and have the functions of CO_2 absorption, fixation, and conversion. They are an important technical means for assisting energy optimization and environmental protection to realize carbon neutralization. However, many scientific and technical problems still need to be studied and solved steadily.

Author Contributions: Investigation, D.M. and W.L.; writing—original draft preparation, D.M. and Z.H.; writing—review and editing, Q.Z.; supervision, Q.Z.; funding acquisition, Q.Z. All authors have read and agreed to the published version of the manuscript.

Funding: This research was funded by the National Natural Science Foundation of China (22105047, 52161033), the Guangxi Special Program for Young Talents (GuiKeAD20159066), the Guangxi Natural Science Foundation (2021GXNSFBA196029) and the Liuzhou Science and Technology Project (2021BDA0102).

Data Availability Statement: Not applicable.

Acknowledgments: We acknowledge the support of the Guangxi University of Science and Technology Doctoral Fund (19Z10) and the Innovation and Entrepreneurship Program for College Students (Guangxi).

Conflicts of Interest: The authors declare no conflict of interest.

References

1. The IMBIE team. Mass balance of the Antarctic ice sheet from 1992 to 2017. *Nature* **2018**, *558*, 219–222. [CrossRef] [PubMed]
2. Mikkelsen, M.; Jørgensen, M.; Krebs, F.C. The teraton challenge. A review of fixation and transformation of carbon dioxide. *Energy Environ. Sci.* **2010**, *3*, 43–81. [CrossRef]
3. Rochelle, G.T. Amine scrubbing for CO_2 capture. *Science* **2009**, *325*, 1652–1654. [CrossRef] [PubMed]
4. Matter, J.M.; Stute, M.; Snaebjornsdottir, S.O.; Oelkers, E.H.; Gislason, S.R.; Aradottir, E.S.; Sigfusson, B.; Gunnarsson, I.; Sigurdardottir, H.; Gunnlaugsson, E.; et al. Rapid carbon mineralization for permanent disposal of anthropogenic carbon dioxide emissions. *Science* **2016**, *352*, 1312–1314. [CrossRef] [PubMed]
5. Kwak, W.J.; Rosy; Sharon, D.; Xia, C.; Kim, H.; Johnson, L.R.; Bruce, P.G.; Nazar, L.F.; Sun, Y.K.; Frimer, A.A.; et al. Lithium–oxygen batteries and related systems: Potential, status, and future. *Chem. Rev.* **2020**, *120*, 6626–6683. [CrossRef]
6. Jung, J.W.; Cho, S.H.; Nam, J.S.; Kim, I.D. Current and future cathode materials for non-aqueous Li–air (O_2) battery technology—A focused review. *Energy Storage Mater.* **2020**, *24*, 512–528. [CrossRef]
7. Wang, D.; Mu, X.W.; He, P.; Zhou, H.S. Materials for advanced Li–O_2 batteries: Explorations, challenges and prospects. *Mater. Today* **2019**, *26*, 87–99. [CrossRef]
8. Zou, X.H.; Lu, Q.; Liao, K.M.; Shao, Z.P. Towards practically accessible aprotic Li–air batteries: Progress and challenges related to oxygen-permeable membranes and cathodes. *Energy Storage Mater.* **2022**, *45*, 869–902. [CrossRef]
9. Mao, Y.J.; Tang, C.; Tang, Z.C.; Xie, J.; Chen, Z.; Tu, J.; Cao, G.S.; Zhao, X.B. Long-life Li–CO_2 cells with ultrafine IrO_2-decorated few-layered δ-MnO_2 enabling amorphous Li_2CO_3 growth. *Energy Storage Mater.* **2019**, *18*, 405–413. [CrossRef]
10. Prehal, C.; Freunberger, S.A. Li–O_2 cell-scale energy densities. *Joule* **2019**, *3*, 321–323. [CrossRef]

11. Takechi, K.; Shiga, T.; Asaoka, T. A Li–O_2/CO_2 battery. *Chem. Commun.* **2011**, *47*, 3463–3465. [CrossRef]
12. Jiao, Y.N.; Qin, J.; Sari, H.M.K.; Li, D.J.; Li, X.F.; Sun, X.L. Recent progress and prospects of Li–CO_2 batteries: Mechanisms, catalysts and electrolytes. *Energy Storage Mater.* **2021**, *34*, 148–170. [CrossRef]
13. Li, J.X.; Wang, L.; Zhao, Y.; Li, S.Y.; Fu, X.M.; Wang, B.J.; Peng, H.S. Li–CO_2 batteries efficiently working at ultra-low temperatures. *Adv. Funct. Mater.* **2020**, *30*, 2001619–2001628. [CrossRef]
14. Zhang, Z.; Zhang, Z.W.; Liu, P.F.; Xie, Y.P.; Cao, K.Z.; Zhou, Z. Identification of cathode stability in Li–CO_2 batteries with Cu nanoparticles highly dispersed on N-doped graphene. *J. Mater. Chem. A* **2018**, *6*, 3218–3223. [CrossRef]
15. Zhou, J.W.; Li, X.L.; Yang, C.; Li, Y.C.; Guo, K.K.; Cheng, J.L.; Yuan, D.W.; Song, C.H.; Lu, J.; Wang, B. A quasi-solid-state flexible fiber-shaped Li–CO_2 battery with low overpotential and high energy efficiency. *Adv. Mater.* **2019**, *31*, 1804439–1804448. [CrossRef] [PubMed]
16. Zhang, Z.; Bai, W.L.; Wang, K.X.; Chen, J.S. Electrocatalyst design for aprotic Li–CO_2 batteries. *Energy Environ. Sci.* **2020**, *13*, 4717–4737. [CrossRef]
17. Sun, X.Y.; Hou, Z.P.; He, P.; Zhou, H.S. Recent advances in rechargeable Li–CO_2 batteries. *Energy Fuels* **2021**, *35*, 9165–9186. [CrossRef]
18. Qiao, Y.; Yi, J.; Wu, S.C.; Liu, Y.; Yang, S.X.; He, P.; Zhou, H.S. Li–CO_2 electrochemistry: A new strategy for CO_2 fixation and energy storage. *Joule* **2017**, *1*, 359–370. [CrossRef]
19. Yang, T.T.; Li, H.; Chen, J.Z.; Ye, H.J.; Yao, J.M.; Su, Y.W.; Guo, B.Y.; Peng, Z.Q.; Shen, T.D.; Tang, Y.F.; et al. In situ imaging electrocatalytic CO_2 reduction and evolution reactions in all-solid-state Li–CO_2 nanobatteries. *Nanoscale* **2020**, *12*, 23967–23974. [CrossRef]
20. Xiao, X.; Yu, W.T.; Shang, W.X.; Tan, P.; Dai, Y.W.; Cheng, C.; Ni, M. Investigation on the strategies for discharge capacity improvement of aprotic Li–CO_2 batteries. *Energy Fuels* **2020**, *34*, 16870–16878. [CrossRef]
21. McCloskey, B.D.; Valery, A.; Luntz, A.C.; Gowda, S.R.; Wallraff, G.M.; Garcia, J.M.; Mori, T.; Krupp, L.E. Combining accurate O_2 and Li_2O_2 assays to separate discharge and charge stability limitations in nonaqueous Li–O_2 batteries. *J. Phys. Chem. Lett.* **2013**, *4*, 2989–2993. [CrossRef] [PubMed]
22. Gowda, S.R.; Brunet, A.; Wallraff, G.M.; McCloskey, B.D. Implications of CO_2 contamination in rechargeable nonaqueous Li–O_2 batteries. *J. Phys. Chem. Lett.* **2013**, *4*, 276–279. [CrossRef] [PubMed]
23. Ogasawara, T.; Débart, A.; Holzapfel, M.; Novák, P.; Bruce, P.G. Rechargeable Li_2O_2 electrode for lithium batteries. *J. Am. Chem. Soc.* **2006**, *128*, 1390–1393. [CrossRef]
24. Peng, Z.Q.; Freunberger, S.A.; Chen, Y.H.; Bruce, P.G. A reversible and higher-rate Li–O_2 battery. *Science* **2012**, *337*, 563–566. [CrossRef] [PubMed]
25. Xie, Z.J.; Zhang, X.; Zhang, Z.; Zhou, Z. Metal–CO_2 batteries on the road: CO_2 from contamination gas to energy source. *Adv. Mater.* **2017**, *29*, 1605891–1605899. [CrossRef] [PubMed]
26. Zhao, Z.W.; Pang, L.; Su, Y.W.; Liu, T.F.; Wang, G.X.; Liu, C.T.; Wang, J.W.; Peng, Z.Q. Deciphering CO_2 reduction reaction mechanism in aprotic Li–CO_2 batteries using in situ vibrational spectroscopy coupled with theoretical calculations. *ACS Energy Lett.* **2022**, *7*, 624–631. [CrossRef]
27. Zhu, Q.C.; Du, F.H.; Xu, S.M.; Wang, Z.K.; Wang, K.X.; Chen, J.S. Hydroquinone resin induced carbon nanotubes on Ni foam as binder-free cathode for Li–O_2 batteries. *ACS Appl. Mater. Inter.* **2016**, *8*, 3868–3873. [CrossRef]
28. Zhang, X.; Zhang, Q.; Zhang, Z.; Chen, Y.N.; Xie, Z.J.; Wei, J.P.; Zhou, Z. Rechargeable Li–CO_2 batteries with carbon nanotubes as air cathodes. *Chem. Commun.* **2015**, *51*, 14636–14639. [CrossRef]
29. Yoon, K.R.; Kim, D.S.; Ryu, W.H.; Song, S.H.; Youn, D.Y.; Jung, J.W.; Jeon, S.; Park, Y.J.; Kim, I.D. Tailored combination of low dimensional catalysts for efficient oxygen reduction and evolution in Li–O_2 batteries. *ChemSusChem* **2016**, *9*, 2080–2088. [CrossRef]
30. Lin, J.F.; Ding, J.N.; Wang, H.Z.; Yang, X.Y.; Zheng, X.R.; Huang, Z.C.; Song, W.Q.; Ding, J.; Han, X.P.; Hu, W.B. Boosting energy efficiency and stability of Li–CO_2 batteries via synergy between Ru atom clusters and single-atom Ru-N_4 sites in the electrocatalyst cathode. *Adv. Mater.* **2022**, *34*, 2200559–2200569. [CrossRef]
31. Savunthari, K.V.; Chen, C.H.; Chen, Y.R.; Tong, Z.Z.; Iputera, K.; Wang, F.M.; Hsu, C.C.; Wei, D.H.; Hu, S.F.; Liu, R.S. Effective Ru/CNT cathode for rechargeable solid-state Li–CO_2 batteries. *ACS Appl. Mater. Inter.* **2021**, *13*, 44266–44273. [CrossRef] [PubMed]
32. Kong, Y.L.; Gong, H.; Song, L.; Jiang, C.; Wang, T.; He, J.P. Nano-sized Au particle-modified carbon nanotubes as an effective and stable cathode for Li–CO_2 batteries. *Eur. J. Inorg. Chem.* **2021**, *2021*, 590–596. [CrossRef]
33. Qiao, Y.; Xu, S.M.; Liu, Y.; Dai, J.Q.; Xie, H.; Yao, Y.G.; Mu, X.W.; Chen, C.; Kline, D.J.; Hitz, E.M.; et al. Transient, in situ synthesis of ultrafine ruthenium nanoparticles for a high-rate Li–CO_2 battery. *Energy Environ. Sci.* **2019**, *12*, 1100–1107. [CrossRef]
34. Chen, B.; Wang, D.S.; Tan, J.Y.; Liu, Y.Q.; Jiao, M.L.; Liu, B.L.; Zhao, N.Q.; Zou, X.L.; Zhou, G.M.; Cheng, H.M. Designing electrophilic and nucleophilic dual centers in the ReS_2 plane toward efficient bifunctional catalysts for Li–CO_2 batteries. *J. Am. Chem. Soc.* **2022**, *144*, 3106–3116. [CrossRef] [PubMed]
35. Chen, C.J.; Yang, J.J.; Chen, C.U.; Wei, D.H.; Hu, S.F.; Liu, R.S. Improvement of lithium anode deterioration for ameliorating cyclabilities of non-aqueous Li–CO_2 batteries. *Nanoscale* **2020**, *12*, 8385–8396. [CrossRef]
36. Thoka, S.; Tsai, C.M.; Tong, Z.Z.; Jena, A.; Wang, F.M.; Hsu, C.C.; Chang, H.; Hu, S.F.; Liu, R.S. Comparative study of Li–CO_2 and Na–CO_2 batteries with Ru@CNT as a cathode catalyst. *ACS Appl. Mater. Inter.* **2021**, *13*, 480–490. [CrossRef]

37. Li, X.; Wang, H.; Chen, Z.X.; Xu, H.S.; Yu, W.; Liu, C.B.; Wang, X.W.; Zhang, K.; Xie, K.Y.; Loh, K.P. Covalent-organic-framework-based Li–CO_2 batteries. *Adv. Mater.* **2019**, *31*, 1905879–1905887. [CrossRef]
38. Bie, S.Y.; Du, M.L.; He, W.X.; Zhang, H.G.; Yu, Z.T.; Liu, J.G.; Liu, M.; Yan, W.W.; Zhou, L.; Zou, Z.G. Carbon nanotube@RuO_2 as a high performance catalyst for Li–CO_2 batteries. *ACS Appl. Mater. Inter.* **2019**, *11*, 5146–5151. [CrossRef]
39. Jin, Y.C.; Chen, F.Y.; Wang, J.L.; Johnston, R.L. Tuning electronic and composition effects in ruthenium-copper alloy nanoparticles anchored on carbon nanofibers for rechargeable Li–CO_2 batteries. *Chem. Eng. J.* **2019**, *375*, 121978–121987. [CrossRef]
40. Kwak, W.J.; Lau, K.C.; Shin, C.D.; Amine, K.; Curtiss, L.A.; Sun, Y.K. A Mo_2C/carbon nanotube composite cathode for lithium–oxygen batteries with high energy efficiency and long cycle life. *ACS Nano* **2015**, *9*, 4129–4137. [CrossRef]
41. Zhu, Q.C.; Xu, S.M.; Harris, M.M.; Ma, C.; Liu, Y.S.; Wei, X.; Xu, H.S.; Zhou, Y.X.; Cao, Y.C.; Wang, K.X.; et al. A composite of carbon-wrapped Mo_2C nanoparticle and carbon nanotube formed directly on Ni foam as a high-performance binder-free cathode for Li–O_2 batteries. *Adv. Funct. Mater.* **2016**, *26*, 8514–8520. [CrossRef]
42. Qi, G.C.; Zhang, J.X.; Chen, L.; Wang, B.; Cheng, J.L. Binder-free MoN nanofibers catalysts for flexible 2-electron oxalate-based Li–CO_2 batteries with high energy efficiency. *Adv. Funct. Mater.* **2022**, *32*, 2112501–2112511. [CrossRef]
43. Hou, Y.Y.; Wang, J.Z.; Liu, L.L.; Liu, Y.Q.; Chou, S.L.; Shi, D.Q.; Liu, H.K.; Wu, Y.P.; Zhang, W.M.; Chen, J. Mo_2C/CNT: An efficient catalyst for rechargeable Li–CO_2 batteries. *Adv. Funct. Mater.* **2017**, *27*, 1700564–1700571. [CrossRef]
44. Yang, C.; Guo, K.K.; Yuan, D.W.; Cheng, J.L.; Wang, B. Unraveling reaction mechanisms of Mo_2C as cathode catalyst in a Li–CO_2 battery. *J. Am. Chem. Soc.* **2020**, *142*, 6983–6990. [CrossRef]
45. Feng, N.N.; Wang, B.L.; Yu, Z.; Gu, Y.M.; Xu, L.L.; Ma, J.; Wang, Y.G.; Xia, Y.Y. Mechanism-of-action elucidation of reversible Li–CO_2 batteries using the water-in-salt electrolyte. *ACS Appl. Mater. Inter.* **2021**, *13*, 7396–7404. [CrossRef]
46. Mao, D.Y.; Yi, S.L.; He, Z.R.; Zhu, Q.C. Non-woven fabrics derived binder-free gas diffusion catalyst cathode for long cycle Li–O_2 batteries. *J. Electroanal. Chem.* **2022**, *915*, 116356–116361. [CrossRef]
47. Zhu, Q.C.; Xu, S.M.; Cai, Z.P.; Harris, M.M.; Wang, K.X.; Chen, J.S. Towards real Li-air batteries: A binder-free cathode with high electrochemical performance in CO_2 and O_2. *Energy Storage Mater.* **2017**, *7*, 209–215. [CrossRef]
48. Chen, M.H.; Liu, Y.; Liang, X.Q.; Wang, F.; Li, Y.; Chen, Q.G. Integrated carbon nanotube/MoO_3 core/shell arrays as freestanding air cathodes for flexible Li–CO_2 batteries. *Energy Technol.* **2021**, *9*, 2100547. [CrossRef]
49. Jin, Y.C.; Liu, Y.; Song, L.; Yu, J.H.; Li, K.R.; Zhang, M.D.; Wang, J.L. Interfacial engineering in hollow NiS_2/FeS_2-NSGA heterostructures with efficient catalytic activity for advanced Li–CO_2 battery. *Chem. Eng. J.* **2022**, *430*, 133029–133038. [CrossRef]
50. Chen, C.J.; Huang, C.S.; Huang, Y.C.; Wang, F.M.; Wang, X.C.; Wu, C.C.; Chang, W.S.; Dong, C.L.; Yin, L.C.; Liu, R.S. Catalytically active site identification of molybdenum disulfide as gas cathode in a nonaqueous Li-CO_2 battery. *ACS Appl. Mater. Inter.* **2021**, *13*, 6156–6167. [CrossRef]
51. He, B.; Li, G.Y.; Li, J.J.; Wang, J.; Tong, H.; Fan, Y.Q.; Wang, W.L.; Sun, S.H.; Dang, F. $MoSe_2$@CNT core-shell nanostructures as grain promoters featuring a direct Li_2O_2 formation/decomposition catalytic capability in lithium-oxygen batteries. *Adv. Energy Mater.* **2021**, *11*, 2003263–2003274. [CrossRef]
52. Xu, Y.Y.; Gong, H.; Song, L.; Kong, Y.L.; Jiang, C.; Xue, H.R.; Li, P.; Huang, X.L.; He, J.P.; Wang, T. A highly efficient and free-standing copper single atoms anchored nitrogen-doped carbon nanofiber cathode toward reliable Li–CO_2 batteries. *Mater. Today Energy* **2022**, *25*, 100967–100975. [CrossRef]
53. Gong, H.; Yu, X.Y.; Xu, Y.Y.; Gao, B.; Xue, H.R.; Fan, X.L.; Guo, H.; Wang, T.; He, J.P. Long-life reversible Li–CO_2 batteries with optimized Li_2CO_3 flakes as discharge products on palladium-copper nanoparticles. *Inorg. Chem. Front.* **2022**, *9*, 1533–1540. [CrossRef]
54. Liu, Y.Q.; Zhao, S.Y.; Wang, D.S.; Chen, B.; Zhang, Z.Y.; Sheng, J.Z.; Zhong, X.W.; Zou, X.L.; Jiang, S.P.; Zhou, G.M.; et al. Toward an understanding of the reversible Li–CO_2 batteries over metal-N_4-functionalized graphene electrocatalysts. *ACS Nano* **2022**, *16*, 1523–1532. [CrossRef] [PubMed]
55. Xu, Y.Y.; Jiang, C.; Gong, H.; Xue, H.R.; Gao, B.; Li, P.; Chang, K.; Huang, X.L.; Wang, T.; He, J.P. Single atom site conjugated copper polyphthalocyanine assisted carbon nanotubes as cathode for reversible Li–CO_2 batteries. *Nano Res.* **2022**, *15*, 4100–4107. [CrossRef]
56. Li, Y.C.; Zhou, J.W.; Zhang, T.B.; Wang, T.S.; Li, X.L.; Jia, Y.F.; Cheng, J.L.; Guan, Q.; Liu, E.Z.; Peng, H.S.; et al. Highly surface-wrinkled and N-doped CNTs anchored on metal Wire: A novel fiber-shaped cathode toward high-performance flexible Li–CO_2 batteries. *Adv. Funct. Mater.* **2019**, *29*, 1808117–1808129. [CrossRef]
57. Kim, H.S.; Lee, J.Y.; Yoo, J.K.; Ryu, W.H. Capillary-driven formation of iron nanoparticles embedded in nanotubes for catalyzed lithium–carbon dioxide reaction. *ACS Mater. Lett.* **2021**, *3*, 815–825. [CrossRef]
58. Thoka, S.; Chen, C.J.; Jena, A.; Wang, F.M.; Wang, X.C.; Chang, H.; Hu, S.F.; Liu, R.S. Spinel zinc cobalt oxide ($ZnCo_2O_4$) porous nanorods as a cathode material for highly durable Li–CO_2 batteries. *ACS Appl. Mater. Inter.* **2020**, *12*, 17353–17363. [CrossRef]
59. Zhang, X.; Wang, C.Y.; Li, H.H.; Wang, X.G.; Chen, Y.N.; Xie, Z.J.; Zhou, Z. High performance Li–CO_2 batteries with NiO-CNT cathodes. *J. Mater. Chem. A* **2018**, *6*, 2792–2796. [CrossRef]
60. Xiao, X.; Zhang, Z.J.; Yu, W.T.; Shang, W.X.; Ma, Y.Y.; Zhu, X.B.; Tan, P. Ultrafine Co-doped NiO nanoparticles decorated on carbon nanotubes improving the electrochemical performance and cycling stability of Li–CO_2 batteries. *ACS Appl. Energy Mater.* **2021**, *4*, 11858–11866. [CrossRef]
61. Lei, D.L.; Ma, S.Y.; Lu, Y.C.; Liu, Q.C.; Li, Z.J. High-performance Li–CO_2 batteries with α-MnO_2/CNT cathodes. *J. Electron. Mater.* **2019**, *48*, 4653–4659. [CrossRef]

62. Liu, Q.N.; Hu, Z.; Li, L.; Li, W.J.; Zou, C.; Jin, H.L.; Wang, S.; Chou, S.L. Facile synthesis of birnessite δ-MnO_2 and carbon nanotube composites as effective catalysts for Li–CO_2 batteries. *ACS Appl. Mater. Inter.* **2021**, *13*, 16585–16593. [CrossRef] [PubMed]
63. Zhang, X.J.; Wang, T.S.; Yang, Y.J.; Zhang, X.; Lu, Z.J.; Wang, J.N.; Sun, C.; Diao, Y.Y.; Wang, X.; Yao, J.N. Breaking the stable triangle of carbonate via W-O bonds for Li–CO_2 batteries with low polarization. *ACS Energy Lett.* **2021**, *6*, 3503–3510. [CrossRef]
64. Gao, J.B.; Liu, Y.D.; Terayama, Y.; Katafuchi, K.; Hoshino, Y.; Inoue, G. Polyamine nanogel particles spray-coated on carbon paper for efficient CO_2 capture in a milli-channel reactor. *Chem. Eng. J.* **2020**, *401*, 126059–126068. [CrossRef]
65. Sun, Z.M.; Wang, D.; Lin, L.; Liu, Y.H.; Yuan, M.W.; Nan, C.Y.; Li, H.F.; Sun, G.B.; Yang, X.J. Ultrathin hexagonal boron nitride as a van der Waals' force initiator activated graphene for engineering efficient non-metal electrocatalysts of Li–CO_2 battery. *Nano Res.* **2022**, *15*, 1171–1177. [CrossRef]
66. Li, X.L.; Zhou, J.W.; Zhang, J.X.; Li, M.; Bi, X.X.; Liu, T.C.; He, T.; Cheng, J.L.; Zhang, F.; Li, Y.P.; et al. Bamboo-like nitrogen-doped carbon nanotube forests as durable metal-free catalysts for self-powered flexible Li–CO_2 batteries. *Adv. Mater.* **2019**, *31*, 1903852–1903860. [CrossRef] [PubMed]
67. Li, X.L.; Zhang, J.X.; Qi, G.C.; Cheng, J.L.; Wang, B. Vertically aligned N-doped carbon nanotubes arrays as efficient binder-free catalysts for flexible Li–CO_2 batteries. *Energy Storage Mater.* **2021**, *35*, 148–156. [CrossRef]
68. Li, Y.J.; Wang, W.Y.; Zhang, B.; Fu, L.; Wan, M.T.; Li, G.C.; Cai, Z.; Tu, S.B.; Duan, X.R.; Seh, Z.W.; et al. Manipulating redox kinetics of sulfur species using Mott-Schottky electrocatalysts for advanced lithium-sulfur batteries. *Nano Lett.* **2021**, *21*, 6656–6663. [CrossRef]
69. Xiao, Y.; Du, F.; Hu, C.G.; Ding, Y.; Wang, Z.L.; Roy, A.; Dai, L.M. High-performance Li–CO_2 batteries from free-standing, binder-free, bifunctional three-dimensional carbon catalysts. *ACS Energy Lett.* **2020**, *5*, 916–921. [CrossRef]
70. Song, L.; Hu, C.G.; Xiao, Y.; He, J.P.; Lin, Y.; Connell, J.W.; Dai, L.M. An ultra-long life, high-performance, flexible Li–CO_2 battery based on multifunctional carbon electrocatalysts. *Nano Energy* **2020**, *71*, 104595–104601. [CrossRef]
71. Na, D.; Jeong, H.; Baek, J.; Yu, H.; Lee, S.M.; Lee, C.R.; Seo, H.K.; Kim, J.K.; Seo, I. Highly safe and stable Li–CO_2 batteries using conducting ceramic solid electrolyte and MWCNT composite cathode. *Electrochim. Acta* **2022**, *419*, 140408–140415. [CrossRef]
72. Xie, H.M.; Zhang, B.; Hu, C.G.; Xiao, N.; Liu, D. Boosting Li–CO_2 battery performances by creating holey structure on CNT cathodes. *Electrochim. Acta* **2022**, *417*, 140310–140316. [CrossRef]
73. Liu, L.L.; Guo, H.P.; Fu, L.J.; Chou, S.L.; Thiele, S.; Wu, Y.P.; Wang, J.Z. Critical advances in ambient air operation of nonaqueous rechargeable Li–Air batteries. *Small* **2021**, *17*, 1903854–1903885. [CrossRef]
74. Li, J.X.; Zhang, K.; Zhao, Y.; Wang, C.; Wang, L.P.; Wang, L.; Liao, M.; Ye, L.; Zhang, Y.; Gao, Y.; et al. High-efficiency and stable Li–CO_2 battery enabled by carbon nanotube/carbon nitride heterostructured photocathode. *Angew Chem. Int. Ed.* **2022**, *61*, e202114612.
75. Zhang, K.; Li, J.X.; Zhai, W.J.; Li, C.F.; Zhu, Z.F.; Kang, X.Y.; Liao, M.; Ye, L.; Kong, T.Y.; Wang, C.; et al. Boosting cycling stability and rate capability of Li–CO_2 batteries via synergistic photoelectric effect and plasmonic interaction. *Angew Chem. Int. Ed.* **2022**, *61*, e202201718.
76. Ma, S.Y.; Lu, Y.C.; Yao, H.C.; Liu, Q.C.; Li, Z.J. Enhancing the process of CO_2 reduction reaction by using CTAB to construct contact ion pair in Li–CO_2 battery. *Chin. Chem. Lett.* **2022**, *33*, 2933–2936. [CrossRef]
77. Wang, L.D.; Lu, Y.C.; Ma, S.Y.; Lian, Z.; Gu, X.L.; Li, J.; Li, Z.J.; Liu, Q.C. Optimizing CO_2 reduction and evolution reaction mediated by o-phenylenediamine toward high performance Li–CO_2 battery. *Electrochim. Acta* **2022**, *419*, 140424–140432. [CrossRef]
78. Lai, J.N.; Xing, Y.; Chen, N.; Li, L.; Wu, F.; Chen, R.J. Electrolytes for rechargeable lithium–air batteries. *Angew Chem. Int. Ed.* **2020**, *59*, 2974–2997. [CrossRef]
79. Zhang, W.Q.; Nie, J.H.; Li, F.; Wang, Z.L.; Sun, C.W. A durable and safe solid-state lithium battery with a hybrid electrolyte membrane. *Nano Energy* **2018**, *45*, 413–419. [CrossRef]
80. Qiu, G.R.; Shi, Y.P.; Huang, B.L. A highly ionic conductive succinonitrile-based composite solid electrolyte for lithium metal batteries. *Nano Res.* **2022**, *15*, 5153–5160. [CrossRef]
81. DeWees, R.; Wang, H. Synthesis and properties of NaSICON-type LATP and LAGP solid electrolytes. *ChemSusChem* **2019**, *12*, 3713–3725. [CrossRef] [PubMed]
82. Li, C.; Guo, Z.Y.; Yang, B.C.; Liu, Y.; Wang, Y.G.; Xia, Y.Y. A rechargeable Li–CO_2 battery with a gel polymer electrolyte. *Angew Chem. Int. Ed.* **2017**, *56*, 9126–9130. [CrossRef] [PubMed]
83. Hu, X.F.; Li, Z.F.; Chen, J. Flexible Li–CO_2 batteries with liquid-free electrolyte. *Angew Chem. Int. Ed.* **2017**, *56*, 5785–5789. [CrossRef]
84. Mu, X.W.; Pan, H.; He, P.; Zhou, H.S. Li–CO_2 and Na–CO_2 batteries: Toward greener and sustainable electrical energy storage. *Adv. Mater.* **2020**, *32*, 1903790–1903811.
85. Tang, M.; Du, J.Y.; Ma, J.L.; Wang, X.D.; Zhang, X.; Shen, Q.Y.; Wang, F.P.; Wang, Y. Cobalt-decorated carbon nanofibers as a low overpotential cathode for nonaqueous Na–CO_2 batteries. *J. Alloys Compd.* **2022**, *911*, 165054–165062. [CrossRef]
86. Mao, Y.J.; Chen, X.; Cheng, H.; Lu, Y.H.; Xie, J.; Zhang, T.; Tu, J.; Xu, X.W.; Zhu, T.J.; Zhao, X.B. Forging ispired processing of sodium-fuorinated graphene composite as dendrite-free anode for long-life Na–CO_2 cells. *Energy Environ. Mater.* **2022**, *5*, 572–581. [CrossRef]
87. Xu, C.F.; Wang, H.W.; Zhan, J.; Kang, Y.; Liang, F. Engineering NH_3-induced 1D self-assembly architecture with conductive polymer for advanced hybrid Na–CO_2 batteries via morphology modulation. *J. Power Sources* **2022**, *520*, 230909–230920. [CrossRef]

88. Hu, X.F.; Li, Z.F.; Zhao, Y.R.; Sun, J.C.; Zhao, Q.; Wang, J.B.; Tao, Z.L.; Chen, J. Quasi-solid state rechargeable Na–CO_2 batteries with reduced graphene oxide Na anodes. *Sci. Adv.* **2017**, *3*, e1602396. [CrossRef]
89. Lu, Y.; Cai, Y.C.; Zhang, Q.; Ni, Y.X.; Zhang, K.; Chen, J. Rechargeable K–CO_2 batteries with a KSn anode and a carboxyl-containing carbon nanotube cathode catalyst. *Angew Chem. Int. Ed.* **2021**, *60*, 9540–9545. [CrossRef]

Article

Porous Carbon Boosted Non-Enzymatic Glutamate Detection with Ultra-High Sensitivity in Broad Range Using Cu Ions

Yifei Ma [1], Jiemin Han [1], Zhaomin Tong [1], Jieling Qin [2,*], Mei Wang [1,*], Jonghwan Suhr [3,4], Jaedo Nam [3], Liantuan Xiao [1], Suotang Jia [1] and Xuyuan Chen [1,5]

[1] State Key Laboratory of Quantum Optics and Quantum Optics Devices, Institute of Laser Spectroscopy, Collaborative Innovation Center of Extreme Optics, Shanxi University, Taiyuan 030006, China; mayifei@sxu.edu.cn (Y.M.); jiemin.han@foxmail.com (J.H.); zhaomin.tong@sxu.edu.cn (Z.T.); xlt@sxu.edu.cn (L.X.); tjia@sxu.edu.cn (S.J.); xuyuan.chen@usn.no (X.C.)
[2] Tongji University Cancer Center, Shanghai Tenth People's Hospital of Tongji University, School of Medicine, Tongji University, Shanghai 200092, China
[3] Department of Polymer Science and Engineering, Sungkyunkwan University, Suwon 16419, Korea; suhr@skku.edu (J.S.); jdnam@skku.edu (J.N.)
[4] School of Mechanical Engineering, Sungkyunkwan University, Suwon 16419, Korea
[5] Department of Micro- and Nanosystem Technology, Faculty of Technology and Maritime Sciences, University College of Southeast Norway, 3184 Borre, Norway
* Correspondence: qinjieling770@hotmail.com (J.Q.); wangmei@sxu.edu.cn (M.W.)

Abstract: A non-enzymatic electrochemical sensor, based on the electrode of a chitosan-derived carbon foam, has been successfully developed for the detection of glutamate. Attributed to the chelation of Cu ions and glutamate molecules, the glutamate could be detected in an amperometric way by means of the redox reactions of chelation compounds, which outperform the traditional enzymatic sensors. Moreover, due to the large electroactive surface area and effective electron transportation of the porous carbon foam, a remarkable electrochemical sensitivity up to 1.9×10^4 μA/mM·cm^2 and a broad-spectrum detection range from nM to mM scale have been achieved, which is two-orders of magnitude higher and one magnitude broader than the best reported values thus far. Furthermore, our reported glutamate detection system also demonstrates a desirable anti-interference ability as well as a durable stability. The experimental revelations show that the Cu ions chelation-assisted electrochemical sensor with carbon foam electrode has significant potential for an easy fabricating, enzyme-free, broad-spectrum, sensitive, anti-interfering, and stable glutamate-sensing platform.

Keywords: carbon foam electrode; copper ion chelation; glutamate detection; high sensitivity

1. Introduction

Huntington's disease (HD) is one of the most prevalent neurodegenerative diseases (NDs), of which the symptoms typically begin in elderly age [1,2]. Similar to other NDs, the symptoms of HD are generally mild at the start yet become worse over time and interfere with daily life [3]. Several researchers have realized that the presence of glutamate in the cerebral cortex is one of the key points for intracellular signal pathways, and the concentration change of glutamate is possibly related to HD [1,4,5]. In addition, glutamate is also an important biomarker for other diseases, such as musculoskeletal pain [6], tumor cells [7], and Alzheimer's disease [8]. Hence, the detection of glutamate can be applied in clinical diagnoses as well as symptom monitoring during the treatments of these diseases [9,10]. The concentrations of glutamate in plasma, serum, cerebral spinal fluid, urine, whole blood and saliva are in the range of 5–100 μM, 97.4 ± 13.2 μM, 0.5–2 μM, 8.5 (3.3–18.4) μM mM^{-1} creatinine, 150–300 μM, and 0.232 ± 0.177 μM, respectively [11]. Therefore, the broader the detection range of glutamate is, the better, and the limit of detection of the non-enzymatic glutamate sensors should be at least lower than 0.05 μM.

Recent studies have reported a number of electrochemical biosensors for the detection of glutamate [12–14]. Although enzymatic glutamate biosensors such as electrochemical sensors using L-glutamate oxidase (GluO$_x$) have demonstrated their capability for glutamate detection, their low sensitivity is problematic due to indirect electron transfer [15]. Moreover, a few other shortcomings of enzymatic detection techniques also need to be overcome, including the complicated enzyme purification procedures, high fabrication costs, instability due to enzyme denaturation, and a narrow detection range. In addition, in terms of the mechanism of glutamate detection in an electrochemical strategy, catalyzing oxidative deamination of glutamate by the enzymes, and the redox reaction of glutamate to oxoglutarate, which is ascribed from the catalytic effect of multivalent cationic metal ions, are the main methods used thus far [16,17]. Unfortunately, the current strategies can only offer a sensitivity of up to ~10^2 μA/mM·cm^2 and a detection linear range in the μM scale, which cannot fulfill the detection requirements. In this regard, undoubtedly, developing a broad-spectrum and highly sensitive glutamate-sensing system is extremely important.

When addressing such formidable challenges of enzymatic detection, researchers have made great efforts to exploit non-enzymatic sensors, which are called the fourth generation of electrochemical glutamate sensors. Of the various factors previously highlighted for establishing a high-performance non-enzymatic glutamate sensor, nanostructured materials as sensing platforms hold an effective strategy for high sensitivity and broad detective concentration ranges which ascribe from their large electrochemically active surface area, as well as a desirable anti-interference, short response time and impressive stability. Therefore, developing nanostructured materials that boosted non-enzymatic glutamate sensors is an irresistible trend that could improve the stability and decrease the cost of sensors. For example, Razeeb et al. firstly developed a non-enzymatic Pt/Ni nanowire array electrode to detect glutamate in 2012 [16]. Disappointingly, despite the complicated and costly synthesis process of the precisely structured nanowire, the sensitivity and linear range of detection were far removed from expectations. Since then, even though the non-enzymatic glutamate sensors have been developed for ten years, there are still a limited number of works of literature published [18–23]. Islam et al. reported RuO$_2$-doped ZnO nanoparticles based on a non-enzymatic glutamate sensor, which reports a high sensitivity of up to 9.6×10^{-5} μA/mM·cm^2 and the lowest detection limit of 0.0001 μM [22]. However, the highest detection limit of this sensor is only 10 μM, which cannot fulfill the detection requirement in many clinical environments, such as plasma, serum, and whole blood.

Owing to its affinitive chelation with Cu ions, glutamate has been employed as a chelation agent to enhance the electrodeposition of copper and prevent the precipitation of copper oxide [24]. This motivated us to develop a Cu ions chelation-assisted system for high-performance glutamate sensing. Along with our recent advance in the synthesis technique of porous carbon foams, we were able to directly detect the concentration of glutamate in an amperometric way. Contributing to the large electroactive surface area and effective electron transportation of the chitosan-derived carbon foam electrode, a high electrochemical sensitivity and a broad-spectrum detection range can be achieved. This study describes the new strategy of a facile and non-enzymatic detection of glutamate, assisted by chelating with Cu ions, and to the best of our knowledge, it reports the highest sensitivity and broadest detection range thus far.

2. Experimental

2.1. Chemicals and Apparatus

L-glutamic acid monosodium salt monohydrate (glutamate, ≥98%), copper chloride (CuCl$_2$, 97%), ascorbic acid (AA), uric acid (UA), dopamine hydrochloride (DA), glucose, 3,4-Dihydroxyphenylacetic acid (DOPAC), chitosan (medium molecule weight), and acetic acid were purchased from Sigma-Aldrich (USA). Phosphate-buffered saline (10 mM of PBS, pH = 7.4) was prepared from NaCl, KCl, Na$_2$HPO$_4$, and KH$_2$PO$_4$. All chemicals were commercially available at analytical grade and were used without further purification.

2.2. Preparation of Electrode Based on a Chitosan-Derived Carbon Foam

The chitosan-derived carbon foam was synthesized from a chitosan foam, which was prepared through a temperature-controlled freeze-casting process [25]. In brief, chitosan powders were dissolved into a 0.3 M acetic acid solution at a chitosan concentration of 10 mg/mL. Subsequently, the solution was frozen at −20 °C and lyophilized in a freeze-dryer at −80 °C for 48 h. Afterwards, the resultant chitosan foam was annealed at 900 °C for 2 h to obtain a cylindrical chitosan-derived carbon foam. The carbon foam was cut with a thickness of 1 mm and attached to a gold plate (as a current collector) with a conductive carbon tape.

2.3. Characterizations

The morphology of the chitosan-derived carbon foam was examined with field emission scanning electron microscopy (FE-SEM, HITACHI, SU8010, Tokyo, Japan). The pyrolysis information was obtained from thermal gravimetric analysis (TGA, Mettler Toledo, TGA 2, Zurich, Switzerland). Focused monochromatized Al Kα radiation (hν = 1486.6 eV) was utilized for the X-ray photoelectron spectroscopy (XPS, ESCALAB, Thermo-Scientific, Brno, Czech). All electrochemical measurements were performed using a VSP potentiostat (Princeton Applied Research, Oak Ridge, TN, USA) at room temperature. A conventional three-electrode system consists of the carbon foam electrode, a platinum plate and Ag/AgCl (saturated KCl solution) as the working, counter and reference electrodes, respectively. The electrochemical performance of the carbon foam electrode on glutamate was studied via cyclic voltammetry (CV) between −0.55 and 0.65 V at a scan rate of 100 mV/s in 10 mM PBS containing 2, 4, and 6 mM $CuCl_2$, respectively. The amperometric responses were operated by chronoamperometry (CA) in 10 mM PBS containing 4 mM $CuCl_2$ at the excited potentials of 0.03 V and 0.31 V, obtained from the previous CV.

3. Results and Discussion

3.1. Chitosan-Derived Carbon Foam Electrode

The chitosan-derived carbon foam was synthesized from a precursor of a chitosan foam made through a freeze-casting process, as illustrated in Figure S1 from Supplementary Materials [25]. The chitosan foam exhibited a porous cellular structure, with the chitosan chains connected with each other, as shown in Figure 1a. After the subsequent pyrolysis process to prepare the carbon foam, the cellular structure kept well while the pore size obviously shrinks (Figure 1b). The change in the pore size was attributed to the weight loss of the chitosan, where the weight after pyrolysis at 900 °C only remained at 22.43% (Figure 1c). The decomposition of chitosan happened at the temperature of ~307 °C and the weight loss occurred steadily at 900 °C, where the porous carbon foam was well synthesized. XPS was checked to precisely demonstrate the N-doping in the carbon foam and chitosan foam, as shown in Figure 1d and Figure S2 from Supplementary Materials. The content of nitrogen in the carbon foam was 5.65%, which was derived from the nitrogen groups in chitosan foam and confirmed from the N 1 s peak in the wide scan of XPS spectra. The deconvoluted N 1 s peak of the N-doped carbon foam shows two distinguished peaks at 398.2 and 401.1 eV, which are attributed to pyridinic N and graphitic N, respectively [26–28], indicating the successful N-doping in the carbon foam during the pyrolysis process.

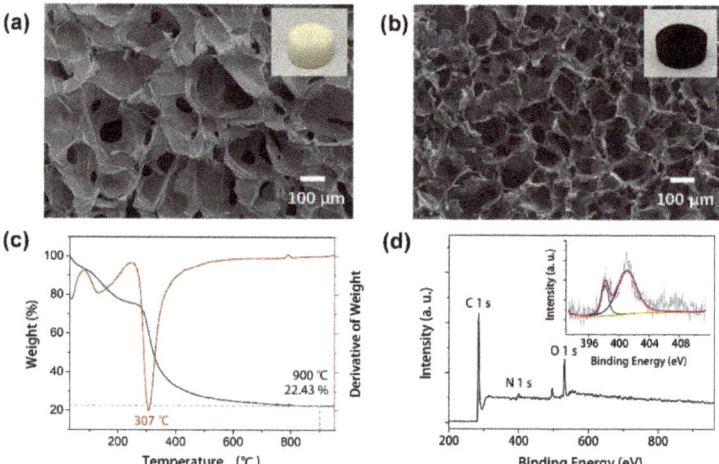

Figure 1. (a) SEM image of chitosan foam; (b) SEM image of chitosan-derived carbon foam; (c) TGA thermogram of weight loss and its derivative of chitosan foam; (d) X-ray photoelectron spectroscopy (XPS) wide scan spectrum and deconvoluted spectra of N 1 s (inset) of carbon foam.

3.2. The Electrochemical Characterization

Porous nanocarbon materials, such as carbon nanotubes, have been widely used as biosensor electrodes for the detection of water-soluble species [29]. In our study, the highly porous chitosan-derived carbon foam was utilized as an electrode for glutamate detection. Figure 2a showed the cyclic voltammograms of the electrode in 10 mM PBS containing 100 µM of glutamate and 4 mM $CuCl_2$ at different potential sweep rates in a wide range of 20–600 mV/s. The dependence of the anodic and cathodic peak currents of glutamate on the scan rates (ν) was depicted in Figure 2b,c. As shown in these figures, the currents of both the oxidation and reduction peaks increased with the increasing scan rates and the peak-to-peak separations also increased simultaneously. The linear regression equations were obtained as follows:

$$\text{Peak 1: } I_{pa} = 7.55 \times 10^{-4}\, \nu + 1.1625,\ R^2 = 0.9938$$

$$I_{pc} = -4.84 \times 10^{-4}\, \nu - 0.3570,\ R^2 = 0.9916$$

$$\text{Peak 2: } I_{pa} = 0.0039\, \nu + 0.3239,\ R^2 = 0.9976$$

$$I_{pc} = -0.0029\, \nu - 0.2542,\ R^2 = 0.9859$$

The perfect linear relationship between the current and scan rates indicates a deposition-controlled process (also called the surface-controlled process), which is ideal for glutamate detection [30–32].

The electrochemical properties of our carbon foam electrode and a gold electrode were evaluated by electrochemical impedance spectroscopy (EIS) in Figure 2d. The electroactive surface areas of these two electrodes were estimated from EIS data and presented in Table S1 [33,34]. The R_s, Z_w, R_{et}, and C in the equivalent circuit represent the solution resistance, the Warburg diffusion resistance, the electron-transfer resistance, and the double-layer capacitance, respectively [35]. Moreover, the electrochemically active specific surface area (S_A) can be calculated from the specific capacitance of the electrochemical double-layer by means of the relationship $S_A = C/C_d$, where C_d is a constant value of 20 µF/cm^2 [36]. As shown in Table S1, the calculated S_A of the carbon-based electrode is 11.66 cm^2/g, which is about 12 times higher than the commonly used gold electrode (0.98 cm^2/g). It indicates that the chitosan-derived carbon foam electrode can effectively provide a large active surface area, and its intrinsic porous structure can enhance the mass transport of glutamate and decrease the diffusion pathway to reach excellent electrochemical performance for glutamate detection.

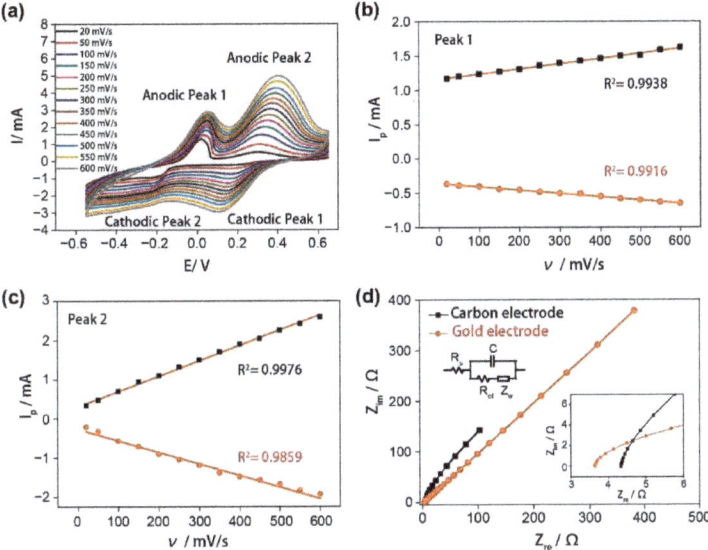

Figure 2. (**a**) Cyclic voltammograms for 4 mM CuCl$_2$ chelation agent in 100 µM glutamate in PBS at scan rates of 20, 50, 100, 150, 200, 250, 300, 350, 400, 450, 500, 550, and 600 mV/s; (**b**) currents of redox peak 1 obtained from (**a**) as functions of the scan rates; (**c**) currents of redox peak 2 obtained from (**a**) as functions of the scan rates; (**d**) Nyquist plots of carbon-based electrode and flat gold electrode in 10 mM PBS containing 100 µM glutamate and 4 mM CuCl$_2$ solution.

3.3. The Sensing Performances

Figure 3a showed the CV response of the carbon-based electrode when detecting glutamate in 10 mM PBS (pH = 7.4) containing 4 mM CuCl$_2$ at the applied potentials between −0.55 and 0.65 V with a scan rate of 100 mV/s. It is worth noting that, without the existence of glutamate, the CV plot of 4 mM CuCl$_2$ presents no obvious peak compared with that of the glutamate solutions with different concentrations. While at the appearance of glutamate, the glutamate would chelate with Cu^{2+}, thereby forming [CuGlu$_2$]$^{2-}$ [24,37]. Hence, the redox peaks are ascribed from the electro-oxidation and electro-reduction of [CuGlu$_2$]$^{2-}$. As shown in Figure 3b,c, the currents are proportional to the logarithmic concentration of glutamate over the range of 0.001 to 1000 µM for peaks 1 and 2 at the potential of 0.03 and 0.31 V, respectively. The linear regression equations of the anodic peaks are I_{pa1} = 0.1545 log C + 0.8805, R^2 = 0.9980 and I_{pa2} = 0.0981 log C + 0.5736, R^2 = 0.9976, with the relative standard deviation (RSD) of 3.15% and 2.32%, respectively. The anodic peaks represent the oxidation of the copper chelate compounds, while the cathodic peaks around 0.17 V and −0.28 V are assigned to the reduction of the chelate compounds.

For confirming the good performance of the carbon foam electrode, the gold plate was undertaken as the electrode for glutamate detection, as shown in Figure S3 (from Supplementary Materials). It can be observed that the current obtained from the porous carbon electrode is 10 times higher than that of using a gold plate electrode at the glutamate concentration of 1 mM, and the glutamate can only be detected under the high glutamate concentration of 0.5–2 mM. The good performance of the carbon electrode is ascribed to the high specific surface area of the carbon foam, which can supply plenty of reaction sites and increase the electrochemical signal. Moreover, the chitosan-derived carbon foams possessed the intrinsic N-doped nature (Figure 1d) [38–40], which can also improve the hydrophilicity of carbon foam electrode and the affinity of the electrode and glutamate, resulting in a broad glutamate detection range using the carbon foam electrode. In addition, Figure S4 (from Supplementary Materials) exhibits the CV curve of the 4 mM $CuCl_2$ aqueous solution using the carbon foam electrode and Figure S5 (from Supplementary Materials) shows the detection of glutamate without the $CuCl_2$. No obvious redox peak can be observed in the CV curve of sole Cu^{2+} and the currents of the anodic peak exhibit no obvious difference with the increasing glutamate concentration without Cu^{2+}, demonstrating the effectiveness of the Cu^{2+} chelation-assisted detection system.

In order to ensure the effectiveness of the detection method at 4 mM $CuCl_2$, glutamate detection was also conducted at 2 and 6 mM $CuCl_2$, as shown in Figure S6 (from Supplementary Materials). Under the condition of 2 mM $CuCl_2$, only peak 2 existed in the plots at various concentrations of glutamate, while both peaks 1 and 2 appeared when the concentration of $CuCl_2$ was 6 mM. The currents in peak 1 show a good linear relation with the log C (R^2 = 0.9926) for the detection of glutamate from 0.01 to 1000 µM, however, the R^2 at 4 mM $CuCl_2$ is higher than that in 6 mM $CuCl_2$, indicating a better performance for the glutamate detection than 2 and 6 mM $CuCl_2$.

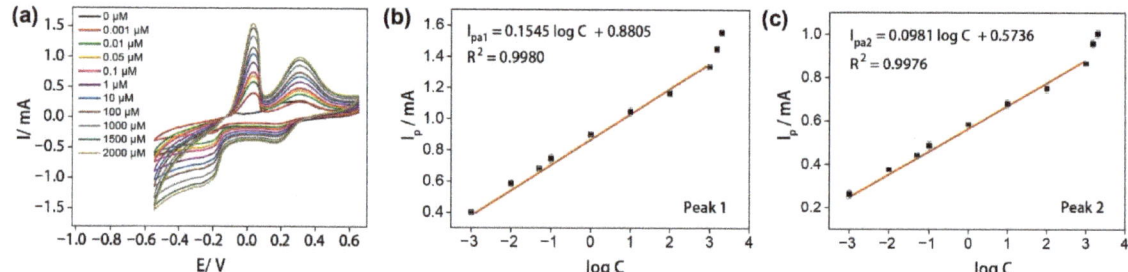

Figure 3. (a) Cyclic voltammograms of carbon-based electrode for the detection of the different concentrations of glutamates (0, 0.001, 0.01, 0.05, 0.1, 1, 10, 100, 1000, 1500, and 2000 µM) in 10 mM PBS containing 4 mM $CuCl_2$; (b) calibration curve of I_p vs. log C of anodic peak 1; (c) calibration curve of I_p vs. log C of anodic peak 2.

3.4. The Sensing Mechanisms

In order to understand the mechanism of the glutamate detection resulting from the complex formation of Cu ions, the electron-transfer mechanism during the electrochemical reactions was investigated. A good linear relationship between the potentials (E_p) of redox peaks and the logarithm of the scan rates (ln ν) were plotted and shown in Figure 4a,b. Laviron derived general expressions for the linear potential scan voltammetric response are as follows [41,42]:

$$E_{pa} = E_0 + A \ln \nu \qquad (1)$$

$$E_{pc} = E_0 + B \ln \nu \qquad (2)$$

where the A = $RT/(1-\alpha)nF$, and the B = $RT/\alpha nF$. E_{pa} and E_{pc} are the anodic and cathodic peak potentials, respectively, and the α, K_s, n and ν are the electron-transfer coefficient, the apparent charge-transfer rate constant, number of electron transfer, and potential sweep rate, respectively. From these expressions, it is possible to determine the α by measuring the variation of the peak potentials with scan rates and the n can be determined for the electron-transfer number between the electrode and the surface-deposited layer by measuring the E_p values (R = 8.314 J/K·mol, T = 298 K, F = 96,485 C/mol). Plots of the E_{pa} and E_{pc} as functions of the ln ν yield two straight lines with slopes equal to $RT/(1-\alpha)nF$ and $RT/\alpha nF$ for the anodic and cathodic peaks, respectively. Figure 4c shows the plot of E_p versus ln ν with slopes equal to 0.0650 and −0.0490 for anodic and cathodic peaks 1, respectively. Using the slopes of plots, the value of α was specified as 0.57 and the electron-transfer number was 1 (0.9194). Figure 4d shows the plot of E_p versus ln ν with slopes equal to

0.0748 and −0.04083 for anodic and cathodic Peaks 2, respectively, thereby the value of α was specified as 0.648 and the electron-transfer number is 1 (0.9722). Hence, all the electron transfers of these two redox peaks are both 1.

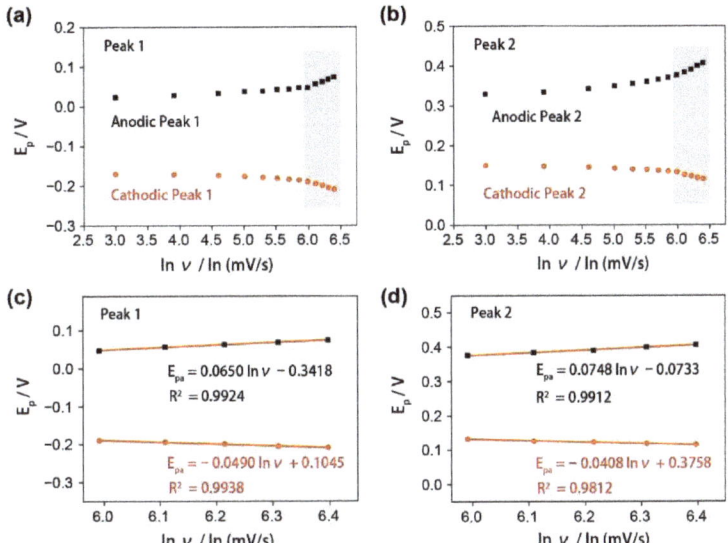

Figure 4. (a) Variations of E_p vs. ln v of the redox peak 1; (b) variations of E_p vs. ln v of the redox peak 2. (c) The calibration curves of E_p vs. ln v of the redox peak 1 when $v > 400$ mV/s. (d) The calibration curves of E_p vs. ln v of the redox peak 2 when $v > 400$ mV/s.

During the cyclic voltammetric process, dark brown copper appears on the carbon foam electrode at the end of the cathodic process. According to this phenomenon and the calculated electron-transfer number, the reaction mechanism could be speculated as follows:

$$\text{Cathodic peak 1: } [Cu^{II}(Glu)_2]^{2-} + e^- \rightarrow [Cu^{I}(Glu)_x]^n \quad (3)$$

$$\text{Cathodic peak 2: } [Cu^{I}(Glu)_x]^n + e^- \rightarrow Cu^0 + Glu^{2-} \quad (4)$$

$$\text{Anodic peak 1: } Cu^0 + Glu^{2-} - e^- \rightarrow [Cu^{I}(Glu)_x]^n \quad (5)$$

$$\text{Anodic peak 2: } [Cu^{I}(Glu)_x]^n - e^- \rightarrow [Cu^{II}(Glu)_2]^{2-} \quad (6)$$

where the $[Cu^{I}(Glu)_x]^n$ stands for the complex formed between Glu^{2-} and Cu^+ [24,43]. Under the appearance of copper ions, chelation compounds of $[Cu^{II}(Glu)_2]^{2-}$ are formed [44,45], and subsequently, the intermediate $[Cu^{I}(Glu)_x]^n$ and final product Cu^0 are synthesized on the carbon foam electrode after the cathodic peak 1 and peak 2, respectively [24,46]. Therefore, in the anodic process afterwards, glutamate interacts with Cu^I or Cu^{II} to form the chelation compounds, in which the glutamate can be detected in an amperometric way due to the redox reactions of Cu.

The amperometric sensing performances of glutamate were carried out under the oxidation potentials of +0.03 V and +0.31 V, respectively. Figure 5a showed typical amperometric response curves of the successive addition of 0.001, 0.01, 1, 5, 50, 100, 200, and 1000 μM of glutamate in 10 mM PBS containing 4 mM $CuCl_2$ for the carbon foam electrode. The current response increased directly after adding the glutamate and achieved a steady-state within 10 s, suggesting the fast rate of electron transfer between glutamate and our proposed electrode. In the calibration curves (Figure 5b), the carbon-based electrode provides a linear range of glutamate from 0.001 to 1000 μM. The linear regression equations are [31,47]: j (mA/cm^2) = 0.0190 C (μM) + 2.6493, R^2 = 0.9943 for 0.03 V; and j (mA/cm^2) = 0.0054 C

(μM) + 0.2106, R^2 = 0.9928 for 0.31 V on carbon foam electrode. Compared with the other reported electrochemical sensors for the detection of glutamate in Table 1, the carbon-based sensor in this study exhibits the highest sensitivity (1.9 × 10^4 μA/mM·cm^2) as well as the comparable detection limit (0.001 μM) and linear range (0.001–1000 μM).

Table 1. Comparison of the performance of different sensor platforms for glutamate detection.

Electrodes	Enzyme	Linear Range (μM)	Limit of Detection (μM)	Sensitivities (μA/mM·cm^2)	Reference
GlutO$_x$/cMWCNT-AuNPs-CHIT/Au	GlutO$_x$	5–500	1.6	155	[17]
GlutO$_x$/Pt-SWCNT/PAA	GlutO$_x$	0.05–1600	0.0046	27.4	[31]
PU/GlutO$_x$/MWCNT/PPy/Pt	GlutO$_x$	0.3–500	0.3	0.384	[48]
GlutO$_x$/APTES/ta-C/Pt	GlutO$_x$	10–500	10	2.9	[49]
GlutO$_x$/BDD/Pt	GlutO$_x$	0.5–50	0.35	24	[50]
Glutamate dehydrogenase/CNT/GCE	Glutamate dehydrogenase	2–225	2	0.71	[51]
GlutO$_x$/ta-C/CNFs	GlutO$_x$	20–500	0.000767	18.8	[12]
GlutO$_x$/ZnO nanorods/PPy/PGE	GlutO$_x$	0.02–500	0.18	N/A	[52]
GlutO$_x$/CeO$_2$/TiO$_2$/CHIT/o-PD/Pt	GlutO$_x$	5–90	0.594	793 pA/μM	[53]
Pt/Ni nanowire array	No enzyme	500–8000	135	65	[16]
NiO/chit/GCE	No enzyme	1000–8000	272	11	[18]
Ni@NC/GCE	No enzyme	0.005–500	135	-	[20]
GluBP/Au NP/SPCE	No enzyme	0.1–0.8	0.15	-	[21]
ZnO/RuO$_2$ NPs/GCE	No enzyme	0.0001–10	9.6 × 10^{-5}	5.42 × 10^3	[22]
MWCNT/Ti-doped ZnO/GCE	No enzyme	100–1000 1000–10,000	11.59	25 4.7	[23]
Cu^{2+} assisted carbon foam	No enzyme	0.001–1000	0.001	1.9 × 10^4	This work

Selective electrochemical detection of glutamate is a challenging task because the oxidizable and electroactive interferents easily interfere with the amperometric measurement of glutamate [51]. To simulate an environment of glutamate in blood, 200 μM of glutamate solution (150–300 μM of glutamate in whole blood) is used to characterize the selectivity and anti-interference ability of the carbon foam-based sensing system. The interference experiment was carried out by the successive addition of 200 μM of glutamate and a high concentration of 50 μM of different interferent species, including 3,4-Dihydroxyphenylacetic acid (DA), ascorbic acid (AA), uric acid (UA), glucose, and dopamine hydrochloride (DH). The current in Figure 5c showed desirable stability under the addition of interferent species. In addition, the current densities obtained at 200 μM and 400 μM of glutamate are 6.84 mA/cm^2 and 9.80 mA/cm^2 (Figure 5c), exhibiting deviations of 6% and 4% from the theoretical values calculated from the equation in Figure 5b, respectively. It indicates satisfactory consistency and repeatability of our glutamate-sensing system. Furthermore, these results demonstrate that the carbon foam-based glutamate sensor possesses satisfactory anti-interference ability and selectivity.

In addition, the stability of our non-enzymatic glutamate detection system was also evaluated from the CA performance for three different electrodes with a relative standard deviation (RSD) of 3.2% in 10 mM PBS containing 4 mM CuCl$_2$ and 100 μM of glutamate over one month. The carbon foam-based electrode was stored at room temperature and tested every five days. The current response toward 100 μM of glutamate retained 98.7% of the initial value after 30 days, as shown in Figure 5d. Hence, our non-enzymatic glutamate sensor exhibited impressive stability, which could be essential for glutamate-sensing applications.

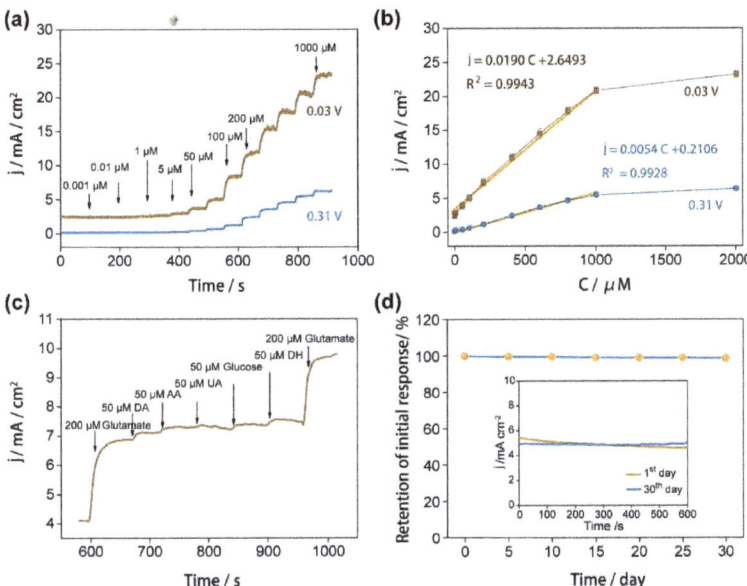

Figure 5. (a) Amperometric responses of carbon-based electrode with stepwise addition of glutamate stock solution at +0.03 V and +0.31 V; (b) plots of the response currents from (a) against the concentration of glutamate; (c) amperometric response of carbon foam electrode in successive addition of 200 µM glutamate, 50 µM of different interference species (3,4-Dihydroxyphenylacetic acid (DA), ascorbic acid (AA), uric acid (UA), glucose, and dopamine hydrochloride (DH) in 10 mM PBS containing 4 mM $CuCl_2$ at 0.03 V; (d) long-term stability of the carbon foam electrode measured in 30 days.

4. Conclusions

In summary, we developed a novel Cu ion chelation-assisted non-enzymatic glutamate detection system on the porous chitosan-derived carbon foam electrode, which was pyrolyzed from a chitosan foam fabricated through a temperature-controlled freeze-drying process. The porous morphology of the electrode provided a large electroactive surface area, which was 12 times larger than the commonly used gold plate electrode, bringing into a low limit of detection (0.001 µM), a broad detection rate of 10^6 µM scale (0.001 to 1000 µM) and a high sensitivity of up to 1.9×10^4 µA/mM·cm^2. The sensing mechanism of the Cu ions chelation-assisted system was finely investigated and proved to be on account of the redox reactions of the chelation compounds of Cu ions and glutamate. Excellent selectivity was also found for glutamate sensing upon various interferent reagents and the sensing performance of our glutamate sensor retains up to 98.7% after 30 days of regular use. We believe our developed non-enzymatic detection system can achieve a low-cost, facile, sensitive, and broad-spectrum glutamate sensor and can also offer new insights into the detection of other reagents.

Supplementary Materials: The following are available online at https://www.mdpi.com/article/10.3390/nano12121987/s1, Figure S1: Scheme of the synthesis process of chitosan-derived graphitic carbon foams, Figure S2: XPS wide scan spectrum (a) and deconvoluted spectra of N 1 s (b) of chitosan foam, Figure S3: (a) Cyclic voltammograms of flat gold electrode for the detection of different concentrations of glutamates (0.5 mM, 1 mM, 1.5 mM, and 2 mM) and carbon electrode for the detection of 1 mM glutamate in 10 mM PBS containing 4 mM $CuCl_2$. The scan rate is 20 mV/s, (b) calibration curve of peak currents of anodic peak in (a) vs. log C, Figure S4: CV curve of the 4 mM $CuCl_2$ aqueous solution measured using the carbon foam electrode, Figure S5: Cyclic voltammograms of carbon foam electrode for the detection of the different concentrations of glutamates (0.5 mM, 1 mM,

1.5 mM, and 2 mM) in 10 mM PBS without $CuCl_2$, Figure S6: (a) Cyclic voltammograms of carbon-based electrode for the detection the different concentrations of glutamates in 10 mM PBS containing 2 mM $CuCl_2$; (b) calibration curve of Ip vs. log C of anodic peak in (a); (c) cyclic voltammograms of carbon-based electrode for the detection of the different concentrations of glutamates in 10 mM PBS containing 6 mM $CuCl_2$; (d) calibration curve of Ip vs. log C of anodic peak 1 in (c), Table S1: EIS data collected from the carbon foam electrode and gold electrode: R_{et}, C, and S_A represent the electron-transfer resistance, the double-layer capacitance, and the surface area, respectively.

Author Contributions: Formal analysis, J.H. and Z.T.; funding acquisition, L.X. and S.J.; investigation, Y.M., J.Q. and M.W.; writing—original draft, Y.M., J.Q. and M.W.; writing—review and editing, J.S., J.N. and X.C. All authors have read and agreed to the published version of the manuscript.

Funding: This research was funded by the National Key R&D Program of China (Grant No. 2017YFA0304203), the National Natural Science Foundation of China (Grants No. 21805174 and 51902190), the Key Research and Development Program of Shanxi Province for International Cooperation (201803D421082), the Scientific and Technological Innovation Programs of Higher Education Institutions in Shanxi (2019L0013 and 2019L0018), the research project was supported by Shanxi Scholarship Council of China (2021-004), the 111 Project (Grant No. D18001), the Changjiang Scholars and Innovative Research Team at the University of Ministry of Education of China (Grant No. IRT_17R70), and the Fund for Shanxi "1331 Project".

Data Availability Statement: The data presented in this study are available on request from the corresponding author.

Conflicts of Interest: The authors declare no conflict of interest.

References

1. Greenamyre, J.T.; Penney, J.B.; Young, A.B.; D'Amato, C.J.; Hicks, S.P.; Shoulson, I. Alterations in L-glutamate binding in Alzheimer's and Huntington's diseases. *Science* **1985**, *227*, 1496–1499. [CrossRef] [PubMed]
2. James, C.M.; Houlihan, G.D.; Snell, R.G.; Cheadle, J.P.; Harper, P.S. Late-onset Huntington's disease: A clinical and molecular study. *Age Ageing* **1994**, *23*, 445–448. [CrossRef]
3. Ross, C.A.; Poirier, M.A. Protein aggregation and neurodegenerative disease. *Nat. Med.* **2004**, *10*, 10–17. [CrossRef] [PubMed]
4. Estrada-Sánchez, A.M.; Montiel, T.; Segovia, J.; Massieu, L. Glutamate toxicity in the striatum of the R6/2 Huntington's disease transgenic mice is age-dependent and correlates with decreased levels of glutamate transporters. *Neurobiol. Dis.* **2009**, *34*, 78–86. [CrossRef] [PubMed]
5. Pépin, J.; Francelle, L.; Sauvage, M.A.C.d.; Longprez, L.d.; Gipchtein, P.; Cambon, K.; Valette, J.; Brouillet, E.; Flament, J. In vivo imaging of brain glutamate defects in a knock-in mouse model of Huntington's disease. *NeuroImage* **2016**, *139*, 53–64. [CrossRef] [PubMed]
6. Baad-Hansen, L.; Cairns, B.; Ernberg, M.; Svensson, P. Effect of systemic monosodium glutamate (MSG) on headache and pericranial muscle sensitivity. *Cephalalgia* **2010**, *30*, 68–76. [CrossRef] [PubMed]
7. Dutta, S.; Ray, S.; Nagarajan, K. Glutamic acid as anticancer agent: An overview. *Saudi Pharm. J.* **2013**, *21*, 337–343. [CrossRef]
8. Maity, D.; Kumar, R.T.R. Highly sensitive amperometric detection of glutamate by glutamic oxidase immobilized Pt nanoparticle decorated multiwalled carbon nanotubes(MWCNTs)/polypyrrole composite. *Biosens. Bioelectron.* **2019**, *130*, 307–314. [CrossRef]
9. Vaquero, J.; Butterworth, R.F. The brain glutamate system in liver failure. *J. Neurochem.* **2006**, *98*, 661–669. [CrossRef]
10. Grace, A.A. Phasic versus tonic dopamine release and the modulation of dopamine system responsivity: A hypothesis for the etiology of schizophrenia. *Neuroscience* **1991**, *41*, 1–24. [CrossRef]
11. Schultz, J.; Uddin, Z.; Singh, G.; Howlader, M.M.R. Glutamate sensing in biofluids: Recent advances and research challenges of electrochemical sensors. *Analyst* **2020**, *145*, 321–347. [CrossRef] [PubMed]
12. Isoaho, N.; Peltola, E.; Sainio, S.; Wester, N.; Protopopova, V.; Wilson, B.P.; Koskinen, J.; Laurila, T. Carbon nanostructure based platform for enzymatic glutamate biosensors. *J. Phys. Chem. C* **2017**, *121*, 4618–4626. [CrossRef]
13. Scoggin, J.L.; Tan, C.; Nguyen, N.H.; Kansakar, U.; Madadi, M.; Siddiqui, S.; Arumugam, P.U.; DeCoster, M.A.; Murray, T.A. An enzyme-based electrochemical biosensor probe with sensitivity to detect astrocytic versus glioma uptake of glutamate in real time in vitro. *Biosens. Bioelectron.* **2019**, *126*, 751–757. [CrossRef] [PubMed]
14. Claussen, J.C.; Artiles, M.S.; McLamore, E.S.; Mohanty, S.; Shi, J.; Rickus, J.L.; Fisher, T.S.; Porterfield, D.M. Electrochemical glutamate biosensing with nanocube and nanosphere augmented single-walled carbon nanotube networks: A comparative study. *J. Mater. Chem.* **2011**, *21*, 11224. [CrossRef]
15. Batra, B.; Kumari, S.; Pundir, C.S. Construction of glutamate biosensor based on covalent immobilization of glutmate oxidase on polypyrrole nanoparticles/polyaniline modified gold electrode. *Enzyme Microb. Tech.* **2014**, *57*, 69–77. [CrossRef]
16. Jamal, M.; Hasan, M.; Mathewson, A.; Razeeb, K.M. Disposable sensor based on enzyme-free Ni nanowire array electrode to detect glutamate. *Biosens. Bioelectron.* **2013**, *40*, 213–218. [CrossRef]

17. Batra, B.; Pundir, C.S. An amperometric glutamate biosensor based on immobilization of glutamate oxidase onto carboxylated multiwalled carbon nanotubes/gold nanoparticles/chitosan composite film modified Au electrode. *Biosens. Bioelectron.* **2013**, *47*, 496–501. [CrossRef]
18. Jamal, M.; Chakrabarty, S.; Shao, H.; McNulty, D.; Yousuf, M.A.; Furukawa, H.; Khosla, A.; Razeeb, K.M. A non-enzymatic glutamate sensor based on nickel oxide nanoparticle. *Microsyst. Technol.* **2018**, *24*, 4217–4223. [CrossRef]
19. Baciu, D.D.; Bîrjega, R.; Mărăscu, V.; Zăvoianu, R.; Matei, A.; Vlad, A.; Cojocaru, A.; Visan, T. Enhanced voltammetric response of monosodium glutamate on screen-printed electrodes modified with NiAl layered double hydroxide films. *Surf. Interfaces* **2021**, *24*, 101055. [CrossRef]
20. Xu, Y.; Zhu, T.; Niu, Y.; Ye, B.C. Electrochemical detection of glutamate by metal–organic frameworks-derived Ni@NC electrocatalysts. *Microchem. J.* **2022**, *175*, 107229. [CrossRef]
21. Zeynaloo, E.; Yang, Y.P.; Dikici, E.; Landgraf, R.; Bachas, L.G.; Daunert, S. Design of a mediator-free, non-enzymatic electrochemical biosensor for glutamate detection. *Nanomed. Nanotech. Biol. Med.* **2021**, *31*, 102305. [CrossRef] [PubMed]
22. Alam, M.M.; Uddin, M.T.; Asiri, A.M.; Awual, M.R.; Fazal, M.A.; Rahman, M.M.; Islam, M.A. Fabrication of selective l-glutamic acid sensor in electrochemical technique from wet-chemically prepared RuO_2 doped ZnO nanoparticles. *Mater. Chem. Phys.* **2020**, *251*, 123029. [CrossRef]
23. Pitiphattharabun, S.; Meesombad, K.; Panomsuwan, G.; Jongprateep, O. MWCNT/Ti-doped ZnO nanocomposite as electrochemical sensor for detecting glutamate and ascorbic acid. *Int. J. Appl. Ceram. Technol.* **2022**, *19*, 467–479. [CrossRef]
24. Pary, P.; Bengoa, L.N.; Egli, W.A. Electrochemical characterization of a Cu(II)-glutamate alkaline solution for copper electrodeposition. *J. Electrochem. Soc.* **2015**, *162*, D275–D282. [CrossRef]
25. Wang, M.; Ma, Y.; Sun, Y.; Hong, S.Y.; Lee, S.K.; Yoon, B.; Chen, L.; Ci, L.; Nam, J.D.; Chen, X.; et al. Hierarchical porous chitosan sponges as robust and recyclable adsorbents for anionic dye adsorption. *Sci. Rep.* **2017**, *7*, 18054. [CrossRef]
26. Zhang, C.; Fu, L.; Liu, N.; Liu, M.; Wang, Y.; Liu, Z. Synthesis of nitrogen-doped graphene using embedded carbon and nitrogen sources. *Adv. Mater.* **2011**, *23*, 1020–1024. [CrossRef]
27. Tang, L.; Ji, R.; Li, X.; Teng, K.S.; Lau, S.P. Energy-level structure of nitrogen-doped graphene quantum dots. *J. Mater. Chem. C* **2013**, *1*, 4908–4915. [CrossRef]
28. Zhang, J.; Xia, Z.; Dai, L. Carbon-based electrocatalysts for advanced energy conversion and storage. *Sci. Adv.* **2015**, *1*, 1500564. [CrossRef]
29. Hughes, G.; Pemberton, R.M.; Fielden, P.R.; Hart, J.P. Development of a novel reagentless, screen-printed amperometric biosensor based on glutamate dehydrogenase and NAD^+, integrated with multi-walled carbon nanotubes for the determination of glutamate in food and clinical applications. *Sens. Actuators B Chem.* **2015**, *216*, 614–621. [CrossRef]
30. Shahdeo, D.; Roberts, A.; Archana, G.J.; Shrikrishna, N.S.; Mahari, S.; Nagamani, K.; Gandhi, S. Label free detection of SARS-CoV-2 Receptor Binding Domain (RBD) protein by fabrication of gold nanorods deposited on electrochemical immunosensor (GDEI). *Biosens. Bioelectron.* **2022**, *212*, 114406. [CrossRef]
31. Yang, J.; Cho, M.; Lee, Y. Synthesis of hierarchical $NiCo_2O_4$ hollow nanorods via sacrificial-template accelerate hydrolysis for electrochemical glucose oxidation. *Biosens. Bioelectron.* **2016**, *75*, 15–22. [CrossRef] [PubMed]
32. Ding, Y.; Wang, Y.; Su, L.; Zhang, H.; Lei, Y. Preparation and characterization of NiO–Ag nanofibers, NiO nanofibers, and porous Ag: Towards the development of a highly sensitive and selective non-enzymatic glucose sensor. *J. Mater. Chem.* **2010**, *20*, 9918–9926. [CrossRef]
33. Lee, S.H.; Yang, J.; Han, Y.J.; Cho, M.; Lee, Y. Rapid and highly sensitive MnO_x nanorods array platform for a glucose analysis. *Sens. Actuators B Chem.* **2015**, *218*, 137–144. [CrossRef]
34. Jung, H.; Lee, S.H.; Yang, J.; Cho, M.; Lee, Y. $Ni(OH)_2$@Cu dendrite structure for highly sensitive glucose determination. *RSC Adv.* **2014**, *4*, 47714–47720. [CrossRef]
35. Kim, S.; Lee, S.H.; Cho, M.; Lee, Y. Solvent-assisted morphology confinement of a nickel sulfide nanostructure and its application for non-enzymatic glucose sensor. *Biosens. Bioelectron.* **2016**, *85*, 587–595. [CrossRef]
36. Chou, S.; Wang, J.; Liu, H.; Dou, S. Electrochemical deposition of porous Co_3O_4 nanostructured thin film for lithium-ion battery. *J. Power Sources* **2008**, *182*, 359–364. [CrossRef]
37. Bukharov, M.S.; Shtyrlin, V.G.; Mukhtarov, A.S.; Mamin, G.V.; Stapf, S.; Mattea, C.; Krutikov, A.A.; Il'in, A.N.; Serov, N.Y. Study of structural and dynamic characteristics of copper(II) amino acid complexes in solutions by combined EPR and NMR relaxation methods. *Phys. Chem. Chem. Phys.* **2014**, *16*, 9411–9421. [CrossRef]
38. Primo, A.; Atienzar, P.; Sanchez, E.; Delgado, J.M.; García, H. From biomass wastes to large-area, high-quality, N-doped graphene: Catalyst-free carbonization of chitosan coatings on arbitrary substrates. *Chem. Comm.* **2012**, *48*, 9254–9256. [CrossRef]
39. Primo, A.; Sánchez, E.; Delgado, J.M.; García, H. High-yield production of N-doped graphitic platelets by aqueous exfoliation of pyrolyzed chitosan. *Carbon* **2014**, *68*, 777–783. [CrossRef]
40. Hao, P.; Zhao, Z.; Leng, Y.; Tian, J.; Sang, Y.; Boughton, R.I.; Wong, C.P.; Liu, H.; Yang, B. Graphene-based nitrogen self-doped hierarchical porous carbon aerogels derived from chitosan for high performance supercapacitors. *Nano Energy* **2015**, *15*, 9–23. [CrossRef]
41. Beitollahi, H.; Ardakani, M.M.; Ganjipour, B.; Naeimi, H. Novel 2,2′-[1,2-ethanediylbis(nitriloethylidyne)]-bis-hydroquinone double-wall carbon nanotube paste electrode for simultaneous determination of epinephrine, uric acid and folic acid. *Biosens. Bioelectron.* **2008**, *24*, 362–368. [CrossRef] [PubMed]

42. Laviron, E. General expression of the linear potential sweep voltammogram in the case of diffusionless electrochemical systems. *J. Electronal. Chem.* **1979**, *101*, 19–28. [CrossRef]
43. Pufahl, R.A.; Singer, C.P.; Peariso, K.L.; Lin, S.J.; Schmidt, P.J.; Fahrni, C.J.; Culotta, V.C.; Penner-Hahn, J.E.; O'Halloran, V.T. Metal ion chaperone function of the soluble Cu(I) receptor atx1. *Science* **1997**, *278*, 853–856. [CrossRef]
44. Fitts, J.P.; Persson, P.; Brown, G.E.; Parks, G.A. Structure and bonding of Cu(II)–glutamate complexes at the γ-Al_2O_3–water interface. *J. Colloid Interface Sci.* **1999**, *220*, 133–147. [CrossRef] [PubMed]
45. Valora, G.; Bonomo, R.P.; Tabbì, G. An EPR and voltammetric study of simple and mixed copper(II) complexes with l- or d-glutamate and l-arginate in aqueous solution. *Inorg. Chim. Acta* **2016**, *453*, 62–68. [CrossRef]
46. Balland, V.; Hureau, C.; Savéant, J.M. Electrochemical and homogeneous electron transfers to the Alzheimer amyloid-β copper complex follow a preorganization mechanism. *Proc. Natl. Acad. Sci. USA* **2010**, *107*, 17113–17118. [CrossRef]
47. Kim, S.; Cho, M.; Lee, Y. Iridium oxide dendrite as a highly efficient dual electro-catalyst for water splitting and sensing of H_2O_2. *J. Electrochem. Soc.* **2017**, *164*, 3029–3035. [CrossRef]
48. Ammam, M.; Fransaer, J. Highly sensitive and selective glutamate microbiosensor based on cast polyurethane/AC-electrophoresis deposited multiwalled carbon nanotubes and then glutamate oxidase/electrosynthesized polypyrrole/Pt electrode. *Biosens. Bioelectron.* **2010**, *25*, 1597–1602. [CrossRef]
49. Kaivosoja, E.; Tujunen, N.; Jokinen, V.; Protopopova, V.; Heinilehto, S.; Koskinen, J.; Laurila, T. Glutamate detection by amino functionalized tetrahedral amorphous carbon surfaces. *Talanta* **2015**, *141*, 175–181. [CrossRef]
50. Hu, J.; Wisetsuwannaphum, S.; Foord, J.S. Glutamate biosensors based on diamond and graphene platforms. *Faraday Discuss.* **2014**, *172*, 457–472. [CrossRef]
51. Chakraborty, S.; Raj, C.R. Amperometric biosensing of glutamate using carbon nanotube based electrode. *Electrochem. Commun.* **2007**, *9*, 1323–1330. [CrossRef]
52. Batra, B.; Yadav, M.; Pundir, C.S. l-Glutamate biosensor based on l-glutamate oxidase immobilized onto ZnO nanorods/polypyrrole modified pencil graphite electrode. *Biochem. Eng. J.* **2016**, *105*, 428–436. [CrossRef]
53. Özel, R.E.; Ispas, C.; Ganesana, M.; Leiter, J.C.; Andreescu, S. Glutamate oxidase biosensor based on mixed ceria and titania nanoparticles for the detection of glutamate in hypoxic environments. *Biosens. Bioelectron.* **2014**, *52*, 397–402. [CrossRef] [PubMed]

MDPI AG
Grosspeteranlage 5
4052 Basel
Switzerland
Tel.: +41 61 683 77 34

Nanomaterials Editorial Office
E-mail: nanomaterials@mdpi.com
www.mdpi.com/journal/nanomaterials

Disclaimer/Publisher's Note: The title and front matter of this reprint are at the discretion of the . The publisher is not responsible for their content or any associated concerns. The statements, opinions and data contained in all individual articles are solely those of the individual Editors and contributors and not of MDPI. MDPI disclaims responsibility for any injury to people or property resulting from any ideas, methods, instructions or products referred to in the content.